ROCKET MEN

ROCKET MEN

The Epic Story of
the First Men on the Moon

CRAIG NELSON

JOHN MURRAY

First published in Great Britain in 2009 by John Murray (Publishers)
An Hachette UK Company

1

© The Craig Nelson Company, Inc., 2009

The right of Craig Nelson to be identified as the Author of the Work has been asserted by him in accordance with the Copyright, Designs and Patents Act 1988.

Grateful acknowledgement is made for permission to reprint an excerpt from *Carrying the Fire: An Astronaut's Journey* by Michael Collins. Copyright © 1974 by Michael Collins. Reprinted by permission of Farrar, Straus and Giroux, LLC.

Photographs from NASA

A CIP catalogue record for this title is available from the British Library

Hardback ISBN 978-0-7195-6948-7
Trade paperback ISBN 978-1-84854-291-4

Printed and bound by Clays Ltd, St Ives plc

John Murray policy is to use papers that are natural, renewable and recyclable products and made from wood grown in sustainable forests. The logging and manufacturing processes are expected to conform to the environmental regulations of the country of origin.

John Murray (Publishers)
338 Euston Road
London NW1 3BH

www.johnmurray.co.uk.

Dedicated to the 400,000 men and women of Apollo.
You made the dream come true.

Then what am I [when flying]—the body substance which I can see with my eyes and feel with my hands? Or am I this realization, this greater understanding which dwells within it, yet expands through the universe outside; a part of all existence, powerless but without the need for power; immersed in solitude, yet in contact with all creation?

—Charles Lindbergh, *The Spirit of St. Louis*

Ad astra per aspera—
A rough path leads to the stars.

—Apollo 1 Memorial, Kennedy Space Center

In 1969, a few months after Apollo 11 landed on the Moon, Rhode Island's Senator John Pastore was interrogating Fermilab physicist Robert Wilson at a Senate hearing on whether the federal government should spend $250 million to build a new collider. The senator wanted to know, would this collider add to "the security of the country?"

"No sir, I don't believe so," Wilson answered.

Senator Pastore: "Nothing at all?"

Mr. Wilson: "Nothing at all."

Senator Pastore: "It has no value in that respect?"

Mr. Wilson: "It only has to do with the respect with which we regard one another, the dignity of men, our love of culture. . . . It has to do with, are we good painters, good sculptors, great poets? I mean all the things we really venerate in our country and are patriotic about. . . . It has nothing to do directly with defending our country, except to make it worth defending."

Contents

PART I

1

Behemoth

On May 20, 1969, at 12:30 p.m. EST, a 363-foot, thirty-story-high black-and-white Saturn V rocket known as AS-506 was painstakingly trundled five miles across the raging heat and searing green of central Florida's eastern coast by an eleven-man Kennedy Space Center crew aboard the world's largest land vehicle, a six-million-pound, tank-wheeled crawler out of NASA's Vehicle Assembly Building, itself a 129-million-cubic-foot edifice so massive that its steel accordion doors were forty-five stories high and, without its ten-thousand-ton air conditioner, interior clouds would form under its 525-foot-high ceiling . . . and it would rain. An enormous creation of white ship and red derrick, the rocket and its mated launch tower, a twelve-million-pound engineering and technological goliath, moved so slowly from where they had been assembled to their destination that progress could not be tracked by human eye. Instead, they would be noted at one point, and then a few hours later at another, the sight accompanied by a small shock, the shock of a great red-white-and-black skyscraper coming to life and advancing, imperceptibly, across the Florida sawgrass swampland.

The crawler at the rocket's base was, in itself, such a dramatic piece of engineering legerdemain that it even impressed Kennedy launch operations director (and former West Point tackle) Rocco Petrone: "Somebody in our shop came up with the idea of using giant tracked machines like those used in strip mining. What evolved was the unique crawler or, more politely, transporter. As built by [Ohio's] Marion Power Shovel Company, the crawler took shape with eight tracks, each seven by forty-one feet, with cleats like a Sherman tank, except that each cleat weighed a ton. Mounted over these eight tracks was the platform, bigger than a baseball diamond, on which the Apollo–Saturn V and its mobile launcher would ride majestically from VAB to pad at one mile per hour. The package weighed nine thousand tons, two-thirds cargo, one-third crawler."

NASA itself manufactures very little of what it flies; instead, for Apollo 11, it relied on twelve thousand American corporations and four hundred thousand employees, almost all of whose output ended up at the Vehicle Assembly Building (VAB), which Petrone called "an intricate machine that assembled the vehicle in its final phases." The Saturn V rocket's bottom stage was built by Boeing at the Michoud Assembly Factory in New Orleans, barged down the Mississippi River, around the Florida peninsula, and through a series of canals to the VAB—the great, shimmering reflecting pool that has graced so many NASA launch photos is in fact a turning basin for its arriving barges. The booster's stage two was built by North American Rockwell in Seal Beach, California, shipped through the Panama Canal to be test-fired at NASA's Mississippi facility, and then shipped to Kennedy. Stage three originated with Douglas in Sacramento, California, and was flown to Kennedy via Aero Spacelines' specially manufactured version of Boeing Stratocruiser: the Super Guppy. The Guppy had been created solely through the efforts and finances of Aero's founder, John Conroy, who against many doubters insisted he could turn a generic cargo jet into a carrier that would fit NASA's stratospheric requirements; the agency had previously tried using zeppelins to transport its enormous rockets, an experiment that did not end well. That same plane carried the Command and Service Modules from North American's Downey, California, plant, while the Lunar Module, as light a ship as any in history, traveled by train and truck from Grumman's factory in Bethpage, New York, to Florida.

After seeing their bucket arrive factory-fresh from manufacturer North American Aviation in its blue packing sheet, the Apollo 9 crew had named their Command Module *Gumdrop*, and their Lunar Module *Spider*, for its arachnid appearance. When Apollo 10 in turn christened their ships *Charlie Brown* and *Snoopy*, assistant administrator for public affairs Julian Scheer wrote Manned Spacecraft Center director George M. Low to suggest that perhaps the Apollo 11 crew might consider being less flippant in naming their craft. Scheer, in fact, would end up suggesting *Columbia* for the CM, while Jim Lovell, Neil Armstrong's backup commander, would recommend *Eagle* for the LM. Both ideas would be adopted.

As each rocket part arrived at Kennedy, it was individually inspected, mated to its nethers in the VAB, and the whole then given a plugs-in test, a compete simulation of every dial, switch, pump, light, fan, valve, and motor that would be used on the mission. The near-sadism that had marked the earliest rounds of psychological and physical testing of the Mercury astronauts now found its match in the torturing of these machines. *Columbia* and

Eagle, and their environmental, communications, electrical, and rocket sub-systems, were subjected to fire, ice, collision, shocks, vibration, dust, rain, and 587,500 forms of inspection.

After reaching the distant loneliness of Pad 39A's octagon-shaped concrete slab built hard against the sea, Saturn V–Apollo 11 underwent a flight readiness test. Then its tanks were pumped with liquid oxygen, liquid hydrogen, jet-grade kerosene, and hydrogen peroxide at ten thousand gallons a minute for four hours and thirty-seven minutes, followed by the five-day, ninety-three-hour countdown to liftoff. At every step, a horde of technicians swarmed over the great leviathan, pumping and probing, tightening and reconnecting, moaning with frustration, or quietly moving on to the next of the tasks that seemed as endless as the stars in the night. NASA's astronauts may have been widely admired for their daring and their courage, but these pad workers, under constant threat from enormous machines and lethal gases, were just as brave. Together they shared a utopian and impossible dream: that they would imminently send, for the first time in history, men to the Moon.

Rocco Petrone: "By the time of Apollo 11, the number of printed pages that were required to check out a space vehicle actually surpassed thirty thousand. In our testing we had a building-block approach, very logical, very methodical; you built each test on the last test, and the whole sequence expanded in the process."

The rough-and-tumble Petrone had been working with the terse and severe Kennedy Space Center director Kurt Debus ever since the very first Redstone rocket was launched in 1953—Petrone was then an army ordnance officer—and the two had risen side-by-side to become heads of NASA's Florida directorate. Kurt Debus was one of the original members of the Wernher von Braun rocket team—at the close of the war, his U.S. Army records described him as "an ardent Nazi" known to have "denounced his colleagues to the Gestapo." At NASA, he was notoriously fastidious, a chief so enamored of cleanliness that he would personally tidy up his underlings' desks, while New Yorker Petrone was just as famous for decimating a worker who didn't give it his all. To get from Alan Shepard's ballistic lob to Neil Armstrong and Buzz Aldrin's moonwalk in less than a decade, the American team needed a front line of just this type of rough management in the theater of war that was the Space Race.

Over the previous seven years, Debus's German émigré colleagues had created a booster for Armstrong's Saturn that was fifty times as powerful as John Glenn's Atlas and, most crucially, would have the remarkable legacy, unlike almost every other rocket in history, of never self-immolating on the pad—heroic engineering at its best. "One wonder to me was that no Saturn

V rocket ever blew up," admitted Apollo 11 Command Module pilot Michael Collins. "I mean, that just surprised the pee-willie [out of] me." NASA researcher Jim Slade remembered: "I heard Neil Armstrong one time say that, today, they're shocked when the shuttle doesn't work every time, but they were always surprised when the Saturn V did."

By the time of Apollo 11, Americans had watched so many rockets lifting off in so many countdowns that NASA's dutiful television broadcasts had become as humdrum as breakfast. The reality, however, was far more dramatic, as deeply alien as anything in Heinlein, Clarke, or Asimov. The superchilled LOX and LH2 fuels inside the Saturn—frozen oxygen and hydrogen, relentlessly difficult materials to manufacture, store, and transport, were used to thrust von Braun's Apollo rockets as they were immensely efficient in the crucial ratio of propellant weight to firing power—were so volatile that even NASA's stringent fueling and insulation technologies couldn't stop a portion of them from boiling over, an alchemy of liquid-into-gas coursing through the missile's plumbing that produced whistles, groans, and wheezes across the entirety of the thirty-story machine, as if it were a living, breathing creature. The cryogenic tanks also froze the dew of Florida hard against the rocket's skin, which then gently floated to the ground as clouds of snow. And, at the Saturn V's very tip, the hypergolic fuels used by *Columbia* and *Eagle* had a pronounced odor: they smelled like trout.

Closeout crew manager—and onetime Luftwaffe flight engineer—Guenter Wendt, a wiry, energetic, and demanding overseer known for his stringent behavior and his thick accent as "the pad fuehrer," who had worked on every flight since the first chimpanzee's, said of the fully assembled Saturn V: "It is a monster, that rocket. It is not a dead animal; it has a life of its own."

"Standing up at night and the lights are on it and all this oxygen and hydrogen are boiling over the sides, it's alive, it's moving because the wind makes it sway, and that's going to take you a quarter of a million miles to another planet . . . and when you get in it, you've got control of it," said Apollo 10 and 14's Gene Cernan. "You remained on the pad while the LOX prechilled, with xenon lights, and the wind blowing, and as those pipes chill, they scream," Kennedy rocket scientist Bob Jones remembered. "This thing is groaning and moaning and the hydraulic pumps are coming on. . . . We would watch that thing ignite a beautiful, absolute, thunderous roar, zillions of horsepower, and you visualize them valves workin' and them turbo pumps goin' ch-ch-ch-ch-ch-ch. The thing is smokin' and ventin' and shakin' and screamin'!"

The missile had six million parts, which meant that, under NASA's rigorous target of 99.9 percent reliability, six thousand of its elements statistically might fail. Petrone's pad crew had in fact spent a frightful thirty-six hours the previous week, when it became clear that something was leaking, somewhere. The problem was traced back to the main LOX (liquid oxygen) tank's helium pressurant manifold. Could it be fixed, or would the manifold need replacing—a four-day job that would mean canceling launch? One tech very carefully tightened a nut to see if that would fix the problem . . . and it did. The pad crew returned to their 1,700-page launch control plan, and the countdown continued.

At the rocket's tip sat Command and Service Module *Columbia,* and beneath it, Lunar Module *Eagle,* each costing $100,000, or ten times the *Spirit of St. Louis,* which had crossed the Atlantic a mere forty-two years earlier. For fuel, the CSM and LM used three hundred pounds of monomethylhydrazine and nitrogen tetroxide, which were hypergolic (self-igniting) when mixed—a useful feature in the vacuum of outer space, when they would be the critical components in bringing the astronauts home. Fully loaded for launch, Apollo 11–Saturn V weighed just under 6.5 million pounds, 6 million of which was its fuel and propellant: liquid oxygen (LOX) and kerosene for stage one; LOX and liquid hydrogen (LH2) for stage two; and hypergolics for the tiny modules that in its final days would be the mission's only spaceships.

When Lunar Module pilot Buzz Aldrin was asked what the most dangerous part of his job would be, he said: "Launch." The pad was where the worst could happen, and sometimes did. Kennedy rocket engineer Bob Jones remembered the early days, when "There was a Juno that went up and turned ninety degrees. It was sitting there, and it came back on the pad and a shock wave came up the flame trench and blew the covers off, and a cigarette machine outside was pierced and the candy bars and cigarettes went everywhere and big chunks of concrete—I thought, 'This is sporty business! This beats the hell out of drawing brackets!'"

Pad catastrophe was such a grave possibility, in fact, that NASA had engineered a number of methods to rescue its crews. The key system was a three-rocket apparatus—the launch escape tower—attached to *Columbia's* nose cone, ready to fire, pull the men from their booster, deploy the chutes, and drift into an Atlantic splashdown. One of two reasons that the American space program used ocean landings to bring its ships home, in fact, originated with a capsule designed to safely bail into the Atlantic in case of pad disaster—the truth was, Mercury, Gemini, and Apollo craft could all equally touch down on land. "The principal reason for not landing on land is that coming down on parachutes, you drift," manned spaceflight associate administrator

George Mueller explained. "At that time, our ability to control the reentry profile was limited, and so no one wanted to take the chance of [targeting] White Sands and ending up in Albuquerque." By Apollo 11, however, NASA reentry techniques had become so sophisticated and its splashdowns so accurate that the navy's recovery fleets were cautioned to wait five miles from a planned recovery site to make sure that a capsule did not crash into one of its carriers.

Apollo's predecessor, Gemini, had ejection seats instead of a launch escape tower, "but the problem is, if you fire the seat while you're sitting on the pad, you get shot into the ground," spacecraft manager Ernie Reyes remembered. "You don't get shot up, and the parachutes open. You get shot straight down and you hit the dirt. Not a very well thought-out process . . . So we took some tractors out there with the disks and we plowed up the dirt. You fluffed up the dirt. What else can you do? My God, what else can you do?"

If a NASA crew managed to escape from the capsule in a launch emergency, they could also use Pad 39A's 600-feet-per-minute high-speed elevators and be met on the ground by armored personnel carriers, or jump into a cab attached to a slide wire that would carry all three men 2,500 feet at 50 miles per hour away from an immolating missile. Michael Collins: "The slide wire had attached to it a small cable car which you entered and then released for the long slide down. At the bottom you got out of the car and jumped into a dark slippery tunnel, sliding along beneath earth and concrete, until finally you were spit out at the 'rubber' room. This chamber was shock-mounted so as to survive the earth tremor caused by an exploding Saturn V, and its interior was literally built out of rubber, including rubber floors and rubber chairs to further protect the occupants from vibration."

Pad leader Guenter Wendt:

> One day I got a call. "Mr. Burke is here and he would like to talk to you." [Walter F. Burke was the vice president of McDonnell Aircraft, Wendt's employer.]
>
> I go in, and Walter said, he says, "I came all the way from St. Louis because something is bothering us in St. Louis very much."
>
> I said, "What is that?"
>
> He said, "There is a rumor going on that you, somewhere in the white room, have stashed away a pipe or something like it, and that you would be willing to kill somebody if they block the exit in an emergency. Is that true or false?"
>
> I said, "Walter, let me give you a background. When they are flight-pressurized, if they spring a major leak [in] the hydrogen [tank or lines and] find

a hydrocarbon . . . we are in a hell of a big flame pit. I have thought many, many nights, long and hard, how can I save people. The elevator is no escape. We have a slide wire, but to the slide wire is only one inward-opening door. It cannot be made an outward-opening door. If somebody panics and blocks that door, and he's bigger than I am, I will remove him by any means, and if you'd like to see the pipe I have stowed away, I'll show you the pipe."

He turned kind of whitish. But he went back to St. Louis. He didn't say a word. . . .

[During Mercury, inspired by oil derrick techniques] we built . . . two slide wires and we designed it all. . . . We had a big trampoline, and put it vertically on some poles, so that you would hit the trampoline and bounce off. One time Shepard came out and Grissom and Cooper, and they came out with the pad safety guy. So Shepard says, "Wait a minute. We are supposed to do this, get in the ring and step over the side, and then let goooooo?" And so Safety says, "He can't do that, he can't do that." So Grissom says, "Ja, you're right. I better tell him that." So he grabs one and goes right up there, and Cooper says, "Now, wait a minute. I belong to these guys." So he went. And Safety says, "That's completely illegal. The Air Force will scream. They will shut it down." I said, "I tell you what. I'll go and tell them about it." Well, I hooked up and I went down. So then it became somewhat of a joy ride for some people.

Now that whole system, we put in for about $10,000 [to] $12,000 because we scrounged most of the material or we had it donated one way or another. But then later on, when we went to Apollo from there, I think the slide-wire system became a $1M project. And we then had a cable car which held nine people. [There was] a blast room down below, you know, and so on. But I once asked the helicopters, I said, "Now, if that happens to us and there's a fire down below that we go up on top of the structure, would you come and get us?" And the major says, "Air Force regulations are: we stay 1400 feet away from you and we can wave at you." . . .

The thing that always scared us was, in Gemini, when the Titan [was] flight-pressurized. And we had that glorious BFRC. You're familiar with that term? It stands for "big fucking red cloud." See, whenever they had a leak, they said, "Don't open the elevator door!" It was big red stuff. You're talking about nitrogen tetroxide all around us. . . . [Laughter] They had problems. It gave you a sour taste on your tongue and so on. We didn't think it was that bad. Until we found out if you parked a car close by or if the Security came up and they walked through the clouds, all their badges turned black and the bumpers turned black on the cars. [And] you always could tell the people that worked on the peroxide systems, because they always had white hair in front, even being twenty years old. Because hydrogen peroxide does something like that. . . .

NASA lifers had spent nearly a decade waiting for this very moment in the summer of 1969. They had endured crushing workloads, heartbreaking failures, and too many nights believing that this was one dream that would never come true. For most of the federal government's Space Race employees, in fact, imagining that Apollo 11 would succeed in taking its first crew to the Moon and back home again took more faith than they could draw. They wanted to believe, but agency history had dashed that hope so many times before.

As the mission's countdown began, a Soviet fishing trawler, bristling with radio antennae, floated peacefully in the ocean nearby, just as it had for every NASA liftoff.

Posted across the campus of Houston's Manned Spacecraft Center, meanwhile, were billboards asking every NASA employee:

"WILL YOU BE READY?"

2

The General's Command

On June 12, 1969, at 12:30 p.m. EST, twenty-three days after Apollo 11–Saturn V had first settled onto its pad, the piercingly thin and brutally direct Apollo program director, Lt. Gen. Samuel C. Phillips, chaired a meeting out of his fifth-floor suite at the new L'Enfant Plaza offices of NASA headquarters in Washington, D.C. A dozen executives were arrayed in Phillips's conference room, with two dozen more on squawk boxes across the country, and representatives listening in from McDonnell Douglas, GE, AC Electronics, MIT, IBM, Boeing, Martin Marietta, North American Rockwell, Philco-Ford, Chrysler, United Aircraft, and Grumman. It was this meeting that would decide the question: Would Apollo 11 launch the following month? And if it did, would it attempt to land on the Moon?

General Sam Phillips: "First, I'd like to hear from Lee James in Huntsville on the state of the launch vehicle, then from George Low on the spacecraft hardware and any other concern he may have. Third from Rocco Petrone at the Cape on launch readiness, and from Gene Kranz on flight operations, and finally from Deke Slayton, who is at the Cape, on the crew and their training."

Saturn V Manager Lee James: "At eight o'clock this morning we went through, office by office, a review of the launch vehicle. We found we are probably in better shape than ever before in getting lessons from previous flights. I have nothing that makes me concerned about the July date. If we delay until August, a couple of extra considerations do arise."

Manned Spaceflight Program Manager George Low: "If things don't get any worse, no problem. We have had a lot of meetings. We don't yet understand all the Apollo 10 anomalies. But if anything we are in better shape on the CSM and the LM than we were a month before Apollo 10. We are doing a little more work on the LM thermal protection against the blast [of the rocket exhaust as it descends to the lunar surface], and on the landing gear. We see no reason not to press on; no reason to prevent us making a July launch."

Phillips: "What does the data analysis show on the staging?"

Low: "There is nothing new. We've continued to look at all the data. We hypothesize mixed wiring plus a malfunction." This was a reference to the Lunar Module's instabilities in separating its two sections on leaving the Moon, instabilities that had been so severe on the Apollo 11 dress rehearsal of Apollo 10 that they caused Gene Cernan to cry out, "Son of a bitch!" NASA and Grumman engineers spent countless hundreds of hours investigating the matter, but it would never be completely understood.

George Low: "The Lunar Receiving Laboratory finally seems to be ready. We did review flight operations and flight crew readiness." The suite of rooms at NASA's Houston campus that would house returning and quarantined astronauts and their samples of lunar rock had gone through a chaotic series of management upheavals.

Kennedy's Rocco Petrone: "I agree with George Low and Lee James. All the open work [unresolved issues] we have seen we can handle. We have four and one-half to five days positive slack in the schedule."

As flight director Chris Kraft couldn't attend, one of his lieutenants, Gene Kranz, had been deputized to represent Mission Control's ground team. Kranz: "Basically we are in a very good posture. We are tight in a couple of spots. Tomorrow we have a telecon set up with the prime flight crew. The major open item is joint training with the crew. But we are in good shape for a July launch."

Phillips: "Which is the most critical item in the simulator area?"

Kranz: "We need a one-day tune-up effort, and we need a LM checkout we want to run through once. Basically, the simulations we've got scheduled with the crew are tight, but we have a good set of procedures and flight plans. The crew training with us is ninety percent effective."

Flight crew operations director Deke Slayton: "Our story hasn't changed appreciably. Training is scheduled up to the sixteenth. We have had to compromise in the CSM-LM area. We should have one hundred more hours, but we'll have to fit the training in only half of that. I think we are comfortable with what we've got. The LLTV [Lunar Landing Training Vehicle, a fussy and erratic contraption of metal crossbars, rocket engine, and thrusters, which simulated piloting the Lunar Module under the Moon's one-sixth gravity] is an open area. Neil will fly the LLTV Saturday, Sunday, and Monday, and maybe Tuesday and the following weekend."

Phillips: "Let's explore it. Go back to early this year in assessing what had to be done. LLTV training justified the risk; we all took that into consideration. Neil had flown the LLTV. We would cancel the LLTV if we felt we couldn't get it in shape to fly; we felt we could go ahead without it. I've thought LLTV was highly desirable but so far haven't concluded that it is mandatory. If we find the LLTV can't be flown, what then?"

Deke Slayton: "If we can't fly it this weekend, we are going to have to bite the bullet. The cutoff date is the weekend of the twenty-first. We don't want to get Armstrong's mind cluttered up. There is one other constraint, the outside input. Everybody thinks they've got to talk to the crew."

Phillips: "Has the crew been asked to do anything that's giving you trouble?"

Slayton: "No, except the press conference."

Phillips: "In my opinion, subject to comment by others, it's essential that the crew feel confidence in regard to maneuvers in normal and emergency situations, and in abnormal situations where reaction has to be immediate. There are an infinite number of normal situations to figure out and act upon. We want to be sure in cutting back the original plan of seven or eight months ago that nothing is cut out that would be considered taking a shortcut."

Slayton: "We are about where we were this time before Apollo 8. We are in better shape than with 9, but not so good as 10. 1 have no reservations about the crew being adequately trained."

In fact at that moment, Apollo 11's Neil Armstrong, Buzz Aldrin, and Michael Collins did not themselves feel adequately trained, but they were afraid to admit this to Slayton. "Neil used to come home with his face drawn white, and I was worried about him," Armstrong's wife, Jan, remembered. "I was worried about all of them. The worst period was in early June. Their morale was down. They were worried about whether there was time enough for them to learn the things they had to learn, to do the things they had to do, if this mission was to work."

Slayton did acknowledge that Aldrin wanted more hours for his geology studies, which, the astronaut said, "opened my eyes to the immensity of time." Collins confessed that, "I hate geology—maybe that's why they won't let me get out on the moon."

Phillips then asked what would be gained by postponing to an August launch.

Slayton: "We'd gain some time in flight plans and trajectory. Basically, we'd do what we have been doing already. We'd be more comfortable. Honest to say, I don't think we'd be all that much better off."

Physician Chuck Berry, however, disagreed: "The days turn out to be long ones, and then you have more long ones. It's hard to put in concrete terms, but I have this feeling that they ought not to fly in July. . . . Deke is being straight, but realistically it's not going to happen that way."

Hearing Berry's comments, public affairs chief Julian Scheer muted his mic and turned to his colleagues: "Chuck wants to make the decision."

Phillips then asked every manager who hadn't spoken for his opinion. All said to move forward, save for one subcontractor executive, who cautioned, "I feel a little uneasy."

Phillips laughed. "I've been uneasy for the five and a half years I've been here."

After the launch, Deke Slayton summed up what had happened: "The toughest job we had to train for on Apollo 11 was landing on the moon. All the rest of the stuff—rendezvous, zero-G, extravehicular activity—had been done and done again. . . . Apollo 7 shook out the command module; the Apollo 9 guys had to bite the bullet on the lunar module, and that was a helluva job. Apollo 8 went to the moon and back, so we knew how to do that. Apollo 10 checked out lunar orbit rendezvous as opposed to rendezvous in earth orbit. . . . The thing we hadn't done was land anything, manned, anyplace on the moon. Training for that was something else." Much of the reason it would turn out to be "something else" is that, because NASA had stacked up its launches so tightly, Aldrin and Armstrong couldn't actually begin to use the Lunar Module simulator until May, a mere two months before their launch, since the Apollo 10 crew was still training in it. It would be June 10 before they could start their contingency training, not even six weeks before liftoff.

Apollo 10 was such a detailed rehearsal for 11 that George Mueller, who had initially thought Apollo 8 was going too far, now believed the agency didn't need to wait—10 might as well be the first lunar landing (a position its crew seconded). The Lunar Module's engineering team, though, hadn't yet found the right mix for a vehicle light enough to ensure a successful liftoff on the return home. Even so, NASA felt the need to take a special precaution to keep the 10 crew from preempting Apollo 11. "A lot of people thought about the kind of people we were: 'Don't give those guys an opportunity to land, 'cause they might!'" said Apollo 10's Gene Cernan. "So the ascent module, the part we lifted off the lunar surface with, was short-fueled. The fuel tanks weren't full. So had we literally tried to land on the moon, we couldn't have gotten off."

The five years of unease which General Phillips admitted to was echoed by MSC program manager George Low's daily progress reports:

3 April 1969.

I have a list of 149 Apollo [9] anomalies. Considering that this was a perfect flight, I wonder what would have happened if we had had a bad flight.

Buzz flew 79 parabolas in the KC-135 ["Vomit Comet"] yesterday—a new record. Everybody got sick except Buzz.

4 April 1969.

Apollo 9 anomalies . . . My overall list . . . is getting longer instead of shorter. . . .

Lunar Exploration Program meeting. I thought things were going well until Dr. Wise from Hq. indicated in effect that none of the things we were proposing would lead to useful lunar science. I felt I had to walk out since otherwise I might have said things I would have regretted later.

21 April 1969.

Apollo 11. Rash of problems, esp. on LM side. Punctured a small hole about 3 weeks ago, but by the time we'd fixed it we had punctured 2 more holes, and we now have patches on patches on patches. Then we allowed water to enter [the] glycol loop—we decided to drain, evacuate and refill entire system. Will take 5 days. Then an arc was struck against a water glycol line, burning a small hole in that line.

23 April 1969.

Apollo 11. Procedural problem ruined all three fuel cells in CSM 107. Need astronaut motivation work among people working on Apollo 11 to re-emphasize importance of doing everything right. [The NASA technique of astronauts' making personal appearances at factories to remind employees that a human being would be relying on their efforts had been found time and again to dramatically boost the quality of workmanship.]

24 April 1969.

Flight Crew Support Activities. Weekly review of Warren North's schedule discouraging. We have found many times where changes that were necessary for one flight never were applied to succeeding missions. We have tried to coax Warren along by showing him how to develop a good systematic approach to what he has to do, just as Chris Kraft has in his operation and we have in ours. Today I somewhat deliberately lost my temper; perhaps I achieved my purpose. . . .

26 May 1969.

Very little time between now and Apollo 11.

CSM: Forward hatch insulation will be removed; fuel cell sorting problem we may never understand; the fuel cell condenser exit problem is being analyzed.

LM: no final answers yet on S-band comm. difficulties, the Gimbal Drive Actuator; and the cabin pressure lost.

To a civilian outsider, the combination of Low's notes and the pessimistic comments at the Phillips teleconference might seem to suggest one answer: Postpone the mission. To NASA executives, however, work was progressing about as smoothly as it ever had. For ninety minutes, the other executives discussed Apollo 11's final, less crucial details. Then, General Phillips made his decision:

"Go."

3

Anything but What He Is

On December 23, 1968, as Frank Borman, Jim Lovell, and Bill Anders were about to circumnavigate the Moon aboard Apollo 8, Deke Slayton asked Borman's command backup, Neil Armstrong, if he wanted to lead Apollo 11 and, if so, did he want to include another Apollo 8 backup crewman, Buzz Aldrin? Armstrong answered yes to both questions. Deke then asked if, for the third crewman, Armstrong would prefer Fred Haise to pilot the Lunar Module, or Michael Collins on Command Module, with Aldrin slotted into the remaining position, having trained for both? Armstrong talked things over with Collins (who'd just been restored to flight status after undergoing a risky spinal operation) and decided he would get the assignment.

The three were happy to be slotted for a mission but at that moment, no one yet believed that this would be *it*. "I suspected that it was highly unlikely that Apollo 11 would in the final analysis be the first lunar landing flight," Armstrong said. "The Lunar Module had not yet flown, and there were a lot of things about the lunar surface we didn't know. We didn't [even] know if Mission Control could communicate with the Lunar Module and the Command Module simultaneously and successfully. . . . We didn't know whether the radar ranging would work. . . .

"I guess there's some thrill to being first to do something and most of our guys in the program so far have been the first at something, just because there's so few of us and so much to be done for the first time. Of course the first to land on the moon—why that's a considerably bigger thing—but I would probably have to agree with those that said in this feat who the person is is sort of happenstance. . . . It's not the same sort of thing as when Lindbergh crossed the ocean. . . . [That was all] based on his own ideas and his own techniques and his own accomplishments. That's not the sort of a thing this is." Collins was then equally skeptical, putting the odds of Apollos 10, 11, or 12 being first on the Moon at 10, 50, and 40 percent, respectively.

When Buzz Aldrin heard he'd been selected for Apollo 11, his reaction was not what anyone would have expected. Aldrin hoped to be part of a mission of scientific breadth, where his MIT Sc.D. might mean something, and was unnerved by the celebrity that would inevitably result from being one of the first men on the Moon. Her husband was always a loner, but Joan Aldrin knew something was up when Buzz grew even more distant than usual in January. When he finally broke the news, she wasn't sure whether to be thrilled or afraid; Mrs. Aldrin finally had to break her own worry and tension through a giant effort of housecleaning and painting.

During Buzz's Gemini mission, Joan had wondered if "our marriage wouldn't be the same, that it would be so much more magical and meaningful and magnificent because he'd done this wonderful thing." That did not happen. "At first I was disappointed, and then it was comforting to think that it hadn't changed him," Joan finally decided. "He's not the same person I married, but I don't believe that I am the same person, either."

"Having experienced one space flight previously," Buzz Aldrin said, "and the hometown parade and the speeches and everything around me, I didn't relish that part of being an astronaut. I knew that would be peanuts compared to what was going to happen after this flight. And that's why I had an inkling of a thought that if there was a way to avoid that postflight stuff, maybe it would be better to go on a later mission. . . .

"Neil Armstrong, one of the quietest, most private guys I'd ever met, has often been described as taciturn, but that's an understatement. Neil was a man from rural Ohio who'd worked his way through a career in aviation and spaceflight, carefully watching everything he did and said. His family was his social life. He was not the hard-drinking, fast-driving 'right stuff-er' the public seemed to think all the astronauts were. Mike Collins was not quite as quiet, but he was hardly what you'd call flashy. None of us was going to have an easy time with the public relations part of our mission."

Buzz told Joan that he was seriously contemplating resigning from Apollo 11 . . . but then, no astronaut had ever turned down a slot, and doing so might mean the end of his career at NASA. Joan wrote in her diary: "Broke out in blotches last night, which still persist today. I'm covered in pancake makeup and jumpy. Nerves. If I'm like this now, what will I be like when it really happens? . . . I wish Buzz was a carpenter, a truck driver, a scientist— anything but what he is."

The decision of which man would be first to set foot on the Moon turned out to be a long and contentious process, repeatedly mishandled in public, and exactly what happened is still hazy. Among a number of press reports, Arthur J. Snider wrote in the February 27, 1969, *New Orleans Times-Picayune* that "the flight plan as now drawn calls for Aldrin to climb down the ladder from the lunar module shortly after touchdown. He will immediately inspect the long-legged spacecraft for any damage and remove equipment from bays on its exterior. Forty-five minutes later the Apollo commander, Neil A. Armstrong, will descend and join him. The disclosure of Aldrin as the choice comes as a surprise to many who had speculated that the top commander would be en-titled to pull rank and take his place in the history books as the first man to set foot on a satellite of the earth. But the space agency official said that the decision is not Armstrong's to make."

The procedure described by Snider is in keeping with a long navy tradi-tion of not having a ship's commander be the first to enter unknown terri-tory; due to the long relationship between the USN and NASA's predecessor, the National Advisory Committee for Aeronautics (NACA), there's a lot of navy in NASA's genetic code. In March, manned spaceflight associate ad-ministrator George Mueller confirmed to a group of reporters that Buzz Aldrin would be first out. At the same time, however, Deke Slayton told Aldrin that he assumed Armstrong would be first, since he had seniority in the corps.

The confusion finally forced a decision on the agency's executives. "I thought back to the intense and private discussions we'd had about who should be the first man on the moon," flight director Chris Kraft said. "In all the early flight plans and timelines, it was the lunar module pilot. Buzz Aldrin desperately wanted that honor and wasn't quiet in letting it be known. Neil Armstrong said nothing. It wasn't his nature to push himself into any spotlight. If the spotlight came, so be it. Otherwise, he was much like [manned spacecraft chief] Bob Gilruth, content to do the job and then go home."

This wasn't the first time Aldrin would suffer for his lack of political fi-nesse. After he'd explained to astronaut chief Slayton in great detail how his MIT doctorate and continuing studies in rendezvous and docking made him crucial to the success of Project Gemini, Slayton decided not to slot Aldrin to fly in the program at all. Later, when Aldrin tried giving Frank Borman some advice, Borman replied, "Goddamn it, Aldrin, you got a reputation for trying to screw up guys' missions. Well, you're not going to screw up mine."

"[Buzz] came flapping into my office at the Manned Spacecraft Center one day like an angry stork, laden with charts and graphs and statistics, arguing . . . that he, the lunar module pilot, and not Neil Armstrong, should be the first down the ladder on Apollo 11," Gene Cernan remembered. "Buzz had pursued this peculiar effort to sneak his way into history, and was met at every turn by angry stares and muttered insults from his fellow astronauts. How Neil put up with such nonsense for so long before ordering Buzz to stop making a fool of himself is beyond me."

"Neil, who can be enigmatic if he wishes, was just that," Aldrin said. "Clearly, the matter was weighing on him as well, but I thought by now we knew and liked each other enough to discuss the matter candidly. Neil equivocated a minute or so, then with a coolness I had not known he possessed he said that the decision was quite historical and he didn't want to rule out the possibility of going first. I was more surprised by the manner of his reply than by what he said, for what he said did have logic. I kept my silence several more days, all the time struggling not to be angry with Neil. After all, he was the commander and, as such, the boss."

"I thought about it," Chris Kraft continued. "The first man on the moon would be a legend, an American hero beyond Lucky Lindbergh, beyond any soldier or politician or inventor. It should be Neil Armstrong. I brought my ideas to Deke, and then to George Low. They thought so, too."

Other executives arrived at the same decision from different paths. Deke Slayton: "We had procedures guys to score a couple of timelines like that. Their logic was trying to split the workload between the LMP [lunar module pilot] and the CDR [commander], and they figured the CDR was going to be overly worked if they got him out there first, and they were working things around in that direction. That didn't sound right to me, based on the configuration of the spacecraft . . . and secondly, just on a pure protocol basis, I figured the commander ought to be the first guy out." Collins remembered that on Apollo 9, LMP Rusty Schweickart "had to crawl all over CDR [Jim] McDivitt" to be first out of the LM.

At the time, NASA's explanation for its decision was based on the Apollo 9 troubles—the Lunar Module's hatch design meant its mission commander would be forced to exit first. Many at the agency, including Michael Collins, believed there was another side to the story: "Originally, some of the early checklists were written to show a copilot first exit, but Neil ignored these and exercised his commander's prerogative to crawl out first," he said. "This had been decided in April, and Buzz's attitude took a noticeable turn in the direction of gloom and introspection shortly thereafter. Once he tentatively approached me about the injustice of the situation, but I quickly turned him off. I had enough problems without getting into the middle of that one." Collins would later acknowledge that he was not completely sure about the details of this account, but others had similar memories. Contamination control officer Mike Reynolds:

> The story told to me was, "Okay, when you land on the surface of the Moon, what are you going to do?"
>
> Armstrong said, "I'll check this. We'll check this," and they had a check-off list.
>
> "Okay. Buzz, what are you doing to do?"
>
> And Aldrin says, "Well, when Armstrong gets done, I'll open up and I'll egress." He said it went absolutely quiet.
>
> Armstrong looked around, and he said, "Buzz, I'll be the first one out." That was the end of the question. What people don't realize is that these guys were all military, all but Armstrong, and they understood the chain of command.

Buzz's father would in fact spend the ensuing decades trying to make his son "first on the moon" in history. If Armstrong and Aldrin had followed Edmund Hillary and Tenzing Norgay's Everest precedent, and arranged to touch down together, it might have made both of their lives in the aftermath of Apollo 11 a lot easier.

Apollo 12/Skylab 3's Alan Bean:

> When you're getting ready to go to the moon, every day's like Christmas and your birthday rolled into one. I mean, can you think of anything better? . . . Things were moving so fast and everything was changing all the time. Nothing ever seemed to go like you thought it was. The flight wouldn't go right some way, some other flight, people would be changed from the crew. Everything was in a state of flux, always. . . .
>
> Another memory is just your not knowing anything about what's going on in the rest of the world. Vietnam took place, the racial unrest and all that stuff. I can remember getting ready to fly out to California one time, and as we're leaving Houston, someone says, "Don't fly around the Watts area when you're making your approach to LAX." "Why not?" "Well, they've got a riot going on down there." "Oh, okay."
>
> We had an astronaut party once a month, and usually it consisted of conversation that you had at work except wives were there at the time, and they probably thought, "These guys never talk to us, or if they do, we don't know what they're talking about," because we were talking about these same things

over and over again, trying to figure them out. And there was always the controversial items. You'd solve them or make a decision, like, "Do we have radar or some other kind of tracking on the lunar module?" Maybe that went on for six months. Some others would be, "Do we do a one-orbit rendezvous or three?" You'd study it. You'd think about it.

Everybody would have an opinion, and then finally it would get decided, and when it did, then you would say, "Okay. We're doing that. Now what do we have to know as a result of this?" "Okay, now we've got to know how many times this should strobe." You see, it was a building. You had to solve these big problems, everybody had to agree, because they had to know their little part and work with the big solutions. Then it was the next layer, then the next layer, and finally, just before you do the mission, you solved the last layer, and you'd do that, and then you'd go do the mission and see if this would work, and if it would, then you could do the next layer. But all these others had to be working along the way. So it was an all-encompassing time of making an impossible dream come true.

As the men of Apollo 11 would soon learn, however, not every day of preparing for so historic a mission would be celebratory. The chore that astronauts hated more than any other began in earnest on July 5, 1969: a fourteen-hour endurance marathon with the general press, the wire services, magazine journalists, and television broadcasters—the very press conference that Deke Slayton had complained about at General Phillips's teleconference.

Mike Collins, Buzz Aldrin, and Neil Armstrong appeared, early that morning, at the Manned Spacecraft Center's visitor theater wearing gas masks before six hundred journalists, rustling and nervous in their rows of orange plastic seats. Like anywhere in the NASA tropics of Houston, Louisiana, Mississippi, Alabama, and Florida, outside, it was very wet and ninety degrees plus; inside, it was dry as a bone, and sixty-two. Flanked by the medallions of their agency, their mission, and an American flag, the crewmen climbed the stage and entered a big plastic box, which had fans blowing air and press germs away from them, part of a quirky and erratic quarantine procedure. Once safely inside this box, they took off their masks.

After the rapt euphoria of John F. Kennedy's Project Mercury, historians and journalists have had an uneven time with NASA and its employees. A central reason for the creation of the federal space agency was to prove to both the citizens of the United States and the world at large that the American way of life was superior to totalitarian communism. Public relations, on a massive and global scale, was at the very heart of the enterprise. NASA's overseas PR, in fact, would reach a hundredfold success in the aftermath of Apollo 11, creating a worldwide surge of admiration for the United States, a dazzling peak of global prestige that has never been equaled.

For most NASA fliers, however, working with the media was simultaneously a glorious ego boost and excruciatingly oppressive, as enjoyable as having one's intimate thoughts and personal feelings examined as if they were insects under dissection. Astronauts would become famous through the press, and famous to the press for their loathing of the press, a sentiment shared by the great majority of the agency's other employees. Many NASA workers were immersed in principles of celestial mechanics, in work so astronomically complicated that explaining it to a layman was more trouble than it was worth. Abetting them was a military infusion of pilots and contractors, men who believed with all their hearts that the less said, the better (an attitude epitomized by Apollo 11's commander). The agency's biggest celebrities, its combat-test-pilot astronauts, were notoriously laconic, epitomized by the story of the fighter pilot on his radio (where the cardinal rule is to only transmit absolutely crucial information) who kept shouting that he was about to be shot down: "I've got a MiG at zero! A MiG at zero!" Another navy man cut in to say: "Shut up and die like an aviator."

Even though the agency was founded on principles of openness and transparency in deliberate contrast with Soviet clandestine belligerence, the men and women of NASA had an almost innate anti-press bias. During the first American launches, "We reporters would watch from the beach, but they wouldn't let us in," space correspondent Jay Barbree recalled. "But we kept banging at the gates and kicking at the fences until they said, 'You're more of a nuisance outside than you would be inside,' so they finally let us in." Other reasons for institutional secrecy were NASA's fears of revealing to Congress and taxpayers just how risky its missions actually were; the astronauts' exclusive contract with the relentlessly anodyne *Life* magazine; the stereotypical engineering attitude of being more interested in things than in people; and the grave hurdle of NASA jargon, a vernacular that combined the studied flair of pocket-protecting engineers with the delicate nuance of Pentagon bureaucrats.

Regardless their lack of interest in providing detailed description and anything that even resembled small talk, NASA's military employees, pilots, and engineers came from long-standing traditions of communication that virtually parodied human speech. Though it was not as explicitly humorous as the United States Air Force's brand of space lingo—which called one program

"Man in Space Soonest" (MiSS) and a spacecraft, Dyna-Soar—novices to the American space agency were lost without a translator:

NASA	Civilian
enable	turn on
disable	turn off
peripheral secondary objectives	other choices
obtaining maximum advantage possible	doing our best
very high confidence level	confident
concentric sequence initiation	orbit
terminal phase initiation	dock
eat cycle	meal
integrated thermal meteoroid garment	space suit
normative	as expected
contingency	astronauts about to die

The July 5, 1969, press conference was a near-perfect display of NASA's fundamental PR conundrum—having to talk, but hating to talk—which the agency's executives tried to address by relying more and more extensively on their universally loved celebrities, the astronauts, in something of a bait and switch. On joining the corps, NASA fliers were required to be the finest test pilot–engineers that America could produce. After taking part in successful missions, however, they were expected to become mediagenic spokesmen, delightful after-dinner speakers, statesmen who always knew exactly what and what not to say, philosophers with profound notions of the cosmos, poets with lyrical talents and extraordinary sense memory for describing their journeys, and all-around agency pitchmen. But because pilots and engineers, especially great pilots and engineers, are not inclined to engage in crowd-pleasing banter and evocative descriptions of earthshine, it's almost shocking that more of them didn't end up as dazed and confused as Buzz Aldrin after his return from the Moon. When Gemini 5/Apollo 12's Pete Conrad, for one example, was asked what traveling to the Moon was like, he said, "Super! Really enjoyed it!" That any of them (most notably John Glenn) succeeded in both capacities is a startling achievement.

Mike Collins tried to explain this conflict: "Being a military test pilot was the best background from a technical point of view, but was probably the worst background from a public relations or emotional point of view. We were trained to transmit vital pieces of information. If someone had said from the ground to me in space, 'Well, how do you feel about that,' I would've said, 'What? Huh? I don't know how I feel about that, you want the temperature, you want the pressure, you want the velocity, you want the altitude, what do you mean, how do I feel about that?' It was not within our ken to share emotions or to utter extraneous information."

"The tension between the astronauts and the public affairs office was a continuous problem," public affairs officer Col. John A. "Shorty" Powers admitted. "I think all seven [Mercury] guys really enjoyed the exposure—they are human and they didn't mind seeing their names in the papers. Yet as test pilots, they instinctively rebel at having to spend time talking to the news media. The story that comes out of an interview does not always suit the man who gave the interview and they didn't understand that. They resented me because I represented the news media. . . . One of the classic comments that several of the guys made during the early flight program was in reply to a question from the news guys as to what the toughest part of the flight was. The astronauts' reply was the press conference."

Matching NASA's deficiencies in working with journalists, however, were the journalists' own array of shortcomings. Those who weren't woefully uneducated, disrespectful, or heedless invaders of privacy (with photographers the most egregious offenders) were perhaps the most difficult of all: the hardcore space cultists. Norman Mailer (who himself had a Harvard degree in aeronautical engineering) described his fellow journalists at Apollo 11's July 5 press event as "a curious mixture of high competence and near imbecility; some assigned to Space for years seemed to know as much as NASA engineers; others, innocents in for the big play on the moon shot, still were not just certain where laxatives ended and physics began. . . . everybody was a little frustrated—the Press because the Press did not know how to push into nitty-gritty for the questions, the astronauts because they were not certain how to begin to explain the complexity of their technique." If journalism is the first draft of history, these disparate forces would collide to form a very rough and erratic initial take.

Michael Collins: "What [the press] really wanted to know was: beyond all that technical crap, what did the crew feel? How did it feel to ride a rocket, what thoughts were racing through your mind as you plummeted toward the sea with the parachutes not yet open? How scared were you, anyway? This is what *Life* paid to find out, and what the others pried to find out without paying, and in truth, neither unearthed very much. . . . As technical people, as test pilots whose bread and butter was the cold, dispassionate analysis of complicated facts . . . it didn't seem right somehow for the press to have this morbid, unhealthy, persistent, prodding, probing preoccupation with the frills, when the silly bastards didn't understand how the machines operated or what they had accomplished. It was like describing what Christiaan Barnard wore while performing the first heart transplant." While understandable, the astronauts' collective inability to describe what they were experiencing in detail would unfortunately prevent the rest of the world from sharing with them the rapturous joy of discovery awaiting them on the Moon and across the heavens.

Neil Armstrong was a test pilot's test pilot, an aeronautical engineer's aeronautical engineer, and a loner's loner, who baffled writers and journalists—not to mention his colleagues—for over four decades. All the journalists at the July 5 prelaunch gathering wanted from him, really, was a nice comment or two that would make a good headline or pull-quote. Armstrong would not, or could not, comply. Buzz Aldrin, a famously snazzy dresser for an astronaut (a group whose collective taste in Ban-Lon had much to do with military pay scales), had worn an incandescent green suit to the conference which, combined with his thorough grasp of the mission's technology, helped him somewhat overcome his fear of public speaking and compensate for Armstrong's glacial distance. Still, Aldrin preferred to communicate in full-blown NASAese. Instead of offering direct quotes for the papers, everything he said ultimately had to be translated for the folks back home: "phasing maneuver . . . descent orbit . . . executing this burn . . . inputs in terms of altitude and velocity updates . . . trajectory conditions."

Michael Collins was, as always, the most comfortably outgoing of the three, but then, as he wasn't going to the Moon, only one journalist at the conference bothered asking him a question: What is it like *not* going to the Moon? Collins: "I'm going 99.9 percent of the way there, and that suits me just fine. . . . I couldn't be happier right where I am." Since that event forty years ago, Collins has expressed this sentiment in different ways, any number of times. No one has ever believed him. An astronaut sharing Collins's fate admitted, "I wanted to go with them so bad I could taste it. . . . I wish the damn thing could hold three people." Perhaps if the press that day had known that Collins had effectively resigned from NASA before liftoff, they would have had a remarkable story to pursue. But no one at the time could have imagined such a thing.

When asked what he would do if the Lunar Module's ascent engine didn't work, Armstrong said (with the long pauses that are a part of his normal speech pattern, that of a man who considers every single word before saying it), "Well . . . that's an unpleasant thing to think about. . . . We've chosen not to think about that . . . up to the present time. We don't think that's at all a likely situation. . . . It's simply a possible one." Actually, Armstrong later admitted that he "had proposed many months earlier—maybe even years earlier—that we just put a big manual valve in there to open those propellant valves rather than, or in addition to, having all the electronic circuitry," as he described it. "But management didn't think that that was up to NASA's standards of sophistication."

Armstrong was then asked, "Will you keep a piece of the moon for yourself?"

"No, that's not a prerogative we have available to us."

What extra item would he take with him, if he could take along anything at all?

"More fuel." Armstrong's dry humor was another of his qualities that few journalists at the time captured.

Did he have any worries about the rocket?

"We're quite sure this girl will go."

Why, really, were they doing this?

"I think we're going to the moon because it's in the nature of the human being to face challenges. It's by the nature of his deep inner soul. Yes, we're required to do these things, just as salmon swim upstream."

All throughout the press conference, all three men projected what they called "very high confidence levels," as though confidence were a form of human propellant. But perhaps for astronauts, it is, and this posture of unmitigated ease—that flying a rocket is nothing special, that going to the Moon is just another job—is a stance historically common to NASA employees facing public scrutiny, an invocation of the agency's shield of prodigious technical competence, along with its corollary: "There's nothing to worry about—NASA's in charge."

In fact, the number of concerns about this first voyage to another celestial body was as infinite as dark matter. The most striking unknown facing Apollo 11 was: What, exactly, was the Moon's surface made of? Though a very few lunar probes had managed to touch down and return pictures and chemical composite details, it was still unclear whether or not the surface would support the weight of a grown man, much less two of them and their spaceship. Was the Moon covered in an immensely deep and light stratum of dust that would swallow up a craft and its passengers whole, as one particularly vocal selenologist (lunar scientist), Thomas Gold, claimed? Were its mountains and rilles as fragile as spun sugar, with a rocket landing certain to trigger landslides and devastation? While most NASA scientists did not subscribe to these exotic speculations, they weren't 100 percent certain that such scenarios weren't possible, either. Those who had to plan, design, and engineer Apollo simply had to move forward with their best hunches. Design engineer Caldwell Johnson: "Owen [Maynard] and I got together one morning and we said, 'It's got to be like Arizona! The moon has just got to be like Arizona! Can't be nothin' else. So let's design a landing gear like it was.'"

Why did we know so little about the Moon before sending two men to land on it? Due to the terrible history of American lunar probes:

August 1961: NASA sends Ranger 3 to survey the Moon. It misses by 20,000 nautical miles.

November 1961: Ranger 4 lands on the Moon, but electrical failure ends the mission before any data can be returned.

Ranger 5: misses by 420 nautical miles.

Ranger 6: TV camera fails.

July 31, 1964: Ranger 7 crashes into the Sea of Clouds, but broadcasts back pictures for the entirety of its descent. Its images of boulders convince many that the surface would support a ship. Many others claim that they mean nothing of the sort.

May 30, 1966: Surveyor 1 lands successfully, producing a panorama landscape of a flat plain littered with craters and rocks.

August 10, 1966: Lunar Orbiter 1's motion compensator anomaly leads to blurred, unusable pictures.

November 18, 1966: Orbiter 2, another success, relays detailed pictures of potential landing sites.

Surveyor 2: navigation system fails, and the satellite vanishes.

April 20, 1967: Surveyor 3 lands in the Ocean of Storms, but shuts down after a single night.

Surveyor 4: radio fails.

September 11, 1967: A mere two years before Apollo 11, Surveyor 5 touches down in the Sea of Tranquility and radios back that the soil is composed of calcium, aluminum, magnesium, silicon, and oxygen, which means basalt originating in lava flow, but unlike the Earth's version, it contains high amounts of iron and titanium. A geologist would use these findings to predict accurately what Armstrong and Aldrin would find on the Moon, but his article wouldn't be published until after they'd already come back home with their own samples.

If there was uncertainty about the lunar surface, there were even greater qualms about the voyage as a whole. If NASA had so much trouble getting a simple drone to touch down on the Moon, how were they ever going to send two men? Though Apollo 11 is commonly believed to have been a "perfect" mission, so many things in fact went wrong that Kennedy Space Center director Jay Honeycutt later admitted: "I'll tell you, it would have been damn easy to abort that mission. Damn easy." Perhaps the best insight into how NASA workers coped with the risks of exploring outer space was revealed by remarks Guenther Wendt made, a few days before John Glenn's Mercury launch, to the astronaut's nervous wife: "Annie, we cannot guarantee you safe return of John. This would be lying. Nobody can guarantee you this—there is too much machinery involved. The one thing I can guarantee you is that when the spacecraft leaves it is in the best possible condition for a launch. If anything should happen to the spacecraft, I would like to be able to come and tell you about the accident and look you straight in the eye and say, 'We did the best we could.' My conscience then is clear and there is where my guideline is."

Saturn V's overseer, Wernher von Braun, however, had a very different attitude toward the dangers of spaceflight: "Going to the moon is a picnic, a trifle, a party trick . . . fifty percent of the risk [to the astronauts] is that they'll die in a car crash here on earth; they drive like madmen." And when asked, "What is the purpose of going to the Moon?" he said: "What is the purpose of a newborn baby? We find out in time."

Neil Armstrong agreed: "I always felt that the risks that we had in the space side of the program were probably less than we [had] back in flying at Edwards [Air Force Base] or the general flight-test community. The reason is that when we were out exploring the frontiers, we were out at the edges of the flight envelope all the time, testing limits. Our knowledge base was probably not as good as it was in the space program. We had less technical insurance, less minds looking, less backup programs, less other analysis going on."

The Apollo 11 crew had been allowed to spend the July Fourth weekend in Houston with their families. Then, after that fourteen-hour press day on the fifth, they were sequestered on the third floor of Kennedy's Manned Spacecraft Operations Building in a windowless suite of rooms equipped with air purifiers and filters to block contaminants. The quarantine served two purposes. A flu, or even a minor head cold, could mean delaying launch for a month, which had happened on Apollo 9. After the flight, if Armstrong or Aldrin returned from space carrying a lunar virus or bacterium, NASA doctors could identify it by comparing their pre- and postflight bloodwork. With all the efforts expended on the quarantine, however, there were two severe lapses that compromised the entire process. The first was that the crew was in daily contact with NASA executives, various technicians, and secretaries erratically restricted by quarantine, a blurring of policy that would, in turn, lead to a political uproar. Director of medical research and operations Chuck Berry:

> I got a phone call from an Associated Press reporter that was one of the guys that I knew that was following the space program all the time, was always at press conferences. He called and he said, "Chuck, I just got word that President Nixon is going to have dinner with the crew the night before launch in the crew quarters." And he said, "What do you think about that?"
>
> I said, "I don't know anything about it."
>
> He said, "Well, how is that possible with your quarantine program?" . . .
>
> So here I am in a very bad spot. I said, "Well, it doesn't fit with the quarantine program."
>
> Well, I arrived at the Cape and there's already newspaper headlines, "Dr. Berry Prevents President from Having Dinner," and all hell broke loose. I guess that's as close as I ever came to getting fired in my life. I had phone calls from everybody. It was embarrassing to the president because he had been led down a primrose path, and he was madder than hell. There isn't any doubt about it, he was mad, because it didn't look very smart.
>
> I had been charged by NASA to say that we were indeed not going to bring back lunar plague. That was President Johnson's decision after we had tried to convince them that we didn't think any organism could survive in that environment, but the Academy of Sciences, in their wisdom, said, "Well, yeah, but you don't have any data to prove that. You haven't anything to really prove that at all."
>
> So we ended up developing a program using a plague model. If any of us were having any contact [with the Apollo 11 crew], we know everything about us. We've had samples continually. We're sampling everything about us. If they came down with anything, whatever it was, a cough, a sniffle, or anything else, we were going to have to prove that it didn't come from the Moon. So I think it would be pretty stupid to let somebody just walk into that situation. It would have been a total breakdown of the program.

On July 4, 1969, the CIA's Corona orbiting spy satellite had returned a series of photos disclosing that the Soviets had brought an enormous rocket to the pad of their Baikonur Cosmodrome. Were they also preparing to go to the Moon? There was so little information on the Soviet program that nearly anyone in a position of authority at NASA or the Department of Defense still considered the Space and Missile Race an extremely close call. If the Soviets suddenly announced they had established a Moon base, after all, it would not have been any more surprising than many of their previous achievements, from Sputnik to Laika and Gagarin.

On one orbit, the Corona's pictures revealed the giant N1 rocket and its spacecraft, the L3, on the pad being tanked with fuel. When the satellite returned in its orbit to take another series of pictures, however, the rocket had vanished, as had the launch pad's lightning towers. Its turning tower gantry had been blown off its rail track and the crossbeams holding the rocket above its flame ducts were missing. Instead, there was a strange blur, and a scar upon the ground.

Later the Soviets would reveal that there had been an electrical short. When fuel in stage three had subsequently ignited, it blew apart the fuel lines of LOX, causing a fire that spread to consume the three thousand tons of propellant. The greatest fear of everyone who works with rockets came true again on July 4 at Baikonur: a never-ending cascade of fire, smoke, and explosion as a giant rocket collapsed upon itself and died.

Had Apollo 11 failed, it turned out, the Russians were planning to immediately use this rocket to send cosmonauts to the Moon.

On July 10, NASA administrator Thomas Paine flew to the Cape to have dinner with the 11 crew. He knew that they were courageous, and that they felt they had something to prove on this mission—an attitude that worried Paine. The dinner, accordingly, was not the typical "We're Counting on You, Boys" ceremony. Paine had something genuinely important to say to them: "If you get into trouble up there, do not hesitate to abort. Come on home. Don't get killed. If you do have to abort, I promise this crew will be slipped ahead in the mission sequence. You'll get another chance. Just don't get killed."

4

The Sons of Galileo

In point of grazing, plunging, oblique, or enfilading, or point-blank firing, the English, French, and Prussians have nothing to learn; but their cannon, howitzers, and mortars are mere pocket-pistols compared with the formidable engines of the American artillery. This fact need surprise no one. The Yankees, the first mechanics in the world, are engineers—just as the Italians are musicians and the Germans metaphysicians—by right of birth.

—Jules Verne, *From the Earth to the Moon*

The founding brotherhood of NASA's astronaut corps was drawn exclusively from military pilots, a directive of President Eisenhower's, since these men already had national security clearances—and, if it turned out that, in fact, no one *wanted* to be an astronaut, they could be drafted. "It was one of the best decisions he ever made," manned spaceflight chief Robert Gilruth said. "It ruled out the matadors, mountain climbers, scuba divers, and race drivers and gave us stable guys." (It also, unfortunately, ruled out a dozen outstanding female pilots who had been invited to apply, and who had already passed NASA's physical tests.) Many agency engineers grew to appreciate military-trained fliers, since one thing they were trained in was how to be perfectly obedient. "I like fighter pilots," said Mike Collins. "They're independent, they say what they mean, they prove who they are by what they do. They have nice parties and they let their ids hang out. They're good people."

Many NASA fliers have gone out of their way to repudiate their *Top Gun–Right Stuff*–daredevil-cowboy reputation, and for good reason. Part of the distaste for journalists shared by almost every astronaut is the stereotype that they have fostered of pilots as being nothing but stick-and-rudder good-time

Charlies. Though there is an element of truth in that characterization—Chuck Yeager did, after all, break the sound barrier while suffering chest pains from ribs broken in a horseback accident the day before his historic flight, while Pete Conrad lost his life in a motorcycle accident—the press seemed to willfully ignore the fact that the symbol for where that first sonic boom took place—NASA's High Altitude Flight Test program at Edwards AFB, home to both Yeager's X-1 and Armstrong's X-15—was a slide rule floating above a blue horizon. What did both Neil Armstrong and Buzz Aldrin take with them on Apollo 11? Slide rules.

What most journalists at the time and, in turn, the general public also did not understand was that NASA did not just hire military pilots for its astronauts: they were military *test* pilots. "Fighter pilots can afford to be irresponsible and impetuous, and test pilots can't," Mike Collins said. "Test pilots have to be older, smarter, steadier or they'll make a wrong judgment on an airplane and someone will kill himself later. The old thing about, you know, white scarf trailing, put her in a power dive and see if the wings come off—it's not true. The test pilot has to be more of an engineer, studying charts and graphs for the airplane's limits."

Because these airmen commandeered craft that weren't ready to be handled by ordinary fliers, they had a dramatically different attitude from civilians about the risks of space travel. During Mike Collins's eleven weeks of advanced training at Nellis AFB, twenty-two students were killed. Nearly a quarter of noncombat navy pilots (and 56 percent of those who eject) die every year, while the generic occupation "pilot" is ranked as America's second most dangerous job, after fisherman. As Charlie Duke pointed out, "You're flying machines. And machines break."

"When Deke was a test pilot, I was surrounded by widows," Marjorie Slayton remembered. "Most of them spent their time comforting whoever was comforting them." Yet, as one flier explained, "Being a test pilot is more dangerous than going up in the Mercury spacecraft or the Apollo spacecraft. They take infinite precautions with the spacecraft, not many with planes."

To civilians witnessing the opening salvos of the Space and Missile Race, astronauts were living exemplars of American boldness and courage, as they were brave enough to ride what was essentially an enormous, untested bomb. John Glenn tried to explain the immense national euphoria at his matching the historic achievement of the Soviets by being the first American in orbit: "People are always fascinated by anything new, new work, new explorations, especially if one risks losing his life by them. Risk always rouses their imagination. And, for good or ill, space flights are risky. And then there's the fact of having to face the mysterious, the unknown, of experiencing what no one has ever experienced."

Neil Armstrong, however, had the opposite reaction: "For heaven's sake, I loathe danger, especially if it's useless; danger is the most irritating aspect of our job. How can a perfectly normal technological fact be turned into adventure? And why should steering a spacecraft be risking your life? It would be as illogical as risking your life when you use an electric mixer to make yourself a milkshake. There should be nothing dangerous about making a milkshake and there should be nothing dangerous about steering a spacecraft. Once you've granted this concept, you no longer think in terms of adventure, the urge to go up just for the sake of going up. . . .

"I don't understand the ones who are so anxious to be the first. It's all nonsense, kid stuff, just romanticism unworthy of our rational age. I rule out the possibility of agreeing to go up if I thought I might not come back, unless it were technically indispensable. I mean, testing a jet is dangerous but technically indispensable. Dying in space or on the Moon is not technically indispensable and consequently if I had to choose between death while testing a jet and death on the Moon, I'd choose death while testing a jet."

At the start, NASA scoured 508 service records and invited 110 fliers to apply. This group was then subjected to physical and psychological testing (from such endurance tests as treadmill running and balloon blowing to "Write twenty answers to the question, Who am I?") and winnowed down to select the first Mercury astronauts, such iconic figures in American culture at the time that they were globally known by their *Life* magazine honorific: the Original Seven. In 1962, nine more were admitted ("The New Nine," per *Life*), followed in 1963 by an additional fourteen ("The Fourteen"). Average age, 38.6; height, five feet, ten inches; weight, 160. Nineteen brunettes, seven blonds, two redheads, and one black-haired; sixteen blue-eyed, eight brown, five green. Twenty-two were firstborn in their family, and five more were eldest sons; seven were left-handed.

While an agency female employee once said that she could describe them all with one word—*Romeos*—Pete Conrad saw a very different group: "All shy, [and] shy people are usually lonely people." One physician at San Antonio's Brooks AFB School of Aerospace Medicine echoed this assessment: "When NASA began looking for astronauts, we spent a long time discussing the psychological requisites necessary for an astronaut, and the result of our discussion was that we ought to look for them among the priests. . . . A young, healthy priest, qualified in engineering, in chemistry, in·medicine, in

geology, and able to pilot a plane competently." Not all candidates were exactly priestlike, however; one astronaut bluntly offered, "Am I motivated? I
surely wouldn't let somebody stick something twelve centimeters up my ass if
I weren't."

"Most of them fall into the top two percent of the population intellectually," said a military psychiatrist who worked with the corps. "They're aggressive, competitive people. They've got very good opinions of themselves, for
one important reason; the good opinion is based on reality. Indeed, they can
be rather pleasant to work with because they've less to prove than the next
guy. Perhaps they're not very imaginative; you don't get poets up there. Well,
you've only got to listen to them about the view, 'Gee, it's beautiful!'" Manned
Spacecraft Center public affairs chief Paul Haney noted that, "With a few
exceptions, they're extremely disciplined people with an almost tunnel-vision
ability to lock on to a problem and never let go. They become so disciplined
they can almost dial up the emotional commodity that's needed. If they're
going to a cocktail party, they'll dial up to where it says, 'We're going to be
light and gay.' They might even muster up a slightly off-color story. They're
extraordinarily well-ordered people who don't panic easily. On the other
hand, they don't react to much, either."

After Project Mercury's brilliant success, which had the effect of inspiring
millions of Americans to want to be astronauts, NASA raised the bar: All candidates would aditionally now have to have degrees in engineering or science,
even though most of what the agency would be undertaking was not yet
taught in any school, and not much 1960s textbook aeronautical engineering
was relevant in the airlessness of outer space. Instead of the recurring myth
of a great and raging NASA battle between astronauts and engineers, during
the Apollo era, astronauts *were* engineers, and engineers, pushing the program
forward through a risky series of missions, had as much cowboy sensibility
as as any flier. Though Tom Wolfe may have accurately captured something of
the character of the men and women of Project Mercury, by the time of
Apollo, *The Right Stuff* would be wrong.

Few astronaut profiles in the 1960s discuss how intelligent these men really
were, which is a mysterious oversight. *Life* magazine, in its exclusive coverage,
insistently portrayed the agency's fliers, their families, and everyone else at
NASA as being just like the people next door . . . though they were only the
people next door if you happened to live on the campus of MIT. Pete Conrad,
for example, was typically viewed by the press as the wildest of cowboys, but

he was a student at Princeton when Einstein was still a professor there, a coincidence that so thrilled him that he wept at the physicist's death.

Apollo 11's launch date of July 16 coincided with the anniversary of Trinity, the culmination of America's greatest engineering effort before NASA, the $2 billion Manhattan Project, which led to the birth of the atomic bomb. This convergence underscores a little-researched line of American history. Before the 1990s' Silicon Valley entrepreneurs with their Red Bulls, boxed pizza, and Cheetos, there were the short-sleeved-white-shirted denizens of Houston's NASA with pocket protectors, Mexican takeout, evaporating hot-plate coffee, and ashtrays choked with smoldering cigarette butts, and before them were New York and New Mexico's Manhattan Project brain trust of alpha engineers in their fedoras and soft, floppy jackets.

In so many ways, the race to the Moon would turn out to be a sequel to its predecessor's race for atomic mastery. Both were enormous projects that only a great nation, on a federal level, could afford to attempt, and achieve. Both began with Third Reich émigrés, and a shared geography. Trinity and Alamogordo are a mere eighty miles from Apollo 11's ancestral territory, the ramshackle sprinkling of federal shanties at Fort Bliss that housed the von Braun rocket team in the postwar years (and where Buzz Aldrin lost his virginity to a Mexican prostitute), which in turn is 120 miles from High Lonesome and Roswell, where American rocket pioneer Robert Goddard tested the thrust of the same superchilled fuels that would send Apollo 11 to the Moon. And, if the first lunar landing marked the end of the Space Race and one of the beginnings of the end of the Cold War, the Manhattan Project marked both the end of World War II and the Cold War's birth—for it was then that the English and Americans did not tell the Russians of their wonder weapon until it was ready for Hiroshima and Nagasaki, and the Russians in turn repeatedly and successfully spied on their nominal allies. (Stalin's agents were so good, in fact, that the Russian leader knew about the atomic bomb before Harry Truman did.)

Trinity's witnesses responded just as those to Apollo 11 would, as J. Robert Oppenheimer remembered: "We knew the world would not be the same. A few people laughed, a few people cried, most people were silent." Oppenheimer later said that he beheld his radiant, blooming cloud and thought of Hindu scripture: "Now I am become Death, the destroyer of worlds." Aloud, however, the physicist made the ultimate engineer comment: "It worked."

Like its cousin, the Pentagon, NASA is extraordinarily mythic for a federal agency, but for all the reverence accorded to it, there is not nearly enough popular acclaim for its engineering genius. Why is it so difficult to honor this greatness? It should be seen as part of a string of American accomplishments ranging from the inventions of Bell, Carver, Morse, and Edison, to the triumphs of the Erie Canal, the St. Lawrence Seaway, the Transcontinental Railroad, the Empire State Building, the Panama Canal, the Interstate Highway System, the Rural Electrification Administration and Tennessee Valley Authority, the Hoover Dam, telecommunications satellites, the Internet, and the Global Positioning System. Such dazzling achievements, whose dynamism is a key force in American identity, are today more often than not undervalued, taken for granted, or just ignored by civilians. It was engineers who created these foundations of modern life, and our civilization (especially the continent-wide, U.S. variety) could not exist without the big pipes, the vast roads, the power grids, the dams, and the people-and-cargo-carrying vehicles of heroic engineering and big science.

Before modern times, engineering's profound accomplishments were celebrated, and engineers lionized. Thomas Carlyle described the core of the profession as defining humanity itself: "Man is a Tool-using Animal . . . Nowhere do you find him without Tools; without Tools he is nothing, with Tools he is all." After Galileo proved that the speed of a pendulum remains constant, his son Vincenzio created a clock based on a pendulum escapement (the part of a timepiece that creates a steady movement to translate time into clock hands)—the ultimate example of scientific theory transformed into engineering practice. This is not universally the case, however, as many engineers like to point out—the steam engine was devised and perfected first, primarily by engineer James Watt, and only later came Sadi Carnot's theories of thermodynamics, based on it.

In the eighteenth century, the American founding fathers and others of their Enlightenment generation revered scientists and engineers, considering their work both an inquiry into the very stuff of the cosmos, and a contribution to the greater good of all humankind. During the second industrial revolution of 1850–1950 (when the American census began including "engineer" as an occupation), the public worshipped them as heroes for the stream of miracles they seemed to regularly unleash, from building blocks of steel, refined petroleum, aluminum, and plastic to the transforming power of trains, trolleys, buses, subways, cars, planes, and jets, alongside the release from drudgery by way of light bulbs, sewing machines, cameras, telephones, refrigerators, air conditioners, and washing and drying machines. The era was celebrated with global showcases of technology and invention, from London's

Crystal Palace of 1851 and France's Universal Exposition in 1889 (and its signature building, the Eiffel Tower) to New York's 1939 and 1964 World's Fairs (and their signature Futuramae). As one of the forty million marveling at the eighty thousand exhibits of the 1900 Paris Exhibition, Henry Adams came to believe that, as the Virgin Mary had once inspired the great leap forward represented by Mont-Saint-Michel and Chartres, so technology would transform modern civilization. And so it has.

Perhaps, however, the great modern divide between engineers and civilians is not solely the undereducated and unappreciative public's fault. At the same time that they produce what to the uninitiated is so magical as to be overwhelming, engineers commonly alienate civilians with their inability to explain their work in layman's terms, their belief that, as humans are error-prone creatures, looking human is unattractive, and their obsession, to an antisocial level, with inanimate objects. "I knew I didn't want to be any of the human skill-oriented people, because I didn't think I did well with interactions with other people, and I tended toward things rather than people," remembered Kennedy Space Center rocket engineer Bob Jones. "Psychology classes, where you had to sit and talk about yourself, drove me crazy; but a physics class, I loved." Flight controller John Aaron summed up a key desire many engineers share: "Just by my nature, I can't stand to be around anything that I don't know how it works. I'm always intrigued by knowing how it works or why it works. I will tend to dig into anything until I understand it."

The men who made NASA trace their first glimmers of interest in their craft to a fascination with planetariums and museums of science and industry, Erector sets and chemistry kits. (The International Space Station is essentially an Erector set lobbed by NASA's engineers into low-earth orbit, forcing the agency's press office to issue news releases extolling the wonders of the truss.) "We used to subscribe to Doc Savage, and that had some quite far-out scientific things," remembered Marshall Flight Center's Dan O'Neill. "He had chemists, engineers, and doctors, and they'd go out and solve all of these big problems." Bob Jones: "We made gunpowder, you know, kids get into things. We put a ball bearing right through the grammar school cafeteria wall one time. It made a neat, one-inch hole."

The holy grail for NASA designers has always been elegant simplicity, partly because minimalism satisfied the era's engineering ethos, but primarily because a drive to simplicity reduces the number of elements and processes that can fail. "Any intelligent fool can make things bigger and more complex and more violent," said Albert Einstein. "It takes a touch of genius and a lot of courage to move in the opposite direction." And, if elegant simplicity was one lodestone, redundancy was the other. Performance aerodynamics/engineering and development director Maxime Faget: "If I am an engineer, I better damn well understand what reliability and what failure means, otherwise I am not an engineer. . . . We had redundant valves, quad-redundant valves, everything else. I basically said the best way to deal with risk management is in the basic conceptual design, get the damn risk out of it. And I think that is what made the program a success."

Simplicity was also a hallmark of many NASA decisions. The three-man Apollo crew was assumed since in the navy, NASA's closest relative, sailors work four hours on and eight hours off, requiring at minimum three men for a twenty-four-hour shift. The distinctively rounded bottoms of the Command and Service Modules were the result of Hunstville's shaving off stage three's diameter by six inches and needing to make it fit—designer Faget had been content, aeronautically speaking, with a flat-butt cone for his spaceships. The thrusters' fuel system has an emergency reserve tank because Apollo program manager Joe Shea had one in his Karmann Ghia.

NASA is a factory of miracles because it is a nation-state of brilliant tinkerers, who outside their field are often derided as nerds. "After spending a lot of time with different subcultures that I intuitively knew were nerdy, I figured out what they all had in common: a love of rules, a love of hierarchies that were meritocratic and open to everybody, and in some cases the affectation of rationalism (whether computer programming or math)," *American Nerd* author Benjamin Nugent said. "What makes people insiders in high school is their ability to intuitively figure out how the hierarchies work. Some nerds can't follow the hierarchies, don't know how, and sometimes don't even perceive them. Other nerds are unwilling to follow them. But in general most of the people we consider nerds are people who are oblivious to or incompetent at following the hierarchies." There were other identifying characteristics, as Chris Kraft pointed out: "You could identify the houses of the NASA guys who personally kept their old, but well-maintained, cars and motorcycles humming along by the oil stains on their driveway. Smooth-running engines and the harmonic rhythm of the valve train were music to their ears." The triumph of NASA, however, would beget a new American archetype, one that continues today in the persons of Steve Jobs and Sergey Brin—the nerd who is supercool.

Anyone who loves his cell phone, iPod, camera, laptop, automobile, or private plane appreciates the work of engineers, and anyone who has ever successfully fixed a beloved broken gadget has known directly the profession's elemental pleasure. Even its centerpiece—formulae—can be compelling for civilians, under the right circumstances. The top speed that a commercial jet in distress can fly when it strikes an ocean wave to avoid disintegration? It can be reduced to a number. The strength of the metalwork that keeps a lion inside its cage? A computation. The step-by-step process that will begin the age of outer space transportation? The product of many, many formulae. Engineering designer Caldwell Johnson:

> We had gotten to the place we had decided, at this day and time, one should have a numerical expression of the required reliability. I've since [thought] it was the worst thing we ever did, but it seemed like a good idea at the time. So the question came up, "What number should it be? Should 50 percent of the missions be successful and should nine or ten guys come back alive? Or should it be 999 of 1000 times? Or should it be a number of 10,000? Or what should it be?"
>
> Since theoretically the cost of the development of the things are a function of this reliability requirement, if you can afford to lose half of them and half the men, you can build them a damn sight cheaper. On the other hand, in this country, you just don't go around killing people—so we had to pick a number . . . and nobody wanted to pick the number, see.
>
> So one day we walked down to see Dr. Gilruth and we said the time has come to bite the bullet. We have to write a number. He said, you know, certainly if you could make nine out of ten, economically that would seem it could pay off. . . . And, as far as safety was concerned, he didn't know—he hated to even hazard a guess as to what was a reasonable number for that. He said probably 99 out of 100. Lose one man out of 100 out of a mission like that.
>
> About that time, [Mercury operations director] Walt Williams came in. He said, "Oh, that's ridiculous. If you put numbers that low—they're good numbers—but if you put numbers that low, you won't get anything. You ought to make it one in a million." Well, they argued awhile and finally Walt won. So we ended up with a number [for] the crew safety [that] was three nines—999. I think mission success was two nines. And we wrote those numbers down, and they had a most profound effect on the cost of the program. If you took one decimal point off of that thing, in theory you could probably cut the program cost in half. . . . It's the single most important number in the whole program, and it got picked in, I guess, a ten-minute discussion.

Perhaps there will always be a great divide between engineers unable to engagingly describe their work to nonspecialists, and citizens uneducated enough to be interested in the details of science and engineering. "We live in a society exquisitely dependent on science and technology, in which hardly anyone knows anything about science and technology," as Carl Sagan once summed up our modern paradox. Even the least educated, however, would find themselves captivated, and on a global scale, by the Moon landing of Apollo 11. Neil Armstrong himself addressed this magic in a speech to the National Press Club on February 22, 2000:

I am, and ever will be, a white-socks, pocket-protector, nerdy engineer, born under the second law of thermodynamics, steeped in steam tables, in love with free-body diagrams, transformed by Laplace, and propelled by compressible flow. . . .

Science is about what is, and engineering is about what can be. The Greek letter eta, in lowercase, often shows up in engineering documents. Engineers pay a good bit of attention to improving eta because it is a symbol for efficiency—doing an equivalent or better job with less weight, less power, less time, less cost. The entire existence of engineers is dedicated to doing things better and more efficiently. . . .

The twentieth was a century often punctuated with the terror of war and darkened with societal struggles to overcome injustice. But it was also the first century in which technology enabled the tenets and the images of those traumas to reach across the world and touch people in ways that were previously unimagined. John Pierce, the engineer who fathered Telstar, the first satellite to relay television signals across the Atlantic, said that engineering helped create a world in which no injustice could be hidden. . . .

[Arthur C. Clarke's] third law seems particularly apt today: Any sufficiently developed technology is indistinguishable from magic. Truly, it has been a magical century.

5

Mr. Cool Stone

Even those who knew full well the astronauts' background in engineering and test flying often made denigrating generalizations about them as a group. One NASA staff member's remark was typical: "Theirs is not to reason why." In truth these men encompassed a broad range of personalities, from the wealthy, acidic, and testosterone-fueled Alan Shepard, to the more intellectual and socially awkward Buzz Aldrin, to the notably nondrinking, nonexercising, and fervently nonsocializing Neil Armstrong, and even to a few quirky charmers such as Michael Collins, who remembered of his Maryland boyhood: "During the summer, I'd catch crabs for bait to catch fish. Then I'd use the fish in a trap to catch crabs. It was a sort of endless chain of converting crabs to fish and fish back to crabs again." When Mike asked his wife Pat for her hand, she told him he had to write a letter explaining his worthiness to her flinty Boston Roman Catholic lawyer/politician father, Joseph Finnegan. The letter was all of six sentences, and Finnegan immediately called his daughter: "Marry him. Marry him before he gets away."

The least noticed of the Apollo 11 crew, Michael Collins was that combination of competence and sophistication found elsewhere in the 1960s in James Bond's resourceful American agent sidekick, sharp as the edge of town. If he seems at first glance like a regular joe, Collins was the man chosen by Armstrong and Slayton from an entire roster of candidates as being such a strong pilot that he could be entrusted with Armstrong and Aldrin's lives.

Born in Rome, Mike spent his childhood in motion, moving to Oklahoma, then to Governors Island off the coast of New York City, on to Baltimore, to Columbus, Ohio; San Antonio, Texas; and San Juan, Puerto Rico—five homes by the age of ten. The son of a State Department military attaché, Major General James L. Collins, and nephew to the army's famed chief of staff J. Lawton "Lightning Joe" Collins, he in time became a member of that too-rare tribe, the cosmopolitan military man, like a Marshall or a Mountbatten, who travels the world through his years in the service and develops an interest in global cultures, from their food and wine to their music and philosophy. Yet, by choosing the air force instead of the service that was his birthright, Collins took a stand of independence. He graduated from West Point in 1952 (the same class as Ed White, one before Aldrin and two behind Borman), trained with the air force at Columbus, San Marcos, Waco, and Las Vegas, rose to lieutenant colonel, and after service in Europe and Illinois, joined ARPS (the USAF's Aerospace Research Pilot School) at Edwards AFB in 1961, barely overlapping with Neil Armstrong, who left Antelope Valley for Houston in 1962.

Until that point in his life, Collins was decidedly more interested in chess and girls than in planes, but with an engineer's slant on discussing his significant life decisions: "Like most of the early astronauts, I was a test pilot, and it was sort of a stair-step process. I went to the military academy, I went to West Point primarily because it was a free and good education. I emphasize 'free.' My parents were not wealthy. When I graduated from the military academy, there was no Air Force Academy, but we had a choice of going into the army or the air force. The air force seemed like a more interesting choice. Then the question was to fly or not to fly. I decided to fly. To fly little planes or big ones? I became a fighter pilot. To keep flying the same or new ones? I became a test pilot. And so, you see, I've stair-stepped up through five or six increments then, and it was a simple, logical thing to go on to the next increment, which was higher and faster, and become an astronaut." When Collins applied to NASA, an air force buddy suggested, "When they ask you why you want to be an astronaut, tell them it's because of all the money and ass you can get."

Of the three, Collins was "always the easy-going guy who brought levity into things," as Buzz Aldrin recalled. During the run-up to their mission, Collins tried to foster a sense of camaraderie, but could get only so far. After Armstrong's bulwark of shyness proved impenetrable, Collins decided that Neil "never transmits anything but the surface layer, and that only sparingly. I like him, but I don't know what to make of him, or how to get to know him better. He doesn't seem willing to meet anyone halfway. . . . Buzz, on the other hand, is more approachable; in fact, for reasons I cannot fully explain, it is me that seems to be trying to keep him at arm's length. I have the feeling that he would probe me for weaknesses, and that makes me uncomfortable."

Armstrong and Aldrin were both men who spent a large part of their lives deep within their own thoughts. Aldrin focused on mathematics and equations—he was considered by journalists to be the brainiest astronaut—

while Armstrong concentrated on test flight and aeronautical engineering. "But that didn't mean that Neil was all pragmatist or Buzz all theoretician," as Collins observed. "Among the dozen test pilots who had flown the X-15 rocket ship, Neil had been considered one of the weaker stick-and-rudder men [to fly a traditional plane, the control stick moves in four directions to pitch the plane's nose and roll it horizontally, while twin rudder foot pedals yaw the plane left and right like a car's steering wheel], but the very best when it came to understanding the machine's design and how it operated. Buzz, far from being an academic recluse, was an outstanding athlete, a pole vaulter, and a Korean War MiG killer."

While some found Armstrong icy and distant, Buzz Aldrin has at times been too human for his own good. Born on January 20, 1930, in Glen Ridge, New Jersey, Edwin Eugene Aldrin, Jr., was the family's only boy, and nick-named "Brother." Sister Fay Ann pronounced it "Buzzer," and Buzz was rarely called Edwin Eugene ever after; he legally changed his name to Buzz in 1988. As a boy, Aldrin's best friend was Alice the housekeeper, his hero was the Lone Ranger, and his first engineering achievement was in designing and building an elaborate home for his pet mice. He was so athletic, however, that he installed an iron bar over his bed for doing chin-ups first thing in the morning and last thing at night, and spent so much time on football and pole-vaulting that his grades were too poor to be considered for an appoint-ment to West Point or Annapolis. However, in a hallmark of Aldrin's determi-nation, seemingly overnight the teenager leaped from a C to an A student. He is still, in his seventies, fit, solid, and taut.

Aldrin's parents met in the Philippines where Edwin Eugene Senior worked for the visionary Army Air Corps general Billy Mitchell; on Mitchell's court martial, the senior Aldrin resigned from the service as well. Demanding, dif-ficult, opinionated, controlling, and remote, Buzz's father never seemed to be able to be proud of his only son, no matter what he achieved in life.

Buzz was an obedient boy but, at times, as stubborn as the old man. Once his grades had improved, Buzz's father thought he should attend Annapolis, and arranged for a congressional appointment. Instead Aldrin decided on West Point, and insisted his father reverse the congressional favor.

Aldrin spent six months in basic at Bartow, Florida, three months on fighters at Bryan, Texas, and three months on F-86s at Nellis AFB near Las Vegas (Mike Collins would have West Point and Nellis in common with Buzz, and Edwards with Neil). At officer school, a superior told Aldrin "that I was too competitive, too insensitive to others, too determined to be the best, and that if I didn't watch it I'd end up with a reputation as a hot-shot egotist. . . . The truth can sometimes hurt and I had tears streaming down my cheeks. I thanked him." Aldrin flew sixty-six missions in an F-86 for the Fifty-First Fighter Wing over Korea, bringing down two MiGs, and becoming famous when his ship's camera caught the enemy dogfighter's bailout on May 14, 1953, and its pictures were published in *Life*.

By 1955, both Collins and Aldrin were flying nuclear bombers, with Collins in France and Aldrin piloting F-100 SuperSabres for NATO out of Bitburg, West Germany. There, he ran into a track teammate from West Point by the name of Ed White, and over the next two years in Germany, White would become Aldrin's best friend. Ed told Buzz that, after their USAF assignments were finished, he would go on to graduate school in aeronautical engineering; join Flight Test at Edwards; and be one of the first American pilots to fly in outer space. This sounded great to Buzz, who wondered if they would be known as "rocket pilots," even though rockets were primarily controlled by engineers on the ground, and Mercury capsules were so tiny that it was said they weren't even flown. They were worn.

While Ed went to the University of Michigan, Buzz attended his father's alma mater, MIT, in 1959, writing an Sc.D. thesis—"Line of Sight Guidance Techniques for Manned Orbital Rendezvous"—deliberately conceived to have application potential for both the USAF and NASA. Ed White had gone on to Edwards test school, and was applying to the civilian agency, and Buzz did, too. White was accepted, and though Aldrin was rejected, he was already becoming known at NASA. Agena/Lunar Module systems experimenter Mel Brooks:

> One of the guys who was representing the Air Force in meetings [with NASA] was Buzz Aldrin. He wasn't yet an astronaut; he was, I don't know, a major or something like that, a young guy, strange guy, bright guy, very likable fellow. . . . I remember Buzz trying to convince all of us how important gravity gradient stabilization was and why it would work and everything. We'd kind of laugh at him, because, see, Buzz was a strange guy. He had a Phi Beta Kappa key in orbital mechanics back when most people didn't even know what that was. He had this . . . well, it wasn't all his idea, but he was the one selling it to us, trying to get it, he tried selling it everywhere and finally found some ops people that would listen to him . . . that if you had a dumbbell in orbit and you pointed it toward the center of the Earth, lined it up with the center of the Earth, that the equation for gravity is the product of the masses of the two masses, the Earth and the other piece, divided by the square of the distance between their centers. And since these were physically farther apart, the distance was a little bit greater, that the force would be greater on this one

than on that one. We laughed at him, because those forces he was talking about would be like in the fifteenth decimal place or something like that, and nobody ever believed that that would work, but it did. It certainly did work. It's a routine operations plan now.

One time when we were at a meeting at Lockheed, he took us all up to San Francisco for a night of carousing around. He was a carouser. When we came out after dozens of bars, none of us knew where in the heck we were, he was the only one who knew his way around, he couldn't find his car. We didn't know where it was. But he said, "Don't worry, I'll get us." And he'd run out in the street and he was shooting on stars, trying to get a bearing on the stars so he could navigate us to the car. [Laughter] He was quite a guy. I like Buzz.

As an astronaut, Aldrin revealed both his smarts and his simultaneous lack thereof in promulgating his MIT studies so assiduously and his opinions so stubbornly that it annoyed many of his colleagues. "[His] doctoral thesis on space rendezvous . . . made him, in his own eyes, one of the world's leading experts," flight director Chris Kraft said. "Before long, the real experts . . . were calling him, with a touch of sarcasm, 'Dr. Rendezvous.'"

Ed Mitchell remembered, "You wouldn't want to sit near him in a party, because he would start talking about rendezvous, and you would want to be talking about the good-looking girl across the room. He could care less." Drinking bourbon and branch while discussing ecliptic orbits, Aldrin could come across as the squarest of the squares; he referred to himself as a "machine man" during this period in his life. Others saw him as burning with ambition. "If Buzz were a trash man and collected trash, he would be the best trash collector in the United States," his wife Joan remarked.

Many at NASA eventually came to recognize that even if Aldrin didn't always present his views with finesse, they couldn't be ignored. "Buzz would always say, 'Well, now, I'm coming up on terminal phase [docking in rendezvous] and the sun is coming over my right shoulder,'" flight dynamics chief Jerry Bostick recalled. "I'm standing there thinking, 'How does he know where the sun is now?' Because that's an aspect that I wasn't worried a lot about. It was one of the constraints, obviously, because you wanted good lighting for the chase crew to be able to see the target vehicle, but it always bugged me that Buzz knew where the sun was. Usually I would run back to the office after talking with him and I would run that case on the computer, and, lo and behold, he was always right, the sun was coming over his right shoulder just when he said it was. Dad gum, Buzz, you do know some things."

"Buzz knew more than anybody here about rendezvous," Alan Bean insisted. "They didn't want to hear it, but he knew more, and he knew he knew more, and he didn't shove it down your throat, but he stayed on it all the time, and he wouldn't give up. He made one of the great contributions, in my opinion, to the success of Apollo, but everybody made fun of him all the time because they didn't like the fact he knew it and he wouldn't give up and he wouldn't compromise. After it was all over, they did it his way, not because it was him—well, but because his way was the best way."

Buzz Aldrin has "never been great on human relations or public relations, either one," Deke Slayton summed up. "He's a damned good guy technically. That's the reason we had him on that crew. . . . He's an unusual guy." But if Deke found Aldrin unusual, what could he have thought of Neil Armstrong? Though he may be blessed with an all-American name, sandy-haired and freckle-faced boyish good looks, and a remarkable résumé, Armstrong is the polar opposite of "down to earth." Described variously as intense, aloof, enigmatic, impassive, and unknowable even by the people who know him, he is, at the same time, cool and aggressive, egg-headed and hard-nosed, both a fighter pilot and a physics professor. Armstrong even befuddled Norman Mailer, who described Apollo 11's commander as simultaneously innocent and sinister, "extraordinarily remote," "apparently in communion with some string in the universe others did not think to play," and "simply not like other men."

The history of rocketry is a history of solitude. Pete Conrad found astronauts to be shy and lonely people, and the most famously shy and lonely of them all would also be the most famous of them all. Some who worked with Neil Armstrong described him as asocial and hyperrational, to the extent that, in conversation, it wasn't clear whether or not he was listening to anything one said. A friend commented, "If you tell Neil that black is white, he may agree with you just to avoid argument." But another added, "No, most likely he won't say anything at all. He'll smile at you and you'll think he is agreeing. Later on, you'll remember that he didn't say a word." Another story: Armstrong was dating Janet Shearon when NASA's predecessor, the NACA, finally offered him his dream job of test flight at Edwards Air Force Base. He asked for Janet's hand since, as she later remembered, "If I would marry him and come along in the car, he'd get six cents a mile for the trip. If I didn't, he'd only get four."

All the same, this was a man who could be acidly pungent. Asked what it was like to walk on the Moon, he replied: "Pilots take no special joy in walking.

Pilots like flying." When a writer learned that he was fond of books and casually asked what he'd been reading lately, Armstrong snapped, "That can't be
pertinent to Apollo history"; he told another historian: "People are a third-rank
category of things interesting to talk about. . . . Someone once said, 'Great
men talk about ideas, good people talk about things, and everybody else talks
about people.' . . . Consequently, I seldom enjoy talking about people, either
within our own group or outside."

At other times, Armstrong could be as cool as Superman. Asked about his
feelings while in outer space, he said, "There is an obvious kind of order
about the physical world that has to be impressive. I find a good deal of satisfaction in the fact that that particular order seems to exist." This parallels a
noted comment by Albert Einstein: "If something is in me which can be called
religious, then it is the unbounded admiration for the structure of the world so
far as our science can reveal it." Nine years after landing on the Moon, Armstrong was working at his Ohio dairy farm when his wedding ring stuck on a
latch and tore off part of his finger. "Instead of screaming and running for a
doctor, he scooted around until he found his finger," a friend reported, iced
it, drove to the hospital, and had it reattached.

Armstrong would turn out to have much in common with the other two
most famous NASA fliers. Both he and Alan Shepard were known behind
their backs as "Ice Commanders," for both were similarly tough, remote, and
austere. There were benefits to being remote, though, as Charlie Duke explained: "Neil Armstrong was probably the coolest under pressure of anyone
that I ever had the privilege of flying with. He was just Mr. Cool Stone, if you
would." Others at the agency referred to John Glenn as "Mr. Clean" for his
Sunday-school-teacher, loner persona, which could equally be applied to Armstrong, who became an Eagle Scout while a freshman at Purdue, and who always carefully considered each word before speaking. He was so self-controlled,
in fact, that it could be said of him what Alan Shepard said of Glenn: "John
always acts as if he were being watched by an army of Boy Scouts or children,
even when he's scratching his nose or peeing." Armstrong even looked like
Glenn's younger brother, but was quiet and diplomatic, instead of room-filling
and Marine (lacking either Glenn's megawatt charisma or his equally powerful temper).

At the same time, it is almost impossible to find a NASA alumnus who
worked with Armstrong who didn't intensely admire the man. "All through
the preparation for the mission, I was absolutely amazed at how quiet, how
calm he was," Gene Kranz said. "The quiet, absolutely superbly confident as-

surance that Neil had, in retrospect, was pretty inspirational in itself. Here's a guy who knew he was destined to do a job. And I believe that he believed that from the day he was born, this was a job that he was singled out to do. I think every person who ever worked with Neil had such a respect for the very quiet confidence that he exuded; his incredibly professional demeanor. He was literally a man for all ages within Mission Control. And I think every person today has that same respect. Even it's increased. After the mission, the one time that I ever remember Neil talking, almost with boyish glee, was [when] we were just shooting the breeze. And all of a sudden he just says, 'You know, I think this says a lot for American craftsmanship.' And Neil proceeded to elaborate on his feelings about the American craftsmanship and the ability to do something so intensely complex and be successful the first time around, that it was marvelous."

Like Collins, Neil Armstrong had a peripatetic boyhood. While the Collins military family was reassigned from post to post around the world, Armstrong's father, Stephen, was an accountant for the Ohio Bureau of Inspection and Supervision of Public Offices, traveling from town to town across the state and taking his family with him. Neil was born on August 5, 1930, in his grandparents' house outside Wapakoneta, and throughout his childhood the Armstrongs moved to a new Ohio town every year—Lisbon, Warren, Ravenna, Shaker Heights, Cleveland Heights, Jefferson, Moulton, St. Marys, Upper Sandusky—with Neil occupying himself in reading, playing the piano, and building model airplanes.

When he was fourteen, the Armstrong family returned to Wapakoneta, where Neil joined the Boy Scouts—eleven of the twelve Apollo moonwalkers were Scouts—and qualified to be a pilot before he could drive a car, having paid for nine-dollar flying lessons out of his baker boy's salary of forty cents an hour—a job he'd had since he was small enough to fit inside the mixing vats, and strong enough to clean them. He was obsessed with flying, and convinced that he'd lost his chance at greatness with the passing of the Lindbergh era: "I was disappointed by the wrinkle in history that had brought me along one generation late. I had missed all the great times and adventures in flight."

Armstrong was accepted to MIT, but instead entered Purdue with a full scholarship through the Naval Aviation College Program, similar to ROTC and initiated by Wally Schirra's father-in-law (Neil's mother got so excited when she heard the news that she dropped a jar of raspberries and broke her toe). It was, as Armstrong described, a seven-year program: Two years of [university], then go to the navy, go through flight training, get a commission, and then serve in the regular navy for a total then of three years of active duty, after which the plan would be to return to university and finish the last two years. . . .

[In] advanced training, . . . I asked for fighters and got fighters. Then we went to Corpus Christi, Texas, and went through training there in single-occupant aircraft, in my case F8F Bearcat. [His mother, Viola, said that Neil flew solo because, he explained, "I didn't want to be responsible for anybody else. I'd better just watch my own self."] I went to the fleet squadron, was in a standby unit for a while, then assigned to a jet fighter squadron, still was a midshipman making seventy-five bucks a month plus flight pay, 50 percent of seventy-five bucks. . . . I happened to be a day fighter pilot. We had night fighter pilots on the ship I was on, and I thought they were crazy. . . . All my landings on a carrier were in day. I was always happy about that.

All the time we flew off the eastern coast of North Korea, off Wonsan Bay, about a hundred miles out, something like that. Had two kinds of flights. One would be called combat air patrol, which was defense of the fleet, basically. And the other was predominantly interdiction flights, flying against bridges and railroads and trying to find an occasional tank—bombs, bullets, and rockets sometimes, depending on what target it was. We had a combination of two jet fighter squadrons, F4U Corsair squadron, of course, air squadron, and an AD [Skyraider] squadron. [They] could carry the 2,000-pounders and really do some damage.

One navy flight surgeon who served on accident boards concluded, "The main problem is overconfidence. Pilots have difficulty saying, 'I can't.'"

On September 3, 1951, the USS *Essex* catapulted Armstrong's F9F-2 into the air for a bombing run west of Wonsan. After evading clouds of ack-ack, he struck an antiaircraft cable at 350 miles per hour, shearing off a piece of his right wing. He was able to mosey the jet along far enough to eject into a South Korean bay, but the winds were so heavy that he was blown into a neighboring rice field and cracked his tailbone. Given a hitch by a local marine, Armstrong learned that the waters he'd tried parachuting into had been heavily mined, that in fact he'd had a major piece of luck in getting blown ashore. He spent R&R time at an installation outside Mount Fuji, developing a taste for Japanese food and design, and for a good round of golf.

After seventy-eight missions and 121 combat hours—"I'd be lying if I said they'd done me any good"—Armstrong returned to Purdue, read Wernher von Braun's *Conquest of the Moon* (which urged a global effort to reach the

Moon by 1978), joined Robert Goddard's American Rocket Society, met his future wife, Janet, and on graduating in 1955, applied to be a test pilot at the National Advisory Committee for Aeronautics' High-Speed Flight Station at Edwards AFB. He was turned down, but six months later applied again, was accepted, and spent the next seven years flying some of the greatest ships ever built.

Home to Pancho Barnes, Chuck Yeager, khaki pants, bomber jackets, quonset huts, sagebrush, Joshua trees, tumbleweeds, and a forty-year history of aviation science, Antelope Valley's Muroc Field ("Corum," the town's founders, spelled backward) was renamed in 1950 for Glen Edwards, who had died while testing a Northrop YB-49. The motto of the Test Flight Center was *Ad Inexplorata* ("Toward the Unknown"). Neil Armstrong: "They had five pilots, and, if memory serves, seventeen aircraft, pretty much all different. A lot of X-airplanes and fighters, the B-47 [Stratojet] and R4D and a couple of B-29s [Superfortress], all kinds of exotic aircraft. So they let me fly a few of these at first, and as they became more confident in my abilities and as I became more experienced, why, they gave me more and more jobs.

"I did a lot of different test programs in those days. That was the first time I ever flew supersonic. We had two B-29s that we used for dropping rocket aircraft, the X-1s and the Skyrockets. So I, either as the right-seat or the left-seat guy in the B-29, launched over one hundred rocket airplanes in the fifties."

"The general pattern was: first, by classroom lectures, to learn the theoretical aspects of a particular type of test, then to make one or more flights trying out the new technique, and finally to analyze the reams of test data acquired during the flights," said Mike Collins, who flew Delta Darts, Starfighters, and Super Sabres at Edwards for the USAF. "This last part . . . began in flight, when information was handwritten on a kneeboard, or instruments were photographed by a special camera, or an oscillograph traced on graph paper the record of thirty or more specific measurements. After each flight, the developed film and oscillograph paper were delivered to us; and weekends and nights would find us hunched over a desk calculator or peering at a film projector, trying to reduce this overwhelming amount of information into a terse report . . . ah, the glamour of a test pilot's life!" Even so, Armstrong called it "the most fascinating time of my life. I had the opportunity to fly almost every kind of high-performance airplane, and at the same time to do research in aerodynamics."

Eventually Armstrong was promoted to piloting the hottest plane in the world, North American's X-15: a fifty-foot-long and thirteen-foot-high winged missile with a skin of black Iconel X, a nickel and chrome alloy that could withstand the heat of reentry and wind friction up to 1,200 degrees Fahrenheit. The X-15 was aerotowed into the sky, and at the appointed altitude, its Thiokol XLR-99 rocket engine (operating at a peak 70,000 pounds of thrust) would explode the ship 200,000 feet into space at 4,000 miles an hour with 18,775 pounds of fuel (the ship itself weighed 12,500 pounds). The pilot controlled this baby with two systems, a standard aerodynamic for launch and land, and for the great and airless heights, hydrogen-peroxide nose and wingtip thrusters (a system which would also be used on NASA spacecraft). X-15 crews felt zero-g and rose so high into the farthest reaches of the atmosphere they were considered "astronauts," but with only 600,000 horsepower—one-fourth the velocity needed for orbit—there was a big difference in potential between this USAF rocket plane and an actual spaceship. X-15 pilots trained with computer simulations and the Iron Cross, a metal platform with thrusters similar to the Lunar Landing Training Vehicle that would be used for Apollo, and would one day cause Armstrong so much trouble.

What was flying the X-15 like? Armstrong: "There's very little time for wondering, but above two hundred thousand feet you have essentially the same type of view that you have from a spacecraft when you are above the atmosphere. You can't help thinking, 'By George, this is the real thing. Fantastic.' You can see the curvature of the earth. And from the [cold water, no electricity] mountain house where we lived, Jan could use her binoculars to see what was going on. She could see the X-15 drop away from the B-52 mother ship, and she could see the puffs of dust down in the valley as the X-15 landed."

X-15s eventually fell to earth, landing like the Space Shuttle by starting a 360-degree spiral from forty thousand feet into a 35-degree bank at 285 to 345 miles per hour, their touchdown sluicing across the sixty-five-square-mile Rogers and Rosamund dry lakebeds with a nose wheel and rear skis that left two-mile-long skid marks. (The pilots looked for smoke bombs at the lakebeds to know which way the wind was turning.) Every winter, a deluge would immediately inundate Antelope Valley and then just as immediately evaporate, leaving dry lakebeds so smooth you could land a screaming rocket jet right on top of them, their asphalt-flat, leached clay offering generous leeway for experiments gone very wrong.

Though this may sound like a pilot's dream come true, it was a tough territory to live in, if "live" was the word, as service families had to ride out the numerous dust storms by hiding under wet blankets. And there was the fact that you were flying "X" craft, with a mortality rate of one in four. Armstrong:

> I can remember several different system problems in the flights. You almost always had something. . . . [One example:] The X-15 covered such a wide speed range and altitude range, it was impossible to set the gains in the flight control system to a single value that was optimum for all flight conditions. The one and two airplanes you had to continually be changing the gains because at one minute you're at Mach 1, the next minute you're at Mach 5, and the airplane responds quite differently under those two conditions.
>
> We had a limit built into the flight control system that would automatically prevent you from exceeding five g's, and one of the things I wanted to do was demonstrate that this worked. It had never been demonstrated in flight. During the test, I couldn't [quite achieve 5 g's], so I [kept] pulling to try to get the g limiter to work. In the process, I got the nose up above the horizon. I was actually skipping outside the atmosphere. I had no aerodynamic controls. That was not a particular problem, because I still have reaction controls to use, but what I couldn't do is get back down in the atmosphere. . . . I [rolled] over . . . and tried to [drop back into] the atmosphere, but [the aircraft] wasn't going down because there was no air to bite into. So I just had to wait until I [fell low enough] to have aerodynamic control and some lift on the wings, [then] immediately started making a turn back.
>
> By that time it wasn't clear whether I would be able to get back to Edwards. Eventually I landed without incident on the south part of the lake. There's no power on the aircraft, so you're always a glider after the rocket burns out. The rocket only [burned for] a minute and a half.

Armstrong logged over four thousand hours in his X-15 and held, at 6,420 and 6,587 kilometers per hour, two of the plane's top-ten speed records, but he never broke the altitude ceilings held by Joe Walker (killed in a midair collision) or Michael Adams (whose craft fell into a spin on descent, accelerated into 15 g's, and disintegrated across an area of fifty square miles). On March 22, 1956, Armstrong was copiloting a B-29 when it hit thirty thousand feet and engine number four went out. The crew tried feathering the prop—aligning the blade so it would stop spinning—but the gearing wouldn't work, and the blades began to windmill. At this altitude and speed, the propeller could rip apart, turning into life-threatening shrapnel. At the same time, Jack McKay, below in the D-558-2 Skyrocket, radioed that he had a

broken valve, so his towed craft couldn't be released. Pilot Stan Butchart, however, had no other choice; he jettisoned McKay. The B-29's engine four propeller then spun loose, one blade destroying engines two and three, and then
slicing clean through the bay where the Skyrocket and its pilot had been sitting. Butchart and Armstrong had to land a B-29 on one engine . . . and they
did. McKay brought his broken Skyrocket down just fine. Armstrong: "We
were very fortunate. It could have turned ugly."

In his spare time, Armstrong flew sailplanes, soaring above the alfalfa and
Neil and his wife Jan then suffered through the worst agony that any parent can know. Their baby daughter, Karen, developed a brain tumor and died,
at the age of three, on their wedding anniversary. Jan Armstrong: "It was a
terrible time. Somehow he felt responsible for her death, not in a physical way,
but in terms of 'Is there some gene in my body that made the difference?'
When he can't control something, that's when you see the real person. I
thought his heart would break." Neil Armstrong: "I thought the best thing for
me to do in that situation was to continue with my work, keep things as normal as I could, and try as hard as I could not to have it affect my ability to do
useful things."

In his spare time, Armstrong flew sailplanes, soaring above the alfalfa and
onion fields of the Antelope Valley. After being aerotowed up to heaven and
then cable-released—just as in the X-15—gliders cocoon their human passengers in a white-fiberglass bliss of warmth and silence: comfortable, high, and
alone. "Soaring is something that's very easy to do [and] very hard to do well,"
Armstrong said. "It has the magic combination of requiring good equipment,
some luck, and as much skill as you can gather together. Like some other
forms of racing, like automobile racing—the machine is important but the
good guy usually wins. That's certainly true in soaring. . . . The motor skills of
operating the glider are not very important at all compared to mental requirements in trying to outguess the weather, primarily. And then second, knowing
the aerodynamics of sailplanes is the thing that makes you fly them just at the
proper speed in order to get the slightest edge on the competition. . . . You
can't blame the mistakes on anyone but yourself. . . . It's great relaxation. . . .
It's the closest you can come to being a bird." Armstrong had even had a recurring boyhood dream that foretold this avocation: "I could, by holding my
breath, hover over the ground. Nothing much happened; I neither flew nor
fell in those dreams. I just hovered."

"Joining NASA wasn't an easy decision," Armstrong explained. "I was flying the X-15 and I had the understanding or belief that if I continued, I
would be the chief pilot of that project. I was also working on the Dyna-Soar,
and that was still a paper airplane, but was a possibility. Then there was this

other project down at Houston, [the] Apollo program. Gemini hadn't been really much identified yet at that point. It wasn't clear to me which of those paths [would be best]. . . . In those days it wasn't a question of 'Do you want to be an astronaut or do you want to sweep streets?' A whole array of approaches to space was in the works. We were doing some exciting, way-out things in which we were more than just pilots. We were engineers and developers using airplanes merely as tools, the way an astronomer uses a telescope as a tool. We considered the spaceflight task group as babes in the wood. . . . I happened to pick one that was a winning horse. But there would be no way to predict that at the time when it got to that fork in the road. In my case, a three-way fork."

In fact, Armstrong's application to join the corps arrived a week late, but he was so clearly qualified for the program (whose administrators included many Edwards alumni who knew him personally) that it was accepted, even though NACA director Paul Bickle refused to give him a recommendation. He became part of the 1962 group of freshman astronauts known by *Life* as "The New Nine," and if these would-be spacemen were drawn by the stories of such Mercury pioneers as Glenn or Schirra flying across the heavens in solo reverie, they would soon discover that there was, in fact, no more Mercury; the program that had proven that Americans could be shot into outer space, survive weightlessness and accelerated g-forces, eat, sleep, and eliminate successfully, and then be returned safely to Earth was ending in favor of two new ones, Gemini (which would test every step needed to reach the Moon) and Apollo (which would go there). And, if the trainees felt honored at being selected out of hundreds of thousands of candidates—"astronaut" already becoming one of the dream jobs of American boys—they quickly learned their place in the pecking order. The Original Seven—Scott Carpenter, Leroy Gordon Cooper, John Glenn, Jr., Virgil "Gus" Grissom, Walter Schirra, Jr., Alan Shepard, Jr., and Donald "Deke" Slayton—were the most famous and sought-after men in America, so admired that their silver suits inspired Andy Warhol to wallpaper his New York offices in aluminum foil. These modern heroes drag-raced their loaned, matching Corvettes on the freeways of Houston and the tarmacs of Cocoa Beach while lounging in complimentary motel rooms and getting comped at the Astrodome. Gus Grissom finally came right out and told the candidates exactly what they didn't want to hear: you're not really an astronaut until you fly.

At 3:45 a.m. on April 24, 1964, Janet and Neil woke up to the odor of smoke; their home in the NASA suburb of El Lago was on fire. Their air conditioning had broken, so Janet had left the windows open on that hot spring night, a decision that saved her children's lives. Their next-door neighbor,

Buzz Aldrin's best friend from Germany, Ed White, woke up to the smell, and leaped over the six-foot fence separating the properties to help with the rescue. Janet had not been able to get through to the local telephone operator (who had apparently fallen asleep), but Pat White was able to reach the volunteer fire department. While Ed tried to stop the damage with a water hose, Neil pulled the children to safety; he and Ed then remembered to get the cars out of the garage so they wouldn't explode. The house was destroyed, and the Armstrongs had to live with the Whites for a few days until they found a suitable rental. Even with insurance, however, the cost was harsh, financially and emotionally, particularly in the many lost and ruined photographs of their daughter, Karen.

6

Don't Eat Toads

B efore getting to ride their rocket, Armstrong, Aldrin, and Collins each
spent two and a half years amassing over two thousand hours of formal
training. Dave Scott, who served with Neil Armstrong on Gemini VIII, kept
a detailed journal of his own Apollo regimen for the year 1966: 270 hours on
operations, 300 hours on systems, 350 on equipment design, 370 on spacecraft tests, and 680 in module simulations. Besides working in NASA's own
trainers in Houston, Kennedy, and Edwards, Apollo crews tested the Command Module at North American's Downey, California, plant and the Lunar
Module at Grumman in Bethpage, Long Island. Armstrong's direct training
for Apollo 11 totaled 959 hours, with 285 (30 percent) spent in LM simulators; Aldrin logged 1,017 hours of training, with 332 hours (33 percent) in the
LM. But these numbers did not include the time spent in camera lessons,
astronomy class, the gym, giving press interviews, or the hundreds of hours
of travel time in T-38s between Downey and Grumman and Kennedy and
Houston, or to Panama for survival training, or to the Gulf of Mexico for
water landings, or to the Texas Big Bend, the Grand Canyon, Arizona's Meteor and Sunset craters, and an Oregon lava field for geology field trips, or to
the dozens of other locations housing specialized machinery for education
that only NASA could conceive.

Even so, "training was [only] about one-third of our time and effort," Neil
Armstrong added. "A third had to do with planning, figuring out techniques
and methods that would allow us to achieve the trajectories and the sequence of events and the ways of picking from the available strategies the one
that might work the best. The last part was testing, and that's probably equal
to thousands of hours in the labs and in the spacecraft and running systems
tests, all kinds of stuff, seeing whether it would work and getting to know the
systems very well. So the one third that was training is training in a different
sense than most people think of training, because, after all, there wasn't

anybody that had done this and could tell us how to do it, because nobody had the experience."

This is the striking irony of astronaut life: men who were once the nation's best combat and test pilots, men who once lived for cars, and girls, but mostly for commandeering state-of-the-art flying machines, now lived in simulators and classrooms and never flew a damn thing. Like most of their fellow astronauts, Armstrong, Aldrin, and Collins would ride Apollo once and only once, and would each fly a rocket twice in a period of six years.

The intensity of training also turned wives and children into an immaterial set of voices on the telephone, for husbands and fathers who rarely made it home. Some families were destroyed. One of Apollo 11's CapComs, Charlie Duke, nearly lost his wife, Dotty, to suicide: "I considered divorce but I wondered if any man could love me the way I wanted to be loved. Was there even such a thing as the perfect marriage? It didn't look like it anymore. So I began to look for other things to fulfill me. I tried a career. I tried church work. I volunteered with the Head Start program and helped the needy. Nothing worked.

"I'll admit it—I even tried drugs. When the marijuana didn't work, either, I thought, Maybe there *is* no purpose in life. You just live and die and that's all there is. That's when I began to think about suicide. I lost all hope. Doesn't that seem strange? I was married to a famous man, had a nice home and healthy children, plenty of money, yet I had no hope."

"We work, like slaves we work; I haven't been to a movie for nearly two years," Deke Slayton admitted. "Out of the 365 days in the year, I spend at least 200 away from home." One controller said, "I missed the Vietnam War. I watched no television, read no newspapers, came to work at six in the morning and worked until nightfall, six or seven days a week for years."

Armstrong and Aldrin trained to work in lunar gravity (one-sixth of earth's) with scuba gear in a Texas water tank and with Grumman's Peter Pan rig, which used pulleys and cables to lift and artificially lighten a fully suited astronaut. A Boeing 707 jet, its seats removed and its walls padded, flew parabola arcs (like an aerial roller coaster summiting peak after peak) to simulate twenty seconds of zero-g. The "Vomit Comet" was far from the only sadistic machine in NASA's educational arsenal, but at least the Gemini and Apollo trainees no longer had to undergo training with MASTIF (Multiple Axis Space Test Inertia Facility), an early Mercury simulator described by Gus Grissom: "You rotated faster and faster until you were spinning violently in three different directions at once—head over heels, round and round as if you were on a merry-go-round, sideways as if your arms and legs were tied to the spokes of a wheel. Your vision blurred. Your forehead broke out into a clammy sweat. And unless you could stop all of this with your stick, you could get sick enough to vomit." The head of the Astronaut Office, Deke Slayton, admitted, "We didn't know how to train an astronaut in those days. We would use any training device or method that had even a remote chance of being useful, and we would make the training as difficult as possible so that we would be overtrained rather than undertrained."

"Overtraining" barely describes Apollo 11's punishing workload. Besides all the astronauts' spending fourteen-hour days in the Houston simulators learning everything that could possibly go wrong with machines and their ground controllers, Collins also had a docking simulator at Langley, Virginia, to fly; space suits in Delaware to test; and 10 g's of centrifuge to overcome. This last, generated by "The Wheel," at the navy's Acceleration Laboratory at Johnsville, Pennsylvania, was half carnival ride and half torture device, an enormous motor wielding a steel arm, at the end of which a one-man capsule was positioned with a TV camera capturing the details of the subject's face as it contorted in anguish. Each would-be astronaut was monitored by doctors who tested his response to the stress of hard decelerations and up to 15 g's of force, pressure that flattened eyeballs and burst capillaries. Collins was even training for NASA before he got to NASA, because the USAF, annoyed that of the Original Seven, only three were air force men, launched a "charm school" to teach its prospective astronaut candidates how to impress NASA interviewers.

For Apollo crews, the task ahead was so overwhelming that each astronaut additionally had to develop an area of advanced expertise. Buzz Aldrin, with his interest in rendezvous, signed on for mission planning, while Mike Collins became the tester for a three-way competition between Hamilton Standard, David Clark, and International Latex to design and produce the Apollo space suit. (Its centerpiece was a rubber undersuit, like a tire's inner tube, assembled by senior citizens with glue pots; the pressure, at 3.7 pounds per inch, was a precaution against the fluids of the human body boiling in the vacuum of outer space.)

"I don't think it's gotten nearly enough credit, what Al Shepard and Deke Slayton decided to do with the astronauts who weren't directly involved in flights," Apollo 17's geologist/astronaut Dr. Harrison "Jack" Schmitt said. "In that having astronauts who were going to be at the tip of the spear, part of the review teams, part of the critical design reviews, getting very familiar with certain aspects of the spacecraft, the rockets, and the planning and operations of Apollo, was extremely important in making it real to everybody else that people are actually going to use these things that we're building. The astronaut's role was twofold: both a visible part of that quality control, but also a working part of it, in that as an astronaut thinking about flying, you bring a different perspective to the engineering, planning, the design of equipment. And we ended up having an office full of experts on different things.

"Each Monday morning there was a pilots' meeting in which the people on the various assignments would be back in Houston and report on any anomalies, any problems, anything of that kind that they thought everybody should know about. And [on] big decisions, the office came to some kind of consensus on how we ought to proceed (if Al Shepard and Deke Slayton's decision can be called a consensus). But still they had information, coming in from everywhere, on these various spacecraft activities. An extremely important part, I think, to the overall management of the program."

For his part, Neil Armstrong specialized in crew training and simulator design, notably the extremely difficult contraptions developed at Edwards with Buffalo, New York's, Bell Aerosystems: the Lunar Landing Research and Training Vehicles. Using a GE CF-700-2V turbofan jet, thrusters, and a regulation ejection seat, the LLRV/LLTV would use "earth mode" to rocket itself to five hundred feet, and then be throttled into "lunar mode" to simulate one-sixth gravity (its attitude was controlled, like the LM, by peroxide jets, whose appearance accounted for one of its nicknames: *the Belching Spider*). It was designed to imitate, as closely as possible, the Lunar Module, and it was just as difficult to fly.

Neil Armstrong: "The thing that surprises people on their initial flights in 'lunar mode' is the tendency of the vehicle to float far beyond where you think it is going to go . . . you must start to brake much earlier if you are to stop where you want to stop. Similarly, if you are in hover, and change your mind, it takes a lot of effort to get moving again. . . . We hope to have one-and-a-half to two minutes of fuel essentially in hover when we're landing on the moon, but you can use that up really fast if you change your mind frequently about where you want to go."

Armstrong would train so heavily in the LLTV that its team could predict his future. "Probably in 1968, my boss in the program office came to me with a question: 'Who's going to be the commander of the first lunar landing flight?'" LLTV program manager Charles Haines recalled. "I can't remember why we thought we needed to know that, but we did. I dug into it a bit and came back to him and said, 'It's going to be Armstrong. And the reason is, nobody else can meet all the training requirements by the time.' Mostly, in my mind, it was because the LLTV, I had a very bad opinion of it at that time in regard to its reliability and its ability to function and be operable. And I felt like Neil had a big step up on all the other competitors because it was going to take so much time to get that many flights in the LLTV. Nobody else had flown it. Nobody out of the Astronaut Office had ever flown it."

"All the pilots, to my knowledge, thought [the LLTV] was an extremely important part of their preparation for the lunar landing attempt," Armstrong said. "It was harder to fly than the Lunar Module, more complicated, and subject to the problems that wind and gusts and turbulence and so on introduce, that you don't have on the Moon. The systems were somewhat choppier or less smooth than the actual Lunar Module, both propulsion and attitude control systems were so. The Lunar Module was a pleasant surprise. As you may know, [it] was designed to be able to make an automatic landing, but, to my knowledge, no one ever did."

The use of simulators like the LLTV to train pilots began in 1910, nearly at the dawn of flight itself, with a "penguin system" of reduced-wingspan planes that never left the ground, and served to complement the traditional schooling of taxiing, short hops with elevators, and, finally, takeoff. In 1929 the Link Piano and Organ Company of Binghamton, New York, used organ bellow pneumatics to pitch, roll, and yaw a bench seat in response to a student's stick hand and pedal toes. Fitted with correctly responding gauges, Link Trainers were used in both world wars to teach instrument-only "blind" flying, and the onetime musical instrument concern would come to design Project Mercury's first trainer. Little by little, sims were refined until they reached a new level of verisimilitude in 1960 with the X-15, and then a great leap forward at NASA.

Beyond epochal breakthroughs in engineering science itself, NASA's greatest achievements in the run-up to Apollo 11 were its great strides in project management, and the growing sophistication of its crew training. The agency's remarkably advanced simulators were exact mock-ups of the Apollo spaceships and contained computer and movie screens, alarm sounds and lights that imitated a mission as closely as possible. These were so well engineered that NASA could have made a tidy fortune by developing this technology into video games and amusement-park attractions.

The Apollo simulator training regimes, however, were exhausting on every level of human effort, for both astronauts and their ground controllers. After

learning the basics on a sim flight that went smoothly, both sides were subsequently put through their paces with a series of problems like mechanical
breakdowns, conflicting data streams, and all-out system failures. Following
the dicta of flight operations director Chris Kraft—such as "If you don't
know what to do, don't do anything," and "To err is human, but to do so more
than once is contrary to Flight Operations Directorate policy"—an entire
NASA team was assigned the task of designing catastrophe scenarios (engine, communications, or environmental failures) that both astronaut candidates and the Mission Control operatives who flew with them would have to
overcome. During tests in which they failed and all sim-perished, trainers
and students became genuinely anxiety-ridden and ashen-faced. However
unnerving they could be, "simulators really are good because they create a
sense of confidence in oneself," Alan Shepard insisted. "You go up . . . and
the engine quits and you land safely; or you go up and the rocket goes sideways
and you get out, come back home and do it again. So there's a lot of confidence
created in the simulation business."

This process worked so well, in fact, that in time, many astronauts would
calm themselves during real-world crises by thinking: *This is just like a simulation.* Mission simulation chief Carl Shelley:

> The simulation organization in those days was organized pretty much the
> way it is now. It's a mirror image, if you will, of the rest of the Flight Control
> Division at that time. Well, we did have one discipline the rest of the flight
> control people didn't have. We had to simulate the astronaut. When we did
> those math models, we had to play the role of the astronaut. We called that
> position astro-sim.
>
> We had all the disciplines represented under a guy we called a simulation
> supervisor, who was responsible for developing all the scripting materials and
> things for exercising the flight control team. The Sim Sup, his major interface
> with the flight control team was with the flight director, so he and the flight
> director would sometimes collectively scheme for certain kinds of problems.
>
> I know [flight director Gene] Kranz had ideas. He'd say, "Well, you know,
> I'm not real comfortable with the way these guys are reacting over in this other
> area here. Dream up something you know that will exercise that." So there
> was very much a collusion going on between some of the flight directors and
> the Sim Sup. But there was also a lot of exercising that was done in which the
> flight director was not a party to what was going on, because he had to take
> the exam, too.
>
> Remember, now, you're not trying to define all the "what if" situations.
> You're only trying to define those that the flight control team or the crew can

do something about. For example, an obvious example, something like a *Challenger* accident, we didn't bother to simulate something like that. There's no training value in it, and there's nothing anybody could have done about it.

"We had to develop the flight rules for how do you monitor a lunar descent," flight dynamics officer (FIDO) Jay Greene remembered. "By the time it was over, what we decided was that we would use abortability as the criteria to terminate a landing. The first thing we monitored for was to keep the crew from crashing into the Moon. That would have ruined everything.

"[The simulations for Apollo 11 were the most intense of any mission because] (a) we had a deadline; (b) we were doing something that had never been done before; (c) we anticipated aborting. Nobody, nobody on the team believed that we'd make it down the first time; I don't think anybody did. That became particularly bothersome for a flight dynamics guy, because the first part of the job was monitoring the descent; the other part of the job was if we abort, computing all the rendezvous maneuvers to get the LM back with the Command Module. So, it was intense. Then you had this accident investigation if you had a crash, and, yes, we had crashes. The spotlight was on us; the papers were counting down."

If astronaut training itself seemed, after the thousandth hour, mind-numbing to an unbearable degree, there was also extracurricular coursework to contend with, never-ending and all-encompassing. When NASA became concerned about what might happen should its fliers land off-course back on Earth, the astronauts were enrolled in a very different kind of school. Jungle survival training instructor Morgan Smith:

What we did [in Panama] was pair them off. They were issued a little twelve-inch-blade machete [and] a jungle hammock, because the jungle hammock would keep [a person] from getting wet at night and allow them to sleep. They fished, they hunted, but mostly they cut into heart of palm. The heart of palm was the basic for all of these guys in training. And, after all, you don't have to shoot it and hunt it down. It's there, and if you have a pocketknife, which you should always have, you can scrape one out. We had a plant and animal demonstration area where plants were introduced that they would find, and they were from all over the world's tropics. Most all of them practiced cutting a vine and drinking the water out of the vine, which is the only pure water you could get without boiling it. They were shown where to collect dead wood, where to get the fibers from palm trees at the base of the leaf to start a fire, and that sort of thing. They had to finish a lean-to themselves, and if it leaked, they had to repair it, they had to redo it. Balsa trees were there, and they would cut a balsa tree with their machetes, and you make a donut-type [life preserver]. They had to cut at least one tree to make a clearing that they could signal. Mostly they ate fish, small little minnows and small fish they caught in the streams. For meat, once in a while they'd catch a lizard. If they got an iguana, they were very lucky.

Mostly it was an orientation that the tropical rain forests are friendly places. It's how the human reacts that makes them dangerous. For example, they learned that you cut yourself there, you're going to infect fast. Well, the converse is true. All biology goes faster. Heat, light, moisture, that's the tropics. . . . The basic thing I think they came out of, and in talking to them later, is that, "Hey. If we get there, we're okay. We'll wait for rescue." And that's exactly what we wanted to establish.

I know the Air Force generals were beside [themselves when briefed that the astronauts were out in the jungle without communication]. "You did what with them? They're by themselves and they don't have radio? You can't contact us? What if something happens?"

"Then they won't be astronauts."

John Glenn and Neil Armstrong were together. They were in Choco Indian territory, and John Glenn and Neil Armstrong wrote a sign on their little lean-to and called it the Choco Hilton. They did it with charcoal on the white parachute cloth.

John said he remembered that training, and that's the one thing that he said made him feel more comfortable that no matter where he may have wound up in a remote area, the people were going to be great. They didn't have to know anything about the space program or about competition with the Soviet Union and all that. The fact remains that he thought that experience of meeting and dealing with those people—and we're very proud of that and so are the Choco. They realized that they had done [an important] thing. And then when Pete [Conrad] came back and went down and told them, they believed he had been to the Moon. As you know, a lot of people in the United States believe it was all done on film.

What crucial tip did Michael Collins learn in Panama? *Don't eat toads.*

"I was having to try to organize a brand new training program for the Apollo missions late in 1966," geologist/astronaut Jack Schmitt said. "It seemed to me that we were really off track. We were boring the astronauts. The intent, apparently, was of the people then in charge to create—in their words, 'create astronauts with master's degrees in geology.' Well, that's not what we needed. What we needed were focused, relatively narrowly trained field geologists who primarily were pilots, but people who could select—who could observe and select the widest variety of rocks, could tell us about the context in which each of those samples came from, and so forth. And since simulation training was working so well and all other aspects of our training, I went to Al Shepard with a proposal that we begin to focus our science training on an actual simulation of lunar traverses in areas on Earth where we could learn something about the kinds of problems that we would encounter on the Moon."

"I can remember field trips with the Apollo 11 crew, and relatively late field trips in Arizona where we were practicing them getting out of craters if they got into trouble," Apollo lunar exploration geologist James Head said. "So it wasn't so much geological training as actually real terrain training, basic geological things, but also having somebody go down into the crater they can't get out of and then [using this] tether-like thing that they carried. You'd have dinner with them at crew quarters, which is a real treat. I mean, the food was—I've never had pork chops like that before or after. I don't know where the hell they got them, but just say the word 'astronaut' next time you go to the store in Cocoa Beach and maybe that'll work."

In time, Armstrong's, Aldrin's, and Collins's training grew unbearable. Aldrin got so overworked that, while commuting one day in a T-38 jet, he had to double-check the compass to remember whether he was on his way to Florida or Houston. Armstrong grew so frustrated by the endless sim aborts that he decided, when it came to landing on the Moon, there would be one and only one mission commander. "Until *Eagle* was about ten thousand feet high, its altitude was based on Earth radars, and its guidance system could be off by hundreds or thousands of feet," Chris Kraft explained. "Then the LM's own landing radar was supposed to kick in and provide accurate readings. That led to some heated discussions. Neil worried that an overzealous flight controller would abort a good descent, based on faulty information. 'I'm going to be in a better position to know what's happening than the people back in Houston,' he said over and over. 'And I'm not going to tolerate any unnecessary risks,' I retorted. 'That's why we have mission rules.'

"We argued the specifics of the landing radar, and I insisted that if it failed, an abort was mandatory. I just didn't trust the ability of an astronaut, not even one as tried and tested as Neil Armstrong, to accurately estimate his altitude over a cratered lunar surface. It was unfamiliar terrain, and nobody knew the exact size of the landmarks that would normally be used for reference.

Finally we agreed. That mission rule stayed as written. But I could tell from Neil's frown that he wasn't convinced. Everything in his experience had taught him to trust his own judgment. I wondered then if he'd overrule all of us in lunar orbit and try to land without a radar system."

That worry was not unfounded for Armstrong's only previous experience in outer space, Gemini VIII, had been aborted. If his Apollo flight was similarly terminated, what would that say about Armstrong's reputation as a pilot? Flight dynamics officer John Llewellyn: "What got me, when I got into that lunar thing, did you know that they had more ways to abort than be successful? I couldn't believe that. I said, 'What in the hell? I mean, we got eleven aborts and not one landing. We're not here to abort.' I mean, I would go berserk with that. 'I mean, why are we doing this? We didn't come to the Moon to find out how close you could get and come off. That's not what we're doing here, gentlemen.'"

A few weeks before their liftoff, Aldrin and Armstrong were running through yet another lunar landing simulation when they found themselves imperiled by a stuck thruster. Aldrin watched the gauges spin crazily while, out the window, the TV moon rose up at them in deadly menace. Knowing he should not contradict his commander, Aldrin waited until the last possible moment to whisper, "Neil—hit abort!" Armstrong, wanting to test the ground crew, but not explaining this to Aldrin, let the exercise take its course. Mission Control finally announced, "Apollo 11, we recommend you abort," but by then, it was too late. Their ship had crashed, and Armstrong and Aldrin were sim dead.

That night in quarters, as Aldrin was telling Collins that Armstrong's intransigence would be a black mark against all of their records, the door opened. It was Armstrong in his pajamas. "You guys are making too much damn noise out here," he complained. Aldrin then knew he had heard every word. They talked it over the next day, and the matter seemed settled, but the innate chill between the cerebral Aldrin and the remote Armstrong intensified.

As historian Andrew Chaikin noted, "They had less than seven months to train for the most ambitious space mission in history. So it's no surprise that strain got to them." Jack Schmitt:

> Neil was quoted as saying and it was, I think, an accurate quote, is that he felt that he only had a finite number of heartbeats and he wasn't going to waste any of them on exercise. That was, I think, in a *Life* magazine article.

And I'm pretty sure I heard Neil say that in person. Buzz Aldrin was just the opposite. Buzz is very much an exercise freak, if you will.

And one time at about two weeks before launch, Neil came into breakfast with a handgrip—a rubber handgrip—and he just sat down at the table and was gripping this. And Buzz Aldrin was sitting on the table just getting redder and redder because here it was two weeks before launch and Neil knew exactly what he was doing, you know. He was just teasing Buzz by gripping this.

The thing, though, that most people don't realize is that in training in pressure suits, you get an awful lot of exercise. It's very, very difficult to go through training for a mission and not get in pretty fantastic shape for the kinds of things that you have to do in a pressure suit. So, Neil hadn't really skimped on anything. But he was just giving Buzz a hard time.

While Armstrong and Aldrin spent countless hours in the Lunar Module sim, Collins trained solo in the Command Module version and, in time, came to a decision: Enough was enough. He told his wife that, after Apollo 11 was mission accomplished, he would retire. She suggested he could wait until then to make that decision. Instead, just before the Apollo 11 liftoff, Collins turned down astronaut chief Deke Slayton's offer for a slot as backup commander on Apollo 14 (which would have meant a commander position for 17). "Being an astronaut was the most interesting job I ever expect to have, but I wanted to leave before I became stale in it, and I could tell that after Apollo 11 I could not have prevented myself from sliding downhill, in terms of enthusiasm and concentration," Collins explained. "Hence I find myself in the weird position of saying I'm glad I no longer have one of the most fascinating jobs in the world, but I think that is an honest summation."

Apollo training eventually became a tremendous struggle for everyone involved in the effort. Flight director Gene Kranz:

> Training for the lunar mission was probably the most difficult time of my entire life. . . . The training process for Apollo 11 began very late, because the lunar landing software in the simulators wasn't ready. The first month of training with my team went like a champ, and then Sim Sup looked at the team and decided we need to be taught a lesson, and he started increasing the pressure associated with the descent phase. Now, when you're going down to the Moon, just like landing an airplane, there is essentially a dead man's box. No matter what you do, you can throttle up, you can change your attitude,

you're going to touch the ground before you start moving back off again, and it's the same kind of condition, but it isn't a neatly drawn line; it's a set of variables, and it depends upon the altitude and the speed at which you're descending down how this box is defined.

Well, then, if you add in the effect of the lunar time delays, the fact is that everything we're seeing is about three seconds old, and then we have to figure out what our reaction time is and then voice some kind of instruction to the crew on what to do. You have to start defining a set of boundaries. You're going to have to make up your mind before you're actually into the problem. This gets to be very dicey.

Anyway, we're now in one month prior to launch, and as we're moving into this final month of training, Sim Sup really laid it to us, and it related to this dead man's box and this lunar time delay. We went through a series of scenarios that was almost—it seemed like forever. It was only a couple of weeks, but it seemed a lifetime where we could not do anything right. Everything we would do, we would either wait too long and crash or we would jump the gun and abort when we didn't have to, and the debriefings were absolutely brutal during that period of time.

A bit about Dick Koos as Sim Sup. He was one of the very early pioneers of the Space Task Group. He came in out of the Army Missile Command in Fort Bliss, Texas, because in those days you couldn't hire people with computer degrees. You just went after people who had the experience. Well, the Army was working with computers in their ground-to-air missile program. Koos was one of the guys who ended up in training in the very early days, and training in the early days was incredibly rudimentary, but by the time that we got to the Apollo program, it really had become quite sophisticated. Training in Apollo was about as real—I mean, you would get the sweaty palms—when the pressure was on in a training episode, it no longer was training, it was real, and the same emotions, the same feelings, the same energies, the same adrenaline would flow. Koos was causing all this to happen, but he decided my team wasn't ready, so he kept beating us up and beating us up and beating us up. At times it got so bad that the prime Apollo 11 crew, Armstrong and Aldrin, just didn't want to train with us anymore, and we didn't want to train with them, to the point where they'd go off in a different simulator and we'd work with the backup crew or work with the Apollo 12 crew.

Finally, we went through a bad, bad, bad day. We had crashed, and we had crashed. And then to avoid crashing, we'd become unnecessarily conservative; and we'd abort when we could've landed. And by the end of the day, we felt pretty bad. Over in the offices where George Low and Chris Kraft would sit, they would listen to our training exercises. They had these little squawk boxes, and they had the air-ground loop, and they had the flight director's loop, and every time we'd have a bad day or a bad session, they'd grit their teeth. And about that time, Chris Kraft calls up on the phone. And from his initial comments, I knew he had been listening to these simulations, and I knew he was watching us struggle. And he said, "Is there anything I can do to help you?" And for the first time in this entire process, I felt the pressure that, "Hey, maybe our bosses were starting to lose confidence in this team that they had signed to do the mission."

My response was very straightforward. I put a switch on this phone so it wouldn't ring anymore. So, he could call all day and he'd just get a busy signal. But we proceeded to dig ourselves out of the pit that we had somehow dug for ourselves. We set a different set of parameters in defining this dead man's box. We biased the times that we would use to make the calls. We became more conscious of the clock. But piece by piece by piece, we started putting it back together again until we felt not only were we going to get the job, "Hell, yes, we're going to get the job done!" There was no question that we would get this crew down to the surface of the Moon.

No one wanted to imagine the worst, but for many both inside NASA and outside the agency, it was a necessary part of their job. It was in fact what most of the simulation engineers did for a living, alongside the hardware engineers attempts to fail-safe through redundancy. If, however, disaster struck Apollo 11, William Safire had prepared a speech that President Nixon would have given as Buzz Aldrin and Neil Armstrong lived out their final . hours:

Fate has ordained that the men who went to the moon to explore in peace will stay on the moon to rest in peace.

These brave men, Neil Armstrong and Edwin Aldrin, know that there is no hope for their recovery. But they also know that there is hope for mankind in their sacrifice.

These two men are laying down their lives in mankind's most noble goal: the search for truth and understanding.

They will be mourned by their families and friends; they will be mourned by their nation; they will be mourned by the people of the world; they will be mourned by a Mother Earth that dared send two of her sons into the un-known.

In their exploration, they stirred the people of the world to feel as one; in their sacrifice, they bind more tightly the brotherhood of man.

In ancient days, men looked at stars and saw their heroes in the constella-tions.

In modern times, we do much the same, but our heroes are epic men of flesh and blood.

Others will follow, and surely find their way home. Man's search will not be denied.

But these men were the first, and they will remain the foremost in our hearts.

For every human being who looks up at the moon in the nights to come will know that there is some corner of another world that is forever mankind.

7

A Way to Talk to God

Following the announcement of Apollo 11's launch date, a great swarm
headed for Cocoa Beach, the seaside town directly south of Cape Canaveral's 88,000-acre Kennedy Space Center on Merritt Island. Even by NASA
shot standards, the Apollo 11 draw was a crushing flood—over one million
spectators, hoping for a glimpse of history, descended on the narrow barrier
islands southeast of Orlando in central Florida. Wernher von Braun and his
wife alighted from a helicopter on a nearby golf course; in time, they were
joined by Spiro Agnew, Lyndon and Lady Bird Johnson, Sargent Shriver, Jack
Benny, Cardinal Cooke, Daniel Patrick Moynihan, Barry Goldwater, Johnny
Carson, Gianni Agnelli, Prince Napoleon of Paris, 400 foreign ministers,
275 corporate executives, 19 governors, 40 mayors, half of Congress, 1,000
police and state troopers, 3,000 boats anchored in local waters, 3,497 journalists, and an exaltation of Supreme Court justices.

Neil Armstrong's wife and kids, joined by astronaut Dave Scott and *Life*
magazine reporter Dodie Hamblin, were part of a North American Aviation
yacht party on the Banana River. This was unusual, for the great majority of
astronaut families avoided attending launches out of concern for their children, and the media assault, in the event of a disaster. "I remember that we
did not go to the Cape to watch the launch. I found out later it was because,
#1, they could not afford it (money was always tight with five kids and a
military salary) and, #2, my dad did not want us all out in the grandstands
in case the rocket exploded during the launch," Gayle Anders said after
Apollo 8. Neil Armstrong, in fact, had urged Jan not to come, but she insisted. Before she could fly to the Cape, however, Mrs. Armstrong—a synchronized swimming coach and "as strong as horseradish," according to
college friend Gene Cernan—stood atop a screaming-pink dais erected in
front of the family's El Lago, Texas, home, and endured another press conference:

"Will you let the children stay up and watch the moon walk?"

"I don't care for what they do."

"Is this the greatest moment of your life?"

"No, sir. When I was married, it was the greatest moment of my life."

"Are you pleased with the Sea of Tranquility as a place to land?"

"Yes."

"What are you having for dinner tonight? Space food?"

"No, sir."

The astronaut wives were so flabbergasted by the absurdity of the press's questions that they had developed a skit parodying the entire process:

"We're here in front of the trim, modest suburban home of Squarely Stable, the famous astronaut who has just completed his historic mission, and we have with us his attractive wife, Primly Stable. Primly Stable, you must be happy, proud, and thankful at this moment."

"Yes, Nancy, that's true. I'm happy, proud, and thankful at this moment."

"Tell us, Primly, tell us what you felt during the blast-off, at the very moment when your husband's rocket began to rise from the earth and take him on this historic journey."

"To tell you the truth, Nancy, I missed that part of it. I'd sort of dozed off, because I got up so early this morning and I'd been rushing around a lot taping the shades shut, so the TV people wouldn't come in the windows."

"And finally, Primly, I know that the most important prayer of your life has already been answered: Squarely has returned safely from outer space. But if you could have one other wish at this moment and have it come true, what would that one wish be?"

"Well, Nancy, I'd wish for an Electrolux vacuum cleaner, with all the attachments. . . ."

Armstrong had also told his parents that it would be better not to come to Kennedy, so they had stayed home in Wapakoneta, Ohio. Imagine your son being chosen as the first man to go to the Moon; could there be a prouder moment in any parent's life? Instead, Neil Armstrong's mother and father found their small town and their modest home under attack by a ceaseless horde of reporters and photographers, the broadcast networks even parking an eighty-foot transmission tower in their driveway. When they admitted they only had a black-and-white TV to watch their son's historic moment, a big color set arrived, courtesy of ABC, CBS, and NBC.

By Tuesday, July 15, every room for let within fifty miles of Kennedy's Pad

39A was taken; a thirty-mile swath of highway was quadruple-parked end-to-end with untold thousands of cars and trailers and motorcycles and campers stocked with beer, Pepsi, and bikinis. It was the middle of summer in the middle of Florida, meaning a heat that melted asphalt onto the soles of barefoot children and a humidity that made women sweat like Teamsters, especially that remarkable gaggle of lithe and adventurous females that made their way to Cocoa for every shot, pretty young things on the hunt for astronauts, or their best buddies, or somebody who worked at NASA, or *somebody*, girls who could be counted on to have a swinging time at the Satellite, Vanguard, Polaris, Rocket!, or Space Girls taverns, drinking liftoff martinis or moonlanders (vodka, soda, lime juice, crème de menthe, and crème de cacao). Legend has it that a woman known as "Wickie" would trump them all by sleeping with six of the Original Seven. Cocoa's innkeepers knew their trade; even under this giddy torrent, the town never ran out of liquor, gasoline, or food. It did, however, run out of alarm clocks.

In the last days before liftoff, the crew was given the use of a beachside cottage, one of the many properties that came with NASA's purchase of Merritt Island. For their last supper, instead of a state dinner with President Nixon, they had "broiled sirloin, mashed potatoes, tomato puree, buttered asparagus, combination salad, cottage cheese, fruit bowl, bread, butter, and a beverage," as the NASA press office revealed. Their cook, Lew Hartzell, was an ex-marine who'd honed his kitchen skills on a tugboat. It hadn't taken more than a few nights of Lew's grub before Armstrong confided to Jan, "I'm sick of steak!"

Anyone attending a Cape Kennedy launch who had ever dreamed of the Space Age couldn't be faulted for imagining that it would be a Tomorrowland of PeopleMovers and personal jet packs. Instead, the site was as relentlessly industrial a manufacturing facility as one could imagine: a scattering of utilitarian 1960s office buildings, generic assembly factories, and sheds made from slabs of concrete topped with corrugated metal, set against an outback of pine, scrub, and palm. But, instead of valves, or ball bearings, or Chevelles on the assembly line, there were nose cones, and exhaust nozzles, and flotsam and jetsam of rocket parts—a common industrial park, but one testing LOX engines and LH2 tanks.

As deserted as it looked, the Merritt Island that would become Cape Kennedy was home to over five hundred creatures, including a great array of ibises, gulls, herons, egrets, bald eagles, raccoons that fished the swamps, wild pigs that were regularly struck by cars in the dead of night, great circling flocks of turkey vultures, and a family of big, black alligators that lived right next to

the Vehicle Assembly Building. There is a widely held belief that the Cape
had seen a burst in the neighborhood's lynx population, lured by the ready
supply of deaf rabbits living too close to the launchpads' explosive roar. And,
as the sun fell, Kennedy's most ravenous inhabitant would appear: mosquitoes, in uncountable swarms, perhaps explaining why, until sixty years ago,
few humans lived there.

The day before liftoff, a mule-and-wagon carrying the Reverend Ralph Abernathy (head of the Southern Christian Leadership Conference and successor to Martin Luther King) appeared with a group of protesters before the
gates of KSC. The decade had seen a small but voluble public protest against
Project Apollo, and some of the SCLC's chiefs were its most visible opponents,
its leaders arguing that federal monies would be better spent on Earth-based
needs instead of starry dreams. Medical operations director Charles Berry:

> I went up to [Reverend Abernathy] and I said, "You know, I do not under-
> stand why you would come and try and demonstrate and say that we ought not
> to have this flight to the Moon. Do you have any concept at all about what this
> can mean to the world and to us as a nation, having the capability to do this?"
>
> He said, "It's really not about the capability to do this, it's this money that's
> going to the Moon, this money's going to be on the Moon, and it should be
> being spent on these people down here on the Earth."
>
> And I said, "There isn't a single dollar going on the Moon. Not one dollar
> going to be on the Moon. Every one of those dollars that's gone to this pro-
> gram, and a lot of this nation is involved in that, and every one of those dollars
> is going to somebody down here on the Earth. If some of your people wanted
> to be working on some of that, they could have done it. I'm sure that jobs are
> there. You could work on it, and you could be getting some of that so-called
> moon money, if you want to call it that."
>
> "That's not what I'm saying," he said. "The thing is, that money ought to be
> spent on these people right down here."
>
> I said, "Well, you obviously don't understand what is happening here, and
> it's being done for your good and for everyone's good. If a nation is great, it's
> my view that that nation ought to be able to do both things, and we ought to
> be able to do the things that are necessary here. We need the science and the
> technology on the cutting edge if we're going to be a nation that's going to prog-
> ress. If you don't, you're going to die as a nation and you're not going to solve any
> of the problems here on Earth or anywhere else."

Later, NASA administrator Tom Paine went out to meet with the reverend
and offered to let some of the group watch the launch from the Center.

Abernathy would later admit, "I succumbed to the awe-inspiring launch . . . I was one of the proudest Americans as I stood on this soil; I think it's really holy ground."

At launch all NASA rockets are aimed to the east, partly to take advantage of a boost from the Earth's 915 mph rotation, and partly to soar over the Atlantic away from human inhabitation in case anything goes wrong. Apollo 11's liftoff of 9:32 a.m. EST on July 16, 1969, was chosen based on daylight hours available at both the launch and recovery sites, weather predictions, and the elliptical orbit of the Moon, which can be located between 221,000 and 242,000 miles away from the Earth while traveling at 2,287 miles an hour. A lunar mission targets the position of the Moon at the time of the ship's arrival, and there is a mere four-hour margin of error for the moment when a rocket's engine must hurl it out of Earth's orbit and into a lunar trajectory. As recently as 1965, Soviet probe Luna 6 had missed the Moon by 100,000 miles and zoomed off into deep space.

Across Kennedy Space Center on that July 16 of 1969 rose wave after wave of electrical happiness, like that of young children on the night before Christmas, a mass exhilaration mixed with an equal undercurrent of fears (many NASA employees who later saw *Challenger* disintegrate cannot bear to witness another liftoff to this day), as well as a sensation of profound achievement. Even the lowest subcontractor employee most distant from the work on the pad knew: launch is why they were here.

T minus 5.25 hours: The phone rang in crew quarters at 4:15 a.m. with the voice of Astronaut Office head Deke Slayton: "It's a beautiful morning!" After shaving, showering, and dressing, Slayton (along with NASA artist Paul Calle and backup crewman Bill Anders) joined the three crewmen for a breakfast of steak, eggs, orange juice, and toast (which followed the NASA physicians' strategy of low in fiber, low in waste). The 11 men had earlier autographed a set of commemorative stamp envelopes—one stack to be left behind, as a kind of insurance for their wives and children in case the worst happened, and the other to join them on the flight, alongside a host of other outer space memorabilia.

"On Apollo 11 I carried prayers, poems, medallions, coins, flags, envelopes, brooches, tie pins, insignia, cuff links, rings, and even one diaper pin," Collins said. "Some things belonged to me and some to others. The only criterion was that the object had to be small, but none matched the ingenuity of one gent for whom I carried a small hollow bean, less than a quarter of an inch long. Inside it were fifty elephants, carved from slivers of ivory, which he planned to distribute to his coworkers after the flight."

"We ran a detailed rendezvous meeting of some sort, and afterwards I said, 'You know, Buzz, if you want to guarantee the long-term success of Apollo and that we'll continue to go back to the Moon just forever, all you need to do when you get up there, just take a little pouch filled with gold dust and just spread it out on some of those rocks when you collect them,'" Mission design head Kenneth Young recalled. "I said, 'We'll be going back there for a hundred years trying to find that gold.'

"He said, 'Oh, I couldn't do that. I couldn't do that.'"

Neil Amstrong wanted to play a similar joke on the expectant selenologists: "I was very tempted to sneak a piece of limestone up there with us on Apollo 11 and bring it back as a sample. That would have upset a lot of apple carts! But we didn't do it."

Collins, Aldrin, and Armstrong then went upstairs to get suited. What for Gemini had been a uniform trailer next to the launchpad was now a suite, with facilities more suitable to testing and storing outfits that had risen in cost to $100,000, on average—in Apollo 8's Frank Borman's case, his head was so large that his Apollo 8 helmet cost an additional $45,000. The space suits were incredibly complicated affairs, being, in effect, miniature spaceships, with twenty-five layers of nylon, Teflon, beta silica, jersey, neoprene, Kapton, Mylar, Nomex, and Spandex. Apollo 11's suits came with repair kits of replacement gaskets, bladder repair patches, outer shell patches, and cloth tape. Just like the ships and their various components, the suits had to be as thoroughly tested as possible, and came with their own crew, headed up by the elfin and well-tailored Joe Schmitt, who had overseen the dressing of every American astronaut since Alan Shepard:

> We suit technicians had been working in the suit room since 3:30, turning on the air and oxygen supply, making leak checks on the suit consoles, checking out the communications systems, laying out suit equipment, making sure suit pockets were loaded in correct order with pens, flashlights and so forth. On Neil's suit, a small folding shovel with plastic sample bags were placed in the special pocket. These were to be used in the event that their stay on the Moon was to be cut short for any reason, so at least they would come back with a few lunar soil samples.
>
> Two types of space suits were used. Mike Collins wore an intravehicular suit, which means that these suits were only to be used inside the spacecraft, while Neil and Buzz wore extravehicular space suits. Three were [custom-tailored] for each crewman—one for training, one for flight, and one backup

flight suit. It seems like a lot of money but when you consider that the extrave-hicular suits were designed to operate in a −250 degree Fahrenheit to a +310 degree Fahrenheit temperature range, and that it has ultraviolet radiation and a certain amount of micrometeorite protection, well I guess that was a fair price for a twenty-eight-layer space suit.

If a quarterback wore three uniforms on top of one another, he would be as comfortable and as flexible as were the men of Apollo 11. Mike Collins: "They have to be heavy, insulated, bulky, and then, even worse than that, they have to maintain, as their name implies, a pressure with a vacuum out-side, and you've got to be pumped up like the tube in your car, and that makes them very rigid, and then all kinds of ingenious engineering devices come into play to take something which fundamentally wants to be immobile, wants not to change shape, like the tire on your car, and force it to have joints at the elbows and the wrists and the arms and the knees and the heads and the whatevers, so that you can move around and do whatever you have to do inside it while it still is maintaining this rigid pressure. So to engineer a suit properly is an extremely complicated task."

To prepare themselves for suiting up, the crewmen first applied salve and taped on a diaper, and then donned a condom, tube, and waist-mounted col-lection bag for urine. Silver chloride biosensors were glued to their chests, sending heart and respiration data to a belt studded with cell-phone-sized radio transmitters, which forwarded continuous medical information—including the data from radiation dosimeters—to Mission Control physicians. They then donned long johns (made of cotton and officially referred to at NASA as "constant-wear garments") and then the suits themselves, which sealed and zipped from the back, and so always needed a helping hand to get in and out of. Next came "Snoopy hats," the brown and white skullcaps, which contained a mic and earphones. Omega watches set to Houston time (CST) were strapped to each wrist. Plastic wraps protected their boots from Earth-based micro life forms until they reached the spaceship. Oxygen tubes were connected to ventilator machines, which were shaped like small suit-cases and which they then carried for the rest of the journey to the pad, like businessmen on their way to the future.

In their suits and in the ships while traveling in outer space, the crew breathed pure oxygen; on the pad, the CM was filled with a mix of 40 per-cent nitrogen and 60 percent oxygen, a procedure followed in the aftermath of the Apollo 1 fire. Suit technician Joe Schmitt: "Nylon comfort gloves fol-lowed by the suit gloves were donned and locked to the suit arms. Next the fishbowl helmets were locked into the suit neckrings. At this point, the pre-breathing begins, as we turn off the air and turn on the breathing oxygen supply. Pressurized suit leakage checks are made after which the crew would lounge comfortably in their reclining chairs until we got the go-ahead from the pad leader, Guenter Wendt, to proceed to the spacecraft. Also, a ham on rye sandwich was carried along as a quick snack."

As they walked through the hallway to the transfer point, the astronauts were greeted excitedly by NASA well-wishers, but all they could hear was the whisper of their oxygen pumps and the plod of their galoshes. At that moment, Collins believed their chances of a fully successful mission were 50/50, as did Armstrong: "My gut feeling was that we had a 90 percent chance—or better—of getting back safely, and a 50 percent chance of making a successful landing." They approached the firetruck-red scaffolding and its battleship-blue open-cage elevators. "When you get to the base of this gigantic gantry, it's empty, there's nobody there, it's deserted," Mike Collins remembered. "You're accustomed to scores of workers like ants swarming up and down and around it, you're with a crowd of people, and then suddenly, there's nobody there. And you think, 'God, maybe they know something I don't know.'" Joe Schmitt:

We walked through a sealed compartment painted gray which reminded me of the inside of a navy ship. It was along the walls of these corridors that Guenter Wendt had placed signs. These signs read: "The Key," and then another one, "To the Moon," then another, "Located In," and finally, "The White Room." These signs tipped off the crew that Guenter was up to his old tricks again. Another thing I remember was the fishy smell of the hypergolic fuel. Of course, the crew didn't smell anything but pure oxygen as they were tightly sealed in their space suits.

A few more steps and we came to the base of the high-rise elevator. We all boarded the #1 elevator, pushed one button which is programmed to take us to the 320-foot level where the spacecraft hatch was located. As we walked along the creaky walkway we could hear cracking sounds of gases being burned off at the base of the burn pond located a good distance from the vehicle. Snow was falling from the cryogenic-filled tanks glistening in the floodlights.

After removing Neil's yellow boot protectors, he climbed into the commander's seat on the left. I followed him inside and immediately connected his communications line. Neil was still connected to his portable oxygen ventilator during ingress, so I switched over to the ship's [spacecraft's] oxygen line and turned on the oxygen valve. With his feet positioned in the stirrups, his restraint straps consisting of lap belt and shoulder harness were connected and adjusted. I had no voice communications with the crew so he would give me hand signals that he was comfortable. Mike was strapped in the right seat and Buzz in the center seat using the same procedure as Neil's. I made a quick check of everyone's equipment, asked them if everything was OK, and wished them good luck.

Grabbing an overhead bar, each crewman inserted himself into his bespoke couch, lying flat on his back to cushion against the crush of launch and splashdown. Backup crewman Fred Haise then helped Schmitt connect each man's suit to the ship's oxygen and com ports, adjust the seat harness restraints, and check that the dashboard switches hadn't been accidentally bumped by the men in their bulky suits getting settled in. The dashboard's most significant buttons, switches, dials, and latches were locked or hooded so they could not be accidentally turned, switched, or punched. Before Armstrong, Aldrin, and Collins had arrived, Haise had gone through the cabin with a 417-point checklist of to-dos—one more example of the exciting life of astronauts.

American spaceships were so small—the Apollo Command Module, a copper, silver, and white cone made from iron reinforced with porcelain, was a mere eleven feet high and thirteen feet at its widest—that the press regularly asked the astronauts about claustrophobia. Frank Borman explained why this wasn't a problem: "Here on Earth usually, when you're trapped in something, what's good is on the outside. In a spacecraft, what's good is on the inside and what's outside is death." Wally Schirra confirmed that sentiment. "Mostly it's lousy out there. It's a hostile environment, and it's trying to kill you. The outside temperature goes from a minus 450 degrees to a plus 300 degrees. You sit in a flying Thermos bottle."

Aldrin, the last to enter, watched the sun rise over the flat tidelands of Florida, casting into full relief the harsh and enormous shadow of the great American rocket and its accessory umbilical tower. Below lapped a thousand campfires tended by the huge audience that had made it to the beaches, while across the horizon was written the entire history of American space travel in a line of United States Air Force and NASA launchpads that had boosted them all, from chimpanzees Ham and Enos to Shepard's Pad 5, Glenn's 14, and Gemini's 19, the ruins of the fire at 34, and now, Apollo's 39A and B. It was a Florida dawn in July, 85 degrees and 73 percent humidity, with a light wind and a visibility of ten miles. Into the beach's sand, one group had written in enormous letters:

GOOD LUCK APOLLO 11

The crew finished their load-in procedure. Collins, Aldrin, and Armstrong lay on their backs, each cradled in a form-fitting couch, like an egg. Wendt and Haise silently wished them all the luck in the universe. The hatch was closed.

T minus 43 minutes: There was a barely perceptible jolt as the launch tower's uppermost capsule arm swung away to clear itself from the spacecraft, meaning the closeout crew had left, and that now, the launch would actually go forward. Armstrong: "These things were canceled more often than they were launched." Seated left to right, Armstrong, Aldrin, and Collins faced a cockpit that included five windows (momentarily covered by the emergency launch escape tower), numerous gauges and readout panels, an assortment of dials, and more switches than any pilot had ever before seen, over four hundred of them. NASA had learned in their simulator training sessions that pressure-suited men, working in those tiny cabins, could easily bump into the dashboard and flip any number of instrument controls without ever knowing it. Capsule engineers were forced to design a series of locks and security brackets to keep that from happening.

Everywhere on the walls were squares of Velcro, which mated to squares glued to every piece of equipment so that, in zero gravity, tools could be kept at a convenient arm's length without floating away. Posted all over the dashboard were last-minute to-dos, a wallpaper of handwritten cue cards, attached to the Velcro or stuck in the cracks of the instrument panel, annotations to the training, and the rocket user's guide. The whole of the ship—the Saturn's three stages, combined with *Eagle* and *Columbia*—were powered by engines and steered by thrusters totaling ninety-one rocket motors in all.

To the left of Armstrong were the bins containing the mission's complement of food and water; to the right of Collins was the necessary equipment for urinating and excreting, and directly below Aldrin's upraised feet were the navigation tools and the tunnel that, after a space minuet, would lead to the docked Lunar Module. Right next to Neil Armstrong's chair was the most important tool in the ship at that moment, which looked like a big oven dial, on a pole. It was the abort lever.

Guenter Wendt: "A lot of people ask the question, 'Were the astronauts nervous when they were put in the spacecraft?' And the answer is, 'No, they were not.' . . . If you prepare for your biggest vacation trip you ever made, you're going to fly to, let's say, Australia and so on. You worry about the day of the flight. Is the alarm clock going to go off? Is the car going to have a flat tire? Will the car get you to the airport? Will the airplane be on time? Will I make a connection? But once you're on the final airplane, you go back and you say, 'Ahhh, got it made.' See, and that's about the same as these guys. They have been through so many tests, so many dry runs, so many activities, that finally they say, 'Oh man, just close the hatch and let it go.'"

There was, in fact, an unacknowledged prayer shared by astronaut crew and ground control engineers alike at the start of every NASA mission— "Dear Lord, please don't let me screw up"—a sentiment felt more keenly on this flight than on any other in agency history. "We were our nation's envoys, we three, and it would be a national disgrace if we screwed it up," Michael Collins said. "We would be watched by the world, including the unfriendly parts of it, and we must not fail. . . . I don't know about Neil and Buzz, because we never discussed these things, but I really felt this pressure, this awesome sense of responsibility weighing me down, this completely negative sensation. . . . By flight time, I had tics in both eyelids, which went away as soon as we got airborne."

Neil Armstrong:

I was certainly aware that this was a culmination of the work of 300,000 or 400,000 people over a decade and that the nation's hopes and outward appearance largely rested on how the results came out. With those pressures, it seemed the most important thing to do was focus on our job as best we were able to and try to allow nothing to distract us from doing the very best job we could. And, you know, I have no complaints about the way my colleagues were able to step up to that. . . .

Each of the components of our hardware were designed to certain reliability specifications, and far the majority, to my recollection, had a reliability requirement of 0.99996, which means that you have four failures in 100,000 operations. I've been told that if every component met its reliability specifications precisely, that a typical Apollo flight would have about [1,000] separate identifiable failures. In fact, we had more like 150 failures per flight, better than statistical methods would tell you that you might have. I can only attribute that to the fact that every guy in the project, every guy at the bench building something, every assembler, every inspector, every guy that's setting up the tests, cranking the torque wrench, and so on, is saying, man or woman, "If anything goes wrong here, it's not going to be my fault, because my part is going to be better than I have to make it." And when you have hundreds of thousands of people all doing their job a little better than they have to, you get an improvement in performance. And that's the only reason we could have pulled this whole thing off.

When I was working here at the Manned Spacecraft Center, you could stand across the street and you could not tell when quitting time was, because people didn't leave at quitting time in those days. People just worked, and they worked until whatever their job was [was] done, and if they had to be there until five o'clock or seven o'clock or nine-thirty or whatever it was, they were just there. They did it, and then they went home. So four o'clock or four-thirty, whenever the bell rings, you didn't see anybody leaving. Everybody was still working.

The way that happens and the way that made it different from other sectors of the government to which some people are sometimes properly critical is that this was a project in which everybody involved was, one, interested, two, dedicated, and, three, fascinated by the job they were doing. And whenever you have those ingredients, whether it be government or private industry or a retail store, you're going to win.

The pressure, on all three men, was so enormous that, for the first thirty minutes of countdown, they sat in absolute silence.

By 7:30 a.m., the temperature under the tin-roofed shade of the VIP viewing stand had topped 100 degrees; one NASA contractor handed out paper sun hats. Behind the bleachers were the wooden shacks built by the broadcast networks, a string of beach cabanas on a grassy slope. Down below, stationed before the giant countdown clock and the banks of the barge turnaround basin, hundreds of photographers fussed and parried with their equipment—their historic subject, that machine of fire, smoke, and ice, smoking, snowing, and whistling 3.5 miles away, was blurred in their telephoto sights by the fumy thermals of a Florida morning. Though an uncomfortable day for human beings, at least the weather was cooperating for the needs of launch, which required a cloud ceiling of at least five hundred feet, and a windspeed of less than twenty-eight knots.

T minus 5 minutes: The launch escape tower, a 150,000-pound thruster attached to the Command Module, which the astronauts could use to pull themselves to safety in the event of a pad catastrophe, was armed.

T minus 3 minutes 10 seconds: Humans relinquished control of the countdown to the firing room's computer, which began several hundred last-minute tests.

T minus 50 seconds: The Saturn V's internal power engaged, and all but five scaffold service arms decoupled from the rocket's flank and swung away.

T minus 17 seconds: The firing room's computer gave the final trajectory to the Saturn's guidance systems. At this moment, the men aboard knew that they were actually going to fly. Neil Armstrong: "The reality is, a lot of times you get up and get in the cockpit, and something goes wrong somewhere and you go back down. So, actually, when you actually lift off, it's really a big surprise."

T minus 10 seconds: The launchpad's fifty-eight-foot-wide, forty-two-foot-deep blast trench was flooded with water to counteract the rocket's noise and heat.

T minus 8.9 seconds: The Saturn's five F-1 engines ignited. Producing a thrust that was four times the speed of sound, they created a fiery exhaust that hit a forty-foot-deep, 1.3 million-pound wedge deflector (laminated in volcanic ash and calcium aluminate), which split it in two. The twin flames then roared into the blast pit, vaporizing the water. The pit's nozzles constantly refilled it at fifty thousand gallons a minute, and huge clouds of mist and flame began to inflate and bloom from the sides of the rocket's base.

T minus 2 seconds: The five F-1 turbines reached 90 percent of full power—the thrust of 540 jet fighters—consuming ten thousand pounds of fuel every second.

During fueling, the Saturn's main tanks had been chilled to −423 and −297 degrees Fahrenheit so that they could then be topped with cryogenic (low-temperature) liquid hydrogen and LOX propellants. That freeze would in turn alchemize Florida's robust humidity into condensation frost on the rocket's skin, becoming flurries of snow and shards of ice that floated down to the ground during fueling and countdown, and crashed in great chunks during launch. That ice and snow, combined with the fire, smoke, and vapor of launch, as well as the five rocket motors' overbearing thunder, made for a spectacle that was both primordial and unearthly.

Since the onlookers stood at least 3.5 miles away, they witnessed a curious effect, in that the sound of the rocket ran fifteen seconds late, which meant that the flames of ignition and clouds of exhaust initially appeared in distant and absolute silence. Then the roar arrived, tardy but violent, a series of tremors under the feet and pounds to the chest. It could be heard hundreds of miles away (only a nuclear bomb is louder), and its force shook an immense radius of ground. When Walter Cronkite covered the first Saturn V launch for CBS in 1967, the broadcast trailer, also 3.5 miles from the pad, vibrated so severely that American viewers saw their beloved anchorman being pelted by ceiling tiles and insulation. Those watching that same liftoff from the top of Petrone's Launch Control Center, meanwhile, said the center's roof rolled like an earthquake.

The five F-1 engines burned eighty-five thousand pounds of gas in 8.9 seconds before the craft even began to rise. That delay was one of the great innovations of the von Braun team: a ring of hold-down arms attached to the rocket's waist kept it to ground until the thrust had achieved 7.5 million pounds—force enough to stabilize the grant craft's lift into the air. "The hold-down mechanism would release the rocket only after all five engines of the first stage produced full power," Wernher von Braun explained. "If this condition was not attained within a few seconds, all engines would shut down. In such a situation, unless special provisions were made for reattachment of some swing arms, Launch Control would be unable to 'safe' the vehicle and remove the flight crew from its precarious perch atop a potential bomb."

Then, very slowly, Apollo 11–Saturn V began to rise. At that moment of being airborne, mission chronology changed from T minus to GET or MET (General Elapsed Time or Mission Elapsed Time). The rocket's sound and the force of its liftoff were so profound that they overwhelmed the screams of over a million spectators, feeling that roller-coaster mix of rushing glee and tongue-biting fear, now calling out in unison with tears streaming down their cheeks: "Go! Go! Go! Go!"

At her home in Houston, Joan Aldrin could barely stand the tension. She smoked and fidgeted, her eyes wet with fear, the entire household silent and nervous with her. At the Collins home, though, Mike's wife Pat breezily announced to her guests: "There it goes!".

The rocket at first seemed to lift as slowly as possible, but inch by inch, it gained speed, rising faster, then faster still. "There was this enormous light, and the rocket goes up and up, and then it goes through the first skiff of clouds, and then through the second skiff of clouds, and then you see a puff of smoke—the first burnout—and then the rocket disappears," selenologist Harold C. Urey remembered. "The precision! The accuracy! If only a fraction of this precision and accuracy spins off into industry, it will pay for the whole space program." Norman Mailer was so overwhelmed by the experience that he could only describe the launch of Apollo 11 as mankind having found a way to talk to God.

Wernher von Braun had spent nearly thirty years waiting for this moment, for the mission when his greatest creation would be used for travel to another planet. NASA had completed its end of the space race with enormous speed and energy, even more brilliantly than its cousins at the Pentagon—it had been a mere eight years, after all, since the agency had sent its very first astronaut into the sky. Watching his masterpiece rise to the heavens, von Braun prayed out loud with tears in his eyes:

Our Father, who art in Heaven
Hallowed be thy Name,
Thy kingdom come
Thy will be done
On earth as it is in Heaven.

He then turned to a colleague and offered, "You give me ten billion dollars and ten years, and I'll have a man on Mars."

The author of 2001: A Space Odyssey had a similar reaction: "At liftoff, I cried for the first time in twenty years—and prayed for the first time in forty years," said Arthur C. Clarke. Walter Cronkite predicted: "Everything else that has happened in our time is going to be an asterisk," and von Braun later decided, "I think it is equal in importance to that moment in evolution when aquatic life came crawling up on the land."

The astronauts themselves were having very different thoughts and emotions. Michael Collins: "Shake, rattle, and roll! Noise, yes, lots of it, but mostly motion, as we are thrown left and right against our straps in spasmodic little jerks. It is steering like crazy, like a nervous lady driving a wide car down a narrow alley, and I just hope it knows where it's going, because for the first ten seconds we are perilously close to that umbilical tower." Apollo 8's Bill Anders described it as "like being a rat in the jaws of a giant terrier."

GET 00:00:12: After the tower's scaffold was cleared by the four outer F-1 rocket engines gimbaling at a slight yaw, control and communications moved from Kennedy to Houston. If the Saturn's onboard navigation system failed at this point, however, for the first time in NASA history, Neil Armstrong could fly the 363-foot rocket himself. Guenter Wendt: "[After crew insertion] we would go to a very forward position, what we called the roadblock, where the crew—I had always an emergency crew that in case we need to go in a hurry we would stay there right at the roadblock. I also had my electrical technicians, the mechanical technicians, and the system guys that were at the last minute to be available. Once it clears the tower, then you are ready to party, partially, because you have no longer control over anything. Houston has control over it. And there's nothing you can do anymore. Even if they parachute, they go up in the ocean and it's not any of your problem."

For those on the ground with binoculars, their very last sight of Apollo 11 before it disappeared from sight was of a white pencil of a ship, trailing a glowing red brush of fire.

So many around the world who watched the liftoff in awe could only concur with that morning's Parisian newspaper *Le Figaro*, which announced, "The greatest adventure in the history of humanity has begun."

PART II

8

How the Pyramids Were Built

It is unlikely that any of the hundreds of thousands of men and women working valiantly to make Apollo 11 succeed that day had even one moment of peace to consider the question at the heart of this massive undertaking: Why was the American federal government spending $29.5 billion to send human beings to the Moon? The headlong race to be first to another planet had been going on for so long and with such vigor that the only ones addressing this question were a group of social critics who believed that NASA's money could be better spent on earthly matters. But the history of everything that led up to Apollo 11 in fact is as complex and extraordinary as the story of Apollo 11 itself, a saga of remarkable coincidence, fortuitous happenstance, and everyday heroism, motivated by a complex range of factors: Fear. Discovery. Honor. Prestige. Charisma. Politics. Public relations. The 1960s. Humiliation. Because great nations do great things. Because it is there.

The standard version of the history of NASA has always been that, alarmed and deflated by Sputnik and Yuri Gagarin, the United States created a wholly civilian agency that, through the vital legacy of its youngest president, won the Space Race "in peace for all mankind." Besides the fact that almost all of these assertions are either misleading or expressly false, this account reduces the cumulative triumph of NASA, and Apollo 11 in particular, to an engineering marvel accompanied by a dozen or so scientific discoveries and a handful of trickle-down technologies (such as scratch-resistant eyeglasses and Tang). The actual story is much richer, and the achievement far more profound.

The Space Race, for its part, is traditionally regarded as having ended with "one small step" on July 20, 1969, and begun twelve years earlier, when the Soviets launched the world's first satellite. But Sputnik-to-Armstrong, the public and symbolic battle of two nations across the heavens, was only one leg of the great and overwhelming contest at the heart of the Cold War—the

Space and Missile Race—a contest that pitted America's Pentagon, CIA, and NASA against the USSR's Ministry of Defense, KGB, and Scientific Research Institute/State Commission on Piloted Flights. While the global public of the 1960s was enjoying the celestial olympics of Laika, Gagarin, Shepard, Glenn, spacewalks and moonwalks, twins of the rockets that launched astronauts and cosmonauts into the heavens were mass-produced, topped with nuclear warheads, and aimed at one another. In the wake of an ever-escalating ICBM showdown, both Soviets and Americans also fielded a contest of aerial reconnaissance—rockets in space that spied on rockets on the ground. What is affectionately and nostalgically remembered now as the Space Age might more accurately be perceived as the Rocket Age—with rockets in underground silos, rockets aboard submarines, rockets that were state secrets, and rockets that were public spectacles. The United States alone spent $5.5 trillion on its nuclear missile program.

The Space Race's opening salvo of Russian satellites, dogs, and air force officers orbiting across the planet also marked the onset of the American public's awareness of the Cold War's potential threats, in the form of looming missiles and nuclear warheads, and surprise attacks that could annihilate the world. Even today, examining this history, it is almost miraculous to contemplate that the only direct military confrontation between the United States and the Soviet Union were fighter sorties in the Korean War, and that the only time nuclear weapons were ever used in their history was on August 6 and 9 in 1945. At the very heart of the Cold War is the remarkable fact that, of those tens of thousands and trillions of dollars of missiles, the only ones that were actually fired carried satellites, dogs, chimps, and men and not atomic warheads.

After the USSR became a nuclear power in its own right and there were close calls during both the Korean War and the Cuban Missile Crisis, it became patently clear to most Soviet and American leaders that they could not directly engage each other in battle. Instead, they did so through a string of proxies, often with devastating results, in nations such as Guatemala, El Salvador, Egypt, Iraq, Indonesia, Laos, Nicaragua, Congo, Angola, Mozambique, Afghanistan, Chile, Cuba, Cambodia, and Vietnam. The one noble proxy though, was the battle in the heavens between the astronauts and cosmonauts of the Space Race. Throughout the 1950s and '60s, during the Cold War's tensest moments, the United States and the Union of Soviet Socialist Republics engaged each other through the development of satellites to monitor their respective military installations, while battling each other with cosmonaut and astronaut feats, for the sake of national pride and global public opinion.

As Soviet physicist Roald Sagdeev noted, "The Soviet government always was supportive of [space exploration] from the point of view of propaganda and flexing muscles. Venus was considered by them as one of the battle-grounds of the Cold War," while aerospace technologist Paul Lowman said, "The Apollo program was a non-military one, but it nevertheless involved developments in many technological areas that also are crucial to national defense: rocket propulsion, inertial guidance, computer utilization, electronics, radar, remote sensing, advanced materials, and others. The success of Apollo thus provided . . . a demonstration of the inherent superiority of American technology. It seems safe to suggest that this demonstration is a real contribution to prevention of a global thermonuclear war; no potential aggressor could plan a surprise attack on the United States without taking into account the military strength implied by it." Perhaps the most important reason for going to the Moon, then, is that the Space Race kept the Cold War cold.

The other great point in this history is that, while the threat of global nuclear annihilation lent the Cold War a constant background hum of fear, this was an era when many saw great hope for a glorious if not utopian future. Some dreamed of dramatic changes that would make the United States as a society more like the one envisioned by its Declaration of Independence. Others saw a remarkable future based on a revolution in science and technology. The ongoing arguments over the legacy of the 1960s are essentially debates over which dream came true, even though the answer—that both have come to startling fruition in ways far from their original intent—is clear.

The Space and Missile Race did not begin with the floating silver light of Sputnik, but with the dying gasps of Nazi Germany. On September 8, 1944, the Wehrmacht's 485th Artillery Detachment, a railway caravan of ethyl alcohol canisters, LOX tanks, and forty-five-foot, black-and-white checkerboard missiles, came to a stop and prepared their weapons at the lapping shores of the North Sea, in the German-controlled Netherlands, at the Hague. A group of their colleagues to the south, also traveling by rail and at the moment staged in Houffalize, Belgium, simultanously prepared their own attacks. It took two hours to fill the rockets' upper tanks with 9,180 pounds of ethyl alcohol and water, and the lower ones with 12,170 pounds of LOX. Hydrogen peroxide and sodium permanganate were mixed to create pressurized steam, which drove the fuel and its oxidizer at thirty-three gallons per second through eighteen ports of injection into the combustion chamber, where the massing propellant cloud was touched by a flame and ignited. A whistling cloud of escaping LOX was quickly followed by a tail funnel shooting an explosion of sparks between its skirt of stabilizer fins, which became a yellow-orange flame that grew to twenty-five tons of thrust.

The rocket soared ninety-five kilometers (fifty-nine miles) in an arc across the sky, controlled using a variation of German aviation's gyroscope autopilot and tail fins with carbon rudders and molybdenum vanes. Fifty-eight seconds after liftoff, the crew radioed a fuel cutoff message, which settled their flying bomb into the firm and gentle ballistics of gravity. The target: London's western suburb, Chiswick-on-Thames. There, with a telltale pink exhaust, the nose cone smashed into the ground and detonated 1,627 pounds of ammonium nitrate and amatol. Seconds later, another rocket launched from Holland struck to the north, in Epping, while a Belgian-fired missile exploded in just-liberated Paris, killing six and injuring thirty-six.

The Nazis had engineered a new form of artillery: the ballistic missile. For the first time in history, a series of rockets reliably landed where their designers wanted them to strike. In the final months of World War II, 18 more were launched on Paris; 86 fell on Liège, 1,610 hit Antwerp, and 1,403 targeted England, killing 5,400 in all.

The engineers who created and built this rocket called it "Unit 4," or Aggregat 4 (A4), but the Propaganda Ministry publicly referred to it as Vergeltungswaffe-2—Vengeance Weapon 2—and it is today commonly known as the V-2. The original, with today's hindsight, seems fairly primitive, and in fact the cumulative damage of each Nazi rocket fired—three thousand tons—was about what an American Flying Fortress could rain down on Germany in a single day. It has even been argued that the Nazi leaders' obsession with wonder weapons in the twilight of the Reich, with such determined focus and effort devoted to ballistic missiles and flying bombs instead of antiaircraft artillery, tanks, and submarines, helped usher in their doom.

To any military commander, however, the first guided missile was a great leap forward, if not in reality, then in potential. It traveled faster than the speed of sound, could not be intercepted, and struck with no warning. The only defense against them was to find and destroy their launching sites— which Allied sorties never achieved against the Nazis' V-2 mobile units—or so annihilate the front lines that the missile squads would have to retreat beyond the rockets' range. Just after the war, General Eisenhower admitted, "It seemed likely that if the Germans had succeeded in perfecting these new weapons six months earlier than they did, our invasion of Europe would have proved exceedingly difficult, perhaps impossible. . . . 'Overlord' might have been written off." In time, from this simple flying bomb would arise the dynamos of outer space exploration, aerial reconnaissance, and nuclear holocaust.

The men who made the A4/V-2 were led by a very blond, very blue-eyed, strikingly handsome, and immediately likable man, a Prussian baron descended from nobility on both sides of his family, Wernher Magnus Maximilian Freiherr von Braun. After receiving the confirmation present of a telescope from his astronomer-hobbyist mother, Wernher was immediately smitten: "At night I would stand spellbound looking at the moon and telling myself how near it was, how near." At the age of sixteen, he created his own spaceship by attaching a set of toy rockets to a wagon, and sending it exploding down Berlin's Tiergarten Allee. "It never occurred to me that [innocent bystanders] were not prepared to share the sidewalk with my noble experiment." In high school, he failed his first courses in math and physics, but after reading a book about voyaging to outer space, a book so exciting and yet so filled with formulae, he was determined to learn their underlying mathematics.

Although there are eight-hundred-year-old anecdotes of rockets appearing in Chinese fireworks and "arrows of flying fire" used in Mongol attacks (uncontrollable and inaccurate, but terrifying when employed on a massive scale), and although the British of the 1800s developed accurate, spin-stabilized missiles (which proved no match against the era's ever-evolving, dramatically improved artillery), modern-day guided-rocket science is generally considered to have begun at the turn of the century, a product of the obsessions of American Robert Goddard, Romanian Hermann Oberth, and Russian Konstantin Tsiolkovsky. What these far-flung men shared was a disdain for the chains of gravity, a desire to create ships that could travel to other worlds, and lives that were changed forever by reading the science fiction of H. G. Wells and, more crucially, Jules Verne. Verne's 1865 *From the Earth to the Moon* mixed science with fiction to imagine a Baltimore Gun Club that built an enormous cannon, the Columbiad (inspired by Newton's postulation that if a cannon fired its shot hard enough to reach a velocity of 18,000 miles an hour, it would never fall to the ground but instead be suspended in orbit), in Tampa, Florida, to launch three men into lunar orbit and return in a Pacific Ocean splashdown—almost exactly the mission of Apollo 11's historic predecessor Apollo 8. (It was in honor of Verne's visionary novel, and to honor Christopher Columbus, that NASA PR man Julian Scheer suggested that Apollo 11's Command Module be named *Columbia*.) Save for a reversed countdown, Verne's account of his space vehicle's liftoff was an uncannily prescient description of the launches of a century later:

> The moon advanced upward in a heaven of the purest clearness, outshining in her passage the twinkling light of the stars. She passed over the constellation of the Twins, and was now nearing the halfway point between the horizon and the zenith. A terrible silence weighed upon the entire scene! Not a breath of wind upon the earth! not a sound of breathing from the countless chests of the spectators! Their hearts seemed afraid to beat! All eyes were fixed upon the yawning mouth of the Columbiad. . . .
>
> "Thirty-five!—thirty-six!—thirty-seven!—thirty-eight!—thirty-nine!—forty! FIRE!!!" . . .
>
> An appalling unearthly report followed instantly, such as can be compared to nothing whatever known, not even to the roar of thunder, or the blast of volcanic explosions! No words can convey the slightest idea of the terrific sound! An immense spout of fire shot up from the bowels of the earth as from a crater. The earth heaved up, and with great difficulty some few spectators obtained a momentary glimpse of the projectile victoriously cleaving the air in the midst of the fiery vapors!
>
> At the moment when that pyramid of fire rose to a prodigious height into the air, the glare of flame lit up the whole of Florida; and for a moment day superseded night over a considerable extent of the country. . . .
>
> [Columbiad's creator, T. J. Maston, spoke:] "Those three men . . . have carried into space all the resources of art, science, and industry. With that, one can do anything; and you will see that, some day, they will come out all right."

Novelists can rarely be credited with inspiring wholly new avenues of science and technology, yet all three of rocketry's founding fathers read *From the Earth to the Moon*, and it changed the course of their lives. As Robert Goddard recalled: "On the afternoon of October 19, 1899, I climbed a tall cherry tree and, armed with a saw which I still have, and a hatchet, started to trim the dead limbs from the cherry tree. It was one of the quiet, colorful afternoons of sheer beauty which we have in October in New England, and as I looked towards the fields at the east, I imagined how wonderful it would be to make some device which had even the possibility of ascending to Mars. I was a different boy when I descended the tree from when I ascended for existence at last seemed very purposive."

As an adult, Goddard regularly alarmed his Clark University students by idly discussing the various methods to be used to visit the Moon, and developed hand-held rockets powered by nitroglycerin and nitrocellulose for

American forces in World War I. Peace was declared before they could be employed; instead, the technology was refined into World War II's bazookas. Then, on March 16, 1926, the Rocket Age was born when Goddard success-fully launched the first liquid-fueled missile, "Nell"—which he had wanted to power with liquid hydrogen, but because no one could successfully manu-facture it at the time, he was forced to use liquid oxygen and kerosene instead—from the cabbage field of his Aunt Effie's Auburn, Massachusetts, farm. Journalists of the day were not impressed; after a 1929 experiment, one newspaper headline read: "MOON ROCKET MISSES TARGET BY 238,799½ MILES," while *The New York Times* remarked that Goddard "seems to lack the knowledge ladled out daily in high schools" in contending that a rocket could travel through the vacuum of outer space.

Turning bitter and antisocial, Goddard left Aunt Effie's to test his liquid oxygen and gasoline–fueled red-and-silver rockets outside High Lonesome, New Mexico, where, assisted in the 1930s by Charles Lindbergh and the Smithsonian, he resolved fundamental issues of rocket steering, cooling, and plumbing. Like all great engineers, Goddard harvested progress from failure, calling his many disappointments "valuable negative information."

Away in the desert, Goddard became so secretive and unforthcoming that his fellow engineers almost unanimously considered him an eccentric crank . . . until they learned of his two hundred patents. On July 17, 1969, the day after the launch of Apollo 11, the *Times* published "A Correction" for its 1920 edi-torial ridiculing the founding father of American rocketry: "Further investi-gation and experimentation have confirmed the findings of Isaac Newton in the 17th century and it is now definitely established that a rocket can func-tion in a vacuum as well as in an atmosphere. The Times regrets the error."

If Robert Goddard's 1919 *A Method of Reaching Extreme Altitudes* (which outlined the mathematical theories of rocket flight and the pros and cons of solid and liquid fuels) was relatively ignored in the United States, his ideas were studied eagerly by Berlin-based Romanian schoolteacher Hermann Oberth. Oberth became such a well-known popularizer of the concept of rocket travel in Germany that he was hired as a consultant for Fritz Lang's 1928 science fiction classic *Frau im Mond* (which invented the backwards countdown); wrote his own missile book, *Die Rakete zu den Planetenräu-men* (The Rocket into Outer Space), which theorized the future of explora-tion and discovery through liquid propellants; and began, on July 5, 1927, the Verein für Raumschiffahrt (the Society for Spaceship Travel, or VfR), which two years later included a recent high-school graduate among its members—the nineteen-year-old Freiherr von Braun, who in time would oversee the creation of Apollo 11's Saturn V, inspired by reading Oberth's work as a boy. Journeying across the cosmos "will free man from his remaining chains, the chains of gravity which still tie him to this planet," von Braun said. "It will open to him the gates of heaven." Oberth was one of the Kennedy Space Center's celebrity attendees at the Apollo 11 launch—his greatest protégé's greatest achievement.

When Oberth was born in 1894 in Transylvania, it was part of Germany, but after World War I it was ceded to Romania, and the rocket scientist became an alien, traveling back and forth between Romania, where he was a high school teacher, and Berlin, where, working with the VfR, he successfully designed Europe's first liquid-fueled rocket. Even his most ardent disciples found working with Oberth somewhat daunting, however, as he had no background, or even much interest, in engineering—one colleague complained that "if Oberth wants to drill a hole, first he invents the drill press." Oberth's acolytes were also the first to discover the inherent dangers of their chosen profession, as VfR, Peenemünde, and Marshall Space Center engineer Arthur Rudolph recalled:

> One weekend in 1930 we started this run. A big problem was that the metal casing of the engine would start to oxidize after a few seconds which would be the time to stop. You could tell the time of danger when patches appeared on the metal, but you had to keep a close eye on it.
>
> I was sitting on a slope, Riedel was at the bottom of the slope and Valier was walking round the engine looking for the patches. It went OK for ten seconds and then I found myself flat on my back. When I got up I saw that our rocket motor had disappeared and there was a tremendous cloud shooting up.
>
> Valier was weaving back and forth on his feet and I saw Riedel run round the stand to grab him and hold him. When I got to Valier I noticed he was dying, bleeding profusely from the chest. I put a piece of wood under his head and I said to Riedel, "Valier is dead."
>
> The shed wall behind where I'd been sitting was plastered with holes like a machine-gun. How the pieces missed me was a miracle.

The essential formula for rocketry is simple: combine liquid fuel, oxygen (for added power and to operate in a vacuum), and a flame to trigger an explosion of gases, which expels through a throat and then a tail, just as the force of its own trapped air propels an inflated and unleashed balloon. Yet, getting the details right is nearly impossible. You need a big explosion, but not so big that it blows up the craft. You need an enormous amount of heat, and an engine composed of materials that can withstand such heat. You need compressed oxidizers like LOX and LH2, and the capabilities of managing these tremendously difficult materials. One scientist even characterized rocket engine technology as a black art without rational principles. Attempting to balance all these factors is so mysterious that, across human history and in every culture, the progress of missile engineering has followed essentially the same path: The first ship blows itself up on the pad. The next refuses to ignite. The next lifts off, climbs a few inches, then collapses in on itself. The fourth shoots at a right angle straight across the field instead of rising. Attempt five blows up on the pad. Six refuses to ignite. . . .

"The early history of rocketry reads like an account of the burning of witches," Norman Mailer said. "One sorceress did not burn at all, another died with horrible shrieks, a third left nothing but a circle of ash and it rained for eight days. Rockets with solid fuel had a firing chamber that grew larger as the fuel burned away; rockets with liquid fuel were obligated to react to the fact that the fuel sloshed around in the tanks. A world of instruments, of gyroscopes, radios, telemetric devices, computers and various electric monitors and controls exhibited their own peculiarities, working difficulties, tendency to malfunction, and subtle hints of private psychology. So dread inhabited the technology of rockets."

The only way forward for the missile engineer was to try, and to fail, and to try again, to learn from each mistake and each catastrophe, and to always tinker, tinker, tinker. This is essentially what the von Braun team did, until one day in September 1931, a VfR test launch based on Oberth's theories crashed into the roof of a building next door to the local police station, erupted in flames, and ended amateur rocketry in the city of Berlin. Hitler later banned private missiles from Germany altogether, forcing the country's burgeoning aerospace pioneers and rocketry addicts to look for government work, or find something else entirely to do with their lives. As it happened, because World War I's Versailles Treaty had forbidden the nation from developing heavy artillery but made no mention of missiles, German military leaders were very interested in the men of the VfR. Wehrmacht colonel Walter Dornberger offered its members financial support if they would covertly develop rockets for his service, including subsidizing von Braun's graduate studies in missile science. The core team members accepted, with the Freiherr earning his Ph.D. in physics from the University of Berlin. "I had no illusions whatsoever as to the tremendous amount of money necessary to convert the liquid-fuel rocket from [an] exciting toy . . . to a serious machine," von Braun said. "To me, the Army's money was the only hope for big progress toward space travel." One of von Braun's college roommates, Constantine Generales, joined him in studying centrifuge acceleration, via experiments that involved tying a mouse to a bicycle wheel and then spinning it faster and faster. In time, the rodents of their less-than-successful efforts damaged so much wallpaper that the pair was evicted.

At a test on December 21, 1932, Dornberger was disassembling a rocket motor when a spark ignited the propellant. The device exploded in his face, and a military orderly was reduced to using butter and tweezers to remove the hundreds of tiny particles embedded in his skin. Then, after innumerable failures, finally, on December 19 and 20, 1934, the team successfully launched two missiles, which they named Max and Moritz, after the two mischievous cartoon heroes, the Katzenjammer Kids.

Von Braun became a National Socialist on May 1, 1937, joined the SS on May 1, 1940, and received annual promotions to lieutenant, captain, and major (*Sturmbannführer*)—this last grade, being just below Adolf Eichmann's, was awarded by Himmler himself. In an affidavit of June 18, 1947, von Braun claimed he was more or less coerced into these positions: "My refusal to join the [Nazi] party would have meant that I would have to abandon the work of my life. . . . [SS Colonel] Mueller . . . told me that Reichsfuehrer SS Himmler had sent him with the order to urge me to join the SS. I called immediately on [Dornberger, who] informed me that . . . if I wanted to continue our mutual work, I had no alternative but to join." Colonel Dornberger, however, though von Braun's Wehrmacht superior, never joined the Nazi Party.

An oddly similar chain of events helped shape the career of von Braun's Cold War nemesis, Soviet chief designer Sergei Pavlovich Korolyov (sometimes transliterated from the Cyrillic as Korolev). If the Cold War's Space and Missile Race was on one hand fought between Kennedy and Khrushchev, Johnson and Brezhnev, Glenn and Gagarin, it was more particularly a contest between the battling engineers under the leadership of von Braun and Korolyov, leaders whose strength was less in engineering genius than in politics, salesmanship, charisma, and pitch-perfect management.

After giving up his teenage dream of becoming a gymnast, the Soviets' rocket master began an interest in aviation as a boy when he became obsessed with the seaplane base in his hometown of Odessa. He designed gliders to gain entry to Kiev Polytechnic Institute's aviation school (whose alumni included helicopter pioneer Igor Sikorsky) and then Moscow's Technical School, where he was mentored by Andrei Tupolev, who had been Konstantin

Tsiolkovsky's protégé. Tsiolkovsky, the third father of rocketry, was once a teenager so deaf he needed the aid of a tin ear horn, and then a young man self-taught in calculus, trigonometry, and aeronautics (but who learned chemistry from Dmitri Mendeleyev, inventor of the periodic table). He became a high school math teacher who never had enough money to actually test his theories (which he published in 1903 as *The Exploration of Cosmic Space by Means of Reaction Devices*), but they were crucially influential, notably the mathematics of escaping earth's gravity with a "rocket train," or multistaging, to reach 18,000 miles an hour, and the use of LOX and LH2 for fuels.

Soviet chief designer Korolyov was described by his lifelong colleague and bitter rival, Valentin Glushko (head of the Soviet Gas Dynamics Laboratory and designer of the nation's best rocket engines), as "short of stature, heavily built, with head sitting awkward on his body, with brown eyes glistening with intelligence . . . a skeptic, a cynic and a pessimist who took the gloomiest view of the future. 'We will all vanish without a trace' was his favorite expression." Always dressed in turtleneck sweaters and black leather jackets, Korolyov became the leader of Moscow's GIRD rocket club (the Group for the Study of Reaction Motion), which grew so popular that it generated regional satellite clubs in ninety Russian cities and then was taken over by the Soviet military and merged with its own rocket R&D into the Jet Propulsion Research Institute (RNII).

On June 10, 1937, Stalin's NKVD purged all senior armament officers, including RNII deputy chief Korolyov, who was beaten until he confessed and sentenced to ten years of hard labor in the gold mines of Siberia, where he had his jaw broken, lost all of his teeth, suffered from frostbite, nearly died from scurvy, and then learned that he had been denounced by a competitor—rocket-engine designer Valentin Glushko. In time, mentor Tupolev was able to convince his superiors that Korolyov was absolutely vital to the state's engineering needs, and he was returned to Moscow—though still under internment—working under his benefactor (and then, Glushko) on bombers and surface-to-air missiles. In 1945, Korolyov was released from prison and awarded the Badge of Honor, promoted to a colonel in the Russian army, and sent to Germany to gather intelligence on the Nazi V-2.

Korolyov's strange biography illustrates all too well the volatile and contrary attitudes toward science of Soviet leaders, who were opposed to scientific freedom while at the same time in agreement with Lenin that World War I "taught us much, not only that people suffered, but especially the fact that those who have the best technology, organization, and discipline, and the best machines emerge on top. . . . It is necessary to master the highest technology or be crushed."

In time, the Nazis determined that von Braun's rocket team needed a secret facility to keep their wonder weapon research clandestine. His mother suggested an isolated peninsula on the Baltic Sea, where her father loved to go duck hunting. By 1936, property had been bought and construction begun on what would be a joint Luftwaffe-Wehrmacht rocket lab: Peenemünde, a heathered bracken marshland of coot, grebe, geese, rabbit, pheasant, and Pomeranian deer, which shared with future spaceports in Huntsville, Clear Lake City, and Canaveral a distinct lack of human population and immense swarms of mosquitoes. Its dense forest also offered a signal advantage: camouflage. The labs, engineering shops, testing facilities, launchpads, radar station, LOX plant, wind tunnel, guidance and telemetry studios were built and fully staffed by August of 1939. While von Braun's Wehrmacht group developed the A4/V-2 (which, per Dornberger's specifications, would have a range twice that of the German Paris gun, which shot 220-pound shells eighty miles, while being small enough for transport by the German federal railway system), their neighbors at the Luftwaffe developed their V-1 "buzz bomb." The two services did not get along, an enmity that escalated dramatically when one of von Braun's rockets went awry and destroyed the Luftwaffe's runway.

Radio telemetry—which transmits to ground controllers such crucial data as the temperature of mission-critical parts, the voltage of different sections of the electrical system, the moment in time when any valve opens or closes, and the speed of fuel coursing through the plumbing—was used by Houston's Mission Control to monitor 3,552 individual functions of Apollo 11. On the A4/V-2, Nazi scientists had four elements of telemetry, meaning they had to launch and test and tinker many times to amass any usable quantity of data. Their progress was slow, and the results, disappointing. Merely determining the proper adjustments for the propellant shutoff valves took twenty launches. There was additionally a grave problem in using ethyl alcohol as fuel—pilferage. To keep Peenemünde employees from stealing it and drinking it, a laxative was mixed in, but this only made production schedules collapse as a result of reduced workload. When methyl was then substituted, one technician went blind and another died.

In February 1940, Hitler ordered a termination of rocket R&D, and Dornberger had to keep his operation going with diverted Wehrmacht funds.

The first rocket failed, as did the second. But on October 3, 1942, the third was an all-out success, with even Dornberger exclaiming, "Today, the spaceship was born." Citing this achievement in the wake of the February 2, 1943, German defeat at Stalingrad, Armaments Minister Albert Speer petitioned Hitler to make the A4 a priority. The Führer said, "I have dreamed that the rocket will never be operational against England. I can rely on my inspirations. It is therefore pointless to give more support to the project."

Undaunted, on July 7, 1943, von Braun met with Speer, Dornberger, and Hitler at Wolf's Lair in East Prussia, where, von Braun explained, "The bird will carry a ton of amatol in her nose, but it will hit the ground at a speed of over 1,000 meters per second, and the shattering force of the impact will multiply the destructive effect of the warhead."

Hitler was not convinced: "It seems to me that the sole consequence of that high impact velocity is that you will need an extraordinarily sensitive fuse so that the warhead explodes at the precise instant of impact. Otherwise, the warhead will bury itself in the ground, and the explosive force will merely throw up a lot of dirt."

It turned out that the Führer was right about the fuse, but even so, he was ultimately swayed by von Braun's presentation, telling his armaments minister: "The A4 is a measure that can decide the war. And what encouragement to the home front when we attack England with it! This is the decisive weapon of the war and what is more, it can be produced with relatively small resources. Speer, you push the A4 as hard as you can." Speer himself found von Braun's efforts "exerted a strange fascination upon me. It was like the planning of a miracle."

At about that time, a package was left anonymously on the front steps of the United Kingdom's embassy in Oslo, Norway, describing a revolutionary new weapon being developed on the north coast of Germany. British intelligence sent in aerial survey sorties, and after their pictures were developed, a photo interpreter, one of "the backroom girls," spotted a torpedo with fins. With this evidence, Winston Churchill's son-in-law, Duncan Sandys, Crossbow intelligence committee chairman, requested and received war cabinet approval for Operation Hydra. The RAF ordered 497 bombers with 1,593 tons of explosives to destroy Peenemünde and the engineers who were creating these weapons on August 18–19, 1943.

Though 735 died in the bombing and ensuing firestorm, every member of the von Braun team escaped.

Three days later, on August 22 at an 11:30 a.m. Wolf's Lair meeting to explore the rocket program's future, Himmler told Speer that he "could guarantee secrecy" for the wonder weapons' production "the simplest way. If the entire workforce were concentration-camp prisoners, all contact with the outside world would be eliminated. Such prisoners did not even have mail." Speer agreed, and the production of V-1 buzz bombs and V-2 ballistic missiles was moved to a sodium sulfate mine in the Harz Mountains, seventy miles west of Leipzig, and so far underground as to be impregnable from Allied bombs.

Under Himmler's management, the Dora concentration camp was erected outside the cavern's entrance. Eighty percent of Mittelwerk, the new V-1 and V-2 plant, was carved out of the Harz rock by inmates, notably two main tunnels, each two miles long, 14 yards wide and 10 yards high, as well as six parallel tunnels, 220 yards long and 30 yards high. "The steady stream of convoys from Buchenwald unloaded their human cargo," one camp inmate, French resistance fighter Jean Michel, remembered. "They drilled, expanded and fitted out the first tunnel almost without tools, with their bare hands. They carried rocks and machines in the most shocking conditions. The weight of the machines were so great that the men, walking skeletons at the end of their strength, were often crushed to death beneath their burden. Ammonia dust burnt their lungs. The food was even insufficient for lesser forms of life. The deportees toiled for eighteen hours a day. They slept in the tunnel. Cavities were hollowed out: 1024 prisoners in hollows on four levels which stretch for 100 yards. They only saw daylight once a week at the Sunday roll call. The cubicles were permanently occupied, the day team following the night team and then vice versa."

Mittelwerk was a final assembly station, relying on five thousand individual German contractors to deliver parts and materials. After the factory tunnels were constructed, the concentration camp slaves then manned the production lines, overseen by Kapos who in civilian life had been common criminals. Their life and work were such a misery that one prisoner said, "It was at Dora that I realized how the pyramids were built."

In the wake of the attempted assassination of Hitler by a group of army officers on July 20, 1944, the entirety of the wonder weapons' production was turned over to Himmler's SS, with the V-2 under the supervision of Obergruppenführer Hans Kammler, who had overseen the design and engineering of gas chambers, crematoria, and concentration camps, the building of Auschwitz-Birkenau, and the destruction of the Warsaw Ghetto. Speer called him "a cold, ruthless schemer. . . . a fanatic in pursuit of a goal, and as carefully calculating as he was unscrupulous." He would turn Mittelwerk into a factory of strategic weapons and of even greater barbarity. The slave workers died by the tens of thousands, while the survivors did all they could to sabotage the production lines.

It is this chapter in von Braun's history that has discouraged his widow, children, and other relatives living in the United States from cooperating with historians and biographers, even thirty years after the man's death. Any documentary records directly linking von Braun to the war crimes of Dora and Mittelwerk have all but vanished, save for a Smithsonian curator's discovery of a letter written by the Freiherr to Mittelwerk's production planner, which stated: "You proposed to me that we use the good technical education of detainees available to you at Buchenwald . . . I immediately looked into your proposal by going to Buchenwald . . . to seek out more qualified detainees. I have arranged their transfer to the Mittelwerk. . . ." Overall, there is so little concrete evidence available that von Braun biographer Michael J. Neufeld was reduced to concluding that his subject was guilty of "selective memory" and "moral obtuseness."

Jean Michel: "I do not claim that von Braun, Reidel, Gröttrup or Dornberger personally brutalized deportees. I am even sure that the German scientists would have preferred to see their marvelous missiles made in more civilized factories and by a better treated workforce. They probably deplored the delays that our technical incompetence and our physical condition—not to speak of our sabotage—caused their program. I claim only that Dornberger, von Braun, Gröttrup and all those lumped together conveniently as 'the Peenemünde scientists' knew perfectly well what crimes were being perpetrated at Dora. Many fellow prisoners saw them in the tunnel in the workshops."

The postwar von Braun would maintain he knew little of the horrors of Mittelwerk, and that those that he was aware of, he was not responsible for, or that the situation could not be helped. In his defense, the history of aviation is often financed by a nation's military, as the biographies of Goddard, Korolyov, deHavilland, Curtiss, and the Wright brothers surely prove. However, from August 7 to December 30, 1947, at Dachau, the U.S. Military Court prosecuted twenty-four defendants in the Dora-Nordhausen War Crimes Trials, including one civilian, Georg Rickhey, Mittelwerk's general director. Rickhey's defense lawyer called on von Braun and other members of the German team—now living in Texas and employed by the United States Army—to testify on his behalf. The army refused to allow any of its rocket scientist war bounty to return to Germany, but did permit Rickhey's attorney to submit depositions, which were carefully scrutinized to make sure no émigré would incriminate himself. Even so, von Braun admitted that "I have been [to Mittelwerk] 15 to 20 times, approximately, for discussing technical matters in connection with the technical alterations of the A4. The last time I was there was in February 1945."

In 1970, von Braun discussed this troubling history with a group of British journalists:

> The problem of amorality is very, very old. Just read the history of the Renaissance and see how even the greatest architects like Michelangelo were pressed into building war machines and fortresses.
>
> The situation is not entirely new and I think it's always created problems and scruples. But it's always a question of "What is a man's duty to his country?" You see, whether you build the rocket or an airplane that drops bombs there's not much of a difference, because both are basically a means of transportation. I do know, for example, that the people who built the bombing aircraft have asked themselves (and rightly so) precisely the same question that I, of course, have asked myself. . . .
>
> When Hitler took over in 1933 I was twenty-one years old. I was wrapped up in my rocketry ideas, and as I look back at myself then, I most certainly didn't appreciate the significance of that upheaval or the evil of the rule that would follow.
>
> Even in the summer of 1939 I was absolutely convinced there wouldn't be a war. You can say that this was naïve, because everyone else thought there would be war. But, of course, when the war came you had a somewhat different situation. Suddenly your country is at war and whether it's right or wrong it's still your country. Besides, at that time my work had practically no significance, but only a growth potential. All of a sudden it emerged as something which became important. What really made it important was that the Luftwaffe lost the Battle of Britain. Without a rocket we couldn't get there any more, they said. . . .
>
> The V-2 was a fine rocket. The only thing wrong with it was that it landed on the wrong planet.

With the A4/V-2 in active production by Mittelwerk's slaves, the von Braun team began developing the Wasserfall and Taifun (Waterfall and Typhoon) antiaircraft missiles, a space shuttle, and most important, methods for using submarines to ferry V-2s across the Atlantic to strike Washington, D.C., and New York City. In retrospect, some would be bitter about these projects. "Gigantic effort and expense went into developing and manufacturing long-range rockets which proved to be, when they were at last ready for use in the autumn of 1944, an almost total failure," Albert Speer said. "Our most expensive project was also our most foolish one. . . . I not only went along with this decision on Hitler's part but also supported it. This was probably one of my most serious mistakes."

By January 1945, the regular gunfire of Soviet artillery could be heard nearing Peenemünde. Von Braun and his team were simultaneously ordered by their SS superior, Hans Kammler, to flee to Mittelwerk, and by the local overseer, the Gauleiter of Pomerania, to stay and defend the homeland alongside the province's civilians. On January 31, von Braun called his five top lieutenants to a secret meeting and announced: "Germany has lost the war. But our dream of going to the moon and to other planets isn't dead. The V-2s aren't only war weapons; they can be used for space travel. To one end or another, the Russians and the Americans will want to know what we know. To which of them will it be better to leave our inheritance and our dream? We absolutely must place the baby in the right hands." There was a unanimous decision not to wait for the Russians, who were now a mere fifty miles off, but to flee, en masse, to surrender to the Americans in central Germany.

Of Peenemünde's 4,325 employees, two-thirds agreed to join the evacuation. Using trains, truck convoys, and even barges, the team successfully crossed a dying Third Reich—the exodus one more testament to the von Braun group's engineering brilliance—in what was called, to please the Gestapo with the appearance of following Kammler's orders, "The Vengeance Express." One engineer, meanwhile, arranged to hide fourteen tons of critical documents in an abandoned mine. On March 19, Hitler decreed that his scientists and engineers should destroy their laboratories and their paperwork, an order von Braun and his men ignored. On their way to the American territory, while the Freiherr was traveling late at night, his driver fell asleep, hurtling the car over an embankment. Von Braun's shoulder was crushed, his arm broken in two places, and his upper lip split in half. At the hospital, it was also determined that he'd contracted hepatitis.

The team's core arrived in the Alpine foothill town of Oberammergau that spring, meeting with Kammler at Jesus's House (an inn owned by the ardent Nazi who played Christ in the centuries-old passion play). Kammler was hoping to trade rocket secrets with the Americans in exchange for money, or leniency; von Braun had the same ambitions. Then, on May 1, when Hitler's suicide was publicly revealed, Kammler told von Braun he had urgent business elsewhere, and vanished. What happened to him afterward has never been determined.

It was at this point that the confrontation at the heart of the Space and Missile Race would begin in earnest. The Soviets sent a rocket research agency electrical engineer, Boris Chertok, to gather up the cream of Nazi engineering, while the United States ordered Colonel Holger "Ludi" Toftoy to do the same. On April 10, Toftoy received word that an American private by the name of John M. Galione had discovered a train filled with corpses, and decided to walk one hundred miles back along the tracks to discover their source. It was, as he described by cable, Mittelwerk. Checking his map, Toftoy saw that the V-2 factory mine was in the province of Thuringia, deeded to Soviet occupation. The Americans had very little time.

On May 2, 1945, the first sentry at the U.S.-controlled Austrian border town of Schattwald, PFC Frederick Schneikert, was approached by a German on a bicycle. The cyclist explained that he was Magnus von Braun; that his brother, the inventor of the V-2 wonder weapon, was waiting with four hundred other rocket scientists just over the mountains at Oberammergau; that he had come to the American side to represent them, since he spoke the best English; and that, since the SS had orders to kill or imprison every major Nazi scientist before they could be taken by Allied forces, it was necessary "to see Ike as soon as possible." Schneikert listened to all this, and then asked his fellow soldiers, "Hey, I've got a nut here! What should I do with him?"

American officers knew that Peenemünde had been taken by the Russians on March 9, so Magnus's story was on its face scarcely credible. The next day, however, his brother and six other team members crossed the mountains into Austria and surrendered to the Americans. Ike, in fact, was expressly interested, cabling, "The thinking of the scientific directors of this group is twenty-five years ahead of U.S. Recommend that 100 of the very best men of the organization be evacuated to U.S. immediately."

Through Operation Paperclip, Toftoy smuggled 118 members of the team into the United States along with their tons of documents, one hundred V-2s, launchers, test-firing rigs, an LOX plant, and 360 metric tons of additional components. It was one of the great coups at war's end, and when he learned of it, Stalin was enraged: "This is absolutely intolerable. We defeated Nazi armies, we occupied Berlin and Peenemünde, but the Americans got the rocket engineers. What could be more revolting and more inexcusable? How and why was this allowed to happen?"

The Russians did try to reverse their fortunes. On October 15, 1945, the British tested one of their own captured V-2s at Cuxhaven. Secretly in attendance, disguised as a captain, was Soviet chief designer Sergei Korolyov. Dornberger recounted that, while the team was waiting to leave for the United

States, the Russians tried "to kidnap our leading lights from us. . . . They appeared at nighttime in English uniform; they didn't realize it was the American zone. They came to us and wanted to come in. They had a proper pass. But the Americans were quick to realize it and wouldn't let them in. So they got into cars and drove off again. That's how the people work. Real kidnapping, they don't stick to the boundaries at all."

The Russians alternately ended up with a group of lower-level Peenemünde alumni, the most senior of which was von Braun's deputy for electrics and guidance, Helmut Gröttrup. It would take them eighteen months to collect enough parts and materials across the whole of what would now be East Germany to make their own V-2s. The great historic irony of Operation Paperclip, meanwhile, was that almost every key element that Toftoy brought to the United States was present already. While the army was negotiating with von Braun in Europe, the American navy's director of the Bureau of Aeronautics, Robert Goddard, passed away on August 10, 1945. Five years later, his heirs sued the American government for patent infringement, citing such independently-engineered-by-the-Nazi-team-but-already-patented-by-Goddard rocket elements as gyroscope guidance systems, jet vane path controllers, and fuel-line turbo-pumps. Ten years later, the United States paid the Goddard estate $1 million for these rights.

As the Germans migrated from the Baltic to the Alps to Amsterdam to Long Island's Fort Strong to Maryland's Aberdeen Proving Ground to El Paso's Fort Bliss and the White Sands of New Mexico, they played night-long marathon games of Monopoly, revising the rules to allow such innovations as cartels.

After spending two years in prison, Dornberger himself emigrated and, eight years later, was working for Bell Aerosystems in Buffalo, New York, while Magnus von Braun had a job at Chrysler, and his brother Sigismund (who'd served as a translator during the Nuremberg War Crimes Trials) was appointed West Germany's American ambassador. The von Braun army team, meanwhile, lived in tar-paper shacks at Fort Bliss, Texas, a life not so very different from what their counterparts taken to Russia would eventually find at the new cosmodrome in the desert of Kazakhstan. Fort Bliss, at least, had bowling, swimming, and Jack Benny's radio show, which the Germans found excellent for practicing their English. Driving one day through downtown El Paso, they were mystified as to the purpose of one business: The Venetian Blind Man.

Even though it only offered them six-month contracts, the U.S. Army did a remarkable job in bringing the Germans into the United States with as little notice as possible. They carried a very unusual identification card: "Special War Department Employee—In the event that this card is presented off a military reservation to civilian authorities . . . it is requested that this office be notified immediately . . . and the bearer of this card not be interrogated."

At the same time, the army sealed the paperwork concerning von Braun's war trials depositions, his involvement with the SS, and his Third Reich service records until seven years after his death in 1977. Toftoy and his men did such a thorough job that it would take decades before historians could learn the secrets of the Peenemünde team's service records. Meanwhile, the local El Paso press ran stories that transformed the team from enemy alien Nazis brought to the United States illicitly into hardworking immigrants renouncing their past and becoming upstanding American citizens.

If Operation Paperclip appeared a great triumph, it had little impact on the fact that the United States remained the sole owner of the atomic bomb, and that all its apparent enemies had been vanquished. Many at the Pentagon saw their future in bombers; who needed rockets? Even so, von Braun and the other Germans were kept at Fort Bliss and under contract, testing missiles, just in case. Then, on May 29, 1947, one of the gyroscopes failed, and instead of descending to the north a missile veered south, toward El Paso and the Mexican border. To limit the damage, cutoff safety officer Ernst Steinhoff let the rocket burn through all of its fuel. It crashed near a cemetery 1.5 miles south of Ciudad Juárez, blowing a hole fifty feet around and thirty feet deep. In the wake of this attack, the DoD's Joint Chiefs of Staff decided that perhaps the army's Germans should be testing their rockets elsewhere.

After five years in the American desert, sidelined and barely working, a life he called being a "prisoner of peace," von Braun came to believe that, if he was ever going to be able to build a ship that voyaged to outer space, he had, somehow, to ignite the interest and enthusiasm of the American public. As he told another team member, "We can dream about rockets and the Moon until Hell freezes over. Unless the people understand it and the man who pays the bill is behind it, no dice."

Von Braun wrote a book to drum up just such enthusiasm, *The Mars Project,* and submitted it to eighteen U.S. publishers, all of whom turned it down. Then, while attending a symposium on space medicine in San Antonio, he met Cornelius Ryan, author of *The Longest Day* and associate editor of *Collier's* magazine (circulation three million, readership twelve million), and epically described the immense forthcoming adventure awaiting humanity in the stars. The impressed Ryan immediately converted to the cause of interplanetary travel, and over the next two years, *Collier's* published eight outer-space features, beginning on March 22, 1952, with "Man Will Conquer Space Soon. Top Scientists Tell How in 15 Startling Pages!" In one of his own visionary *Collier's* pieces, von Braun outlined a three-stage rocket with two boosters that fell into the ocean and were then salvaged for future flights, topped by a craft with wings (similar to the Space Shuttle) as well as a rotating and orbiting space station (not unlike the one that would appear in the movie *2001: A Space Odyssey*) and a remote-controlled telescope (similar to the Hubble). Insisting that the nation should commit to space as it had earlier to nuclear arms with the Manhattan Project, he outlined virtually the whole of NASA's future.

When von Braun again tried to get a massive rocket project under way, however, he was stymied. In the period from 1932 to 1952, the annual federal budget had exploded from $4 to $85.5 billion, with 57.2 percent allocated to the Department of Defense. Eisenhower tried to stop this trend by appealing directly to the American public, warning in 1953 that "The cost of one modern heavy bomber is this: a modern brick school in more than thirty cities. It is two electric power plants, each serving a town of sixty thousand population. It is two fine, fully equipped hospitals. It is some fifty miles of concrete highway. We pay for a single fighter plane with a half million bushels of wheat. We pay for a single destroyer with new homes that could have housed more than eight thousand people. . . . This is not a way of life at all, in any true sense. Under the cloud of threatening war, it is humanity hanging from a cross of iron." Even after von Braun's passionate lecture to the Armed Forces Staff College on the immense potential of reconnaissance and weapons-targeting satellites, no one at the Pentagon wanted to spend money on his daydream of rockets bursting forth across the cosmos.

While von Braun was publishing his dreams of space in *Collier's*, Walt Disney became obsessed with the idea of creating a California version of Copenhagen's Tivoli Gardens. Roy, his brother and the Walt Disney Company's business manager, refused to allow the corporation to invest in such a crazy scheme, so Walt convinced fledgling television network ABC to offer $4.5 million in loan guarantees for the amusement park in exchange for a Disney television program, both to be called "Disneyland." The television version became America's most popular TV show, and when Disney and ABC wanted to create three episodes showcasing the "Tomorrowland" section of the park, producer Ward Kimball hired as on-air consultants three of the *Collier's* writers, including von Braun. It was estimated that between 42 and 100 million Americans watched the March 9, 1955, premiere episode,

"Man in Space," which effectively established von Braun as the national symbol of rocket science. He looked like a movie star, said things like, "We can lick gravity, but sometimes the paperwork is overwhelming," and with the help of *Collier's* and Disneyland, became the Space and Missile Race's first American celebrity. Physicist Freeman Dyson concluded: "In the end, the amnesty given to him by the United States did far more than a strict accounting of his misdeeds could have done to redeem his soul and to fulfill his destiny."

For his entire life, Wernher von Braun had one aim: to go to Mars. In the mid-twentieth century, only a wealthy and generous federal government could make such a dream come true, and so he became an exemplar of Realpolitik, of doing what needed to be done. Under the Third Reich, he had been a good German, and in the United States he would be an exemplary American. On first entering the country, he became a born-again Christian, joining the Church of the Nazarene, and on April 14, 1955, forty-one members of the German rocket team, including Wernher Magnus Maximilian Freiherr von Braun, were sworn in as citizens of the United States.

9

Total Cold War

After their inadvertent attack on Mexico, the von Braun team was exiled from Fort Bliss to the wilderness of northern Alabama. While today's Huntsville is a cosmopolitan town of 170,000, home to LG, Siemens, and Boeing, when the 117 Germans arrived—settling into a section now known as "Sauerkraut Hill"—it was little but dairy cows and cotton balls, and thanks to the alkaline chalk streams feeding its bogs, proud to claim the title: "Watercress Capital of the World." Scientist Ernst Stuhlinger observed, "Huntsville was a [beautiful] southern town when we came here. It reminded us very much of what we read in *Gone with the Wind*." At Fort Bliss, due to military security and public relations, the Germans had almost no contact with any Texan or New Mexican locals. In Alabama, that would change. "We knew that people here ran around without shoes," engineer Konrad Dannenberg recalled. "They make their money moonshining and that's what they drink for breakfast, and supper. And so we, in a way, were a little disappointed that it was really not that bad." It took the émigrés a while, however, to get used to local ways. At first, every time a shopgirl in a department store at the end of a transaction said, "Y'all hurry on back now, hear?" a German would immediately turn around and return to the cash register.

The Alabamans, for their part, knew full well that these rocket scientists were Nazis who'd worked on the V-2 attacks against England and France. Memories of Hitler and concentration camps were still strong, but the Germans carried with them a remarkable advantage. NASA administrator Jim Webb said of his booster chief, "He had the instinct and intuition of an animal. He could sense danger [and had] a remarkable sense of what his audience wanted to hear." JPL designer Carl W. Raggio put it more simply: "Von Braun could sell ice to Eskimos."

The team chief made it his job to win over the locals, attending every dinner and giving every speech he could, whether to gentry or to peanut pickers. He needed the town on his side, especially as his engine testing would shatter windowpanes and crockery, and force his new neighbors to stuff their ears with cotton. (Even so, many came to believe that, in the wake of von Braun, Huntsville became America's deafest town.) The German PR campaign, however, was in time so successful that, a mere two years after the team's arrival, local Dixiecrats elected Luftwaffe sergeant Walter Wiesman to the presidency of the Huntsville Junior Chamber of Commerce. When his suitability for the position was questioned, as 70 percent of the Chamber's members were World War II veterans, Wiesman replied, "So am I!" During Apollo, NASA's Alabama operation would grow from 117 employees to 6,000.

On October 21, 1946, the Russian general overseeing the now–East German engineers—those who hadn't emigrated to America with von Braun and were working for the Soviets at Mittelwerk and Peenemünde—threw an elaborate celebration for his employees, lasting until four in the morning. At that point, Soviet recruits with automatic weapons appeared at every German home to begin a forced migration. Ninety-two trains ferried the engineers and their families to Russia.

Sergei Korolyov was now in charge of Soviet long-range missile development. He sent some of the Nazi captives to the guarded island of Gorodomliya, where his newest missile, the R-14, would be developed, while the majority lived and worked directly under him at the Scientific Research Institute and Experimental Factory 88—NII-88—in the Moscow exurb of Kaliningrad (which would in time become the site of the Soviets' Mission Control), surrounded by female armed guards and barbed wire. Each team of workers was composed of Germans and Soviet trainees, and as the local talent gained expertise, the Germans were sent home (with the last returning in 1954). On October 30, 1947, the German-Soviet alliance produced a V-2 successor whose successful launching, apogee of 175 miles, and accurate touchdown inspired Korolyov to grab German team leader Helmut Gröttrup in a bear hug and dance with him.

Strategic Air Command chief Curtis LeMay regularly ordered sorties to practice simulated en masse bombing runs against Soviet targets while preparing for the ultimate drop of 750 nuclear warheads that would kill sixty million Russians. At the time, the Soviets had no way to counterstrike, or to defend themselves, with Russian leaders all too aware that the immense expanse of farmland that had shielded the nation against Napoleon and Hitler was useless in the face of America's nuclear strike force. Unlike LeMay, the Russians had no enormous fleets of well-tanked bombers to ferry atomic strikes to the Western Hemisphere—their pilots could only make one-way, suicide runs—and no overseas bases close enough to American shores for deploying intermediate range ballistic missiles (IRBMs).

To compensate, the Russians furiously used their V-2 inheritance to launch a crash course in missile R&D, straining to design, engineer, and manufacture ICBMs as quickly as possible. As Stalin had told a meeting of military chiefs in April 1947: "Do you realize the tremendous strategic importance of machines of this sort? They could be a very effective straitjacket for that very noisy shopkeeper, Harry Truman. We must go ahead with it, comrades. The problem of the creation of the transatlantic rocket is of extreme importance to us." Soviet leaders began to believe that missiles could not only defend the entire nation, but actually alter the global balance of power.

On August 29, 1949, the USSR exploded its own twenty-two-kiloton atomic bomb in the desert of Kazakhstan, with no public announcement—the United States learned about it through geiger counters aboard an American "weather plane"—and the United States was no longer the world's sole atomic power. Four years later, in August of 1953, the Soviets announced the successful development of their own hydrogen bomb. While the Russians' atomic warhead had been an exact duplicate of the Manhattan Project's, they'd achieved hydrogen all on their own. More important for the course of history, before that moment, missiles were not considered sophisticated weapons, as their margin of error was four miles. With hydrogen, however, so much would be annihilated that the margin was now immaterial.

When the news of Soviet hydrogen deployment reached U.S. intelligence, the Pentagon's historic, ironclad confidence in nuclear bombers wavered; its chiefs realized that they had fallen far behind in missile technology, with squabbling among the DoD's various branches stunting development even further. The situation became so dire that a brief moment of interservice cooperation arose. From June 1954 to January 1955, the navy tried to merge its Vanguard rocket development with the air force's Titan IRBM and Thor ICBM (intercontinental, or over 3,500 miles) efforts, as well as with the army's medium-range Redstone program under von Braun. After six months, however, each went its separate way.

Eisenhower would spend the entirety of his presidency nagged by the fear that he might face another Pearl Harbor. When he received yet another embarrassing report detailing the USAF's reconnaissance failures, the president finally called for an outside investigation. What was commonly known as the Killian Surprise Attack Committee (its chair was the Oval Office science adviser and MIT president James Killian) described an America in March of 1954 that was superior to the USSR in bombers and nuclear warheads, but with no early-warning or sophisticated defense systems. It predicted that, in time, the Soviets would reach parity. American intelligence, nuclear submarine fleets, and ICBMs needed to be immediately and dramatically expanded.

The frugal Ike was disinclined to spend a fortune on rockets and submarines, but he realized that the United States desperately needed solid intelligence on its new, secretive foe, intelligence that would be untainted by the navy and air force's unquenchable yearning for an endless supply of new hardware. Eisenhower first tried to negotiate an "Open Skies" arrangement with the Soviets, to be overseen by the United Nations, allowing both nations to fly through the other's airspace to monitor nuclear weapons without triggering acts of war. When Khrushchev refused, Eisenhower approved Operation Aquatone, with the CIA giving Lockheed SkunkWorks the job of engineering and building a new species of aircraft in two and a half months' time. SkunkWorks created the greatest spy plane ever built—a sixty-foot single-seater with a two-hundred-foot wingspan, a carbon-black fuselage, a Hycon camera, a TRW electronic intelligence (ELINT) system, absolutely no clues as to its identification—even the pilot's underwear labels were knifed out before missions—and one great advantage: a ceiling of seventy thousand feet. No Soviet surface-to-air missile could touch it; no MiG fighter could bring it down. It was known (when it had to be referred to) as the Second Provisional Weather Squadron's Utility Plane Number Two, and its operations chief was the creator of the Space and Missile Race's third leg, the onetime Yale economics professor and CIA director of plans (covert operations), Richard Bissell.

For four years, beginning July 4, 1956, the U-2 program worked; for four years, the Soviets were too embarrassed to admit they could not stop the Americans from spying on their country at will. In this, and in its thousands of other illicit surveillance missions over the Soviet Union during the period, however, the United States lost 138 fliers.

On September 16, 1955, Korolyov successfully launched his R11-FM rocket from a diesel submarine. Its range was less than a hundred miles, but the Soviets now had the ability to strike any American coastal city. Then on February 2, 1956, Korolyov had another profound success when his V-2 grandchild, the R-5, with power enough to thrust an eighty-kiloton warhead eight hundred miles, landed near the shores of the Aral Sea. For the the first time, a ballistic missile had carried the equivalent of a nuclear payload, and it was Soviet.

Three weeks later, on February 27, the entire Presidium visited Chief Designer Korolyov at his NII-88 compound. There, he showed them the future—the R-7, which used five rocket engines to command one million pounds of lift at over 24,000 feet per second, with LOX and kerosene for fuel and twelve gimbaled thrusters for guidance. It could reach American targets in less than thirty minutes, making it impervious to Pentagon tracking.

It was just what Khrushchev wanted to protect his nation and enable it to achieve parity as a global superpower. Immediately afterward, the premier slashed funds for tanks, boats, and guns, cutting his military's payroll by 2,300,000 men, and throwing massive support to Korolyov and his rockets. This decision would lead to a dazzling history of Soviet space triumphs— and the onset of the USSR's military chiefs' turning against Khrushchev.

Of that February 27 meeting, the first secretary remembered, "We gawked at what he showed us as if we were a bunch of sheep seeing a new gate for the first time. When he showed us one of his rockets, we thought it looked like nothing but a huge cigar-shaped tube, and we didn't believe it could fly. Korolyov took us on a tour of a launching pad and tried to explain to us how the rocket worked. We were like peasants in a marketplace. We walked around and around the rocket, touching it, tapping it to see if it was sturdy enough—we did everything but lick it to see how it tasted. . . .

"A model of some kind of apparatus lay on the stand. It looked unusual, to put it mildly. A flying machine should have a smooth surface, flowing shapes and clean-cut angles. But this one had some type of rods protruding on all sides and paneling swollen by projections." Designed by the chief designer's chief designer, Mikhail Tikhonravov, this apparatus had a skin of steel polished until it was as reflective as a mirror, because Korolyov wanted one thing most of all from his new creation—that it be seen.

Korolyov warned his bosses that the Americans were about to launch a vessel that would orbit the Earth—the first artificial moon—and explore the heavens. But if the Politburo desired, the USSR could easily beat the West to this glorious achievement, which would surely amaze the world. Khrushchev himself was dismissive of this man-made moon idea, but finally gave it a half-hearted approval: "If the main task [the development of the R-7 ICBM] doesn't suffer, do it."

In June of 1957 a U-2 pilot conducting clandestine aerial reconnaissance over the Soviet Union's Asian deserts saw something strange off in the distance. He went to investigate, and CIA analysis of his photographs discovered that in the wilds of Kazakhstan, a province once famous along the Silk Road but now wedged between Russia and China and blessed in modern times (not unlike New Mexico) with nuclear testing facilities, the Soviets had used slave labor to construct a massive compound that in time would come to be known as the Baikonur Cosmodrome.

The pilot had uncovered the heart of the Soviets' military and civilian space programs, with separate facilities to design, engineer, build, and launch Korolyov's Semyorka, the R-7. The Cosmodrome included everything a rocket scientist could ever want—even a little house for the chief designer's use, set midway between the assembly building and the launchpad, so he could walk, either way, to work—but it was so remote that even drinking water had to be imported. Temperatures reached 131 degrees; the construction had been so excruciating that the officer in charge ended up in a mental hospital; and in the cool of the evenings, engineers and construction workers entertained themselves by dropping pairs of scorpions into empty vodka bottles and wagering on the outcomes of their death battles.

Korolyov's use of the threat of American competition at his meeting with the Presidium was both canny and overstated—the United States was nowhere close to having a rocket powerful enough to lob a satellite into orbit. At the time the USAF had given up on its Atlas rocket and switched to engineering the kerosene-and-LOX-burning Titan; the navy was developing the submarine-launched, solid-fueled Polaris; and the Pentagon had contracted with Western Development for a missile that could be launched in under sixty seconds—the Minuteman. "Development," however, was the key word. While the Russians appeared to be reaping success after success through the brilliance of their chief designer, the August 1957 test firings of the navy's Vanguard were catastrophic, while two Thors exploded and three von Braun Jupiters zigzagged off course. Even with the bounty of the Nazi missile scientists, these first years of American rocketry were so unproductive that many at the Department of Defense wanted to return to ever-greater bombers and ever-bigger artillery guns. As late as the 1960s, it would remain a serious question among Pentagon brass and their engineers whether the nation's ICBMs would actually operate as planned, since they could not be thoroughly test-fired. For years, SAC chief LeMay would exploit this doubt to keep his bomber budget healthy.

In that same August of 1957, Korolyov's revolutionary R-7 struck its target in Kamchatka after flying across nearly the whole of Russia. However, its nose cone heat shield did not work, and its test warhead was destroyed on reentry, a chronic Soviet problem and a key reason why Korolyov was so assiduously promoting his mirrored-skin apparatus. The United States was unaware that the Soviets were suffering from chronic payload reentry mishaps; they only knew the Russians had launched a successful ICBM flight when the U.S. Air Force couldn't even get its IRBM Thors in working order.

At the same time that NASA's predecessor, the NACA, was helping to design its incredible X-1 and X-15 planes, though, it helped to create for the United States exactly what the Soviets wouldn't have for many years: a solution to reentry. Beginning in 1955, nose cones designed by NACA engineers such as Maxime Faget, a gnomish and energetic Cajun, featuring blunt bodies to slow the return to Earth and and beryllium shingles or ablative heat shields of phenolic resins (which scorched and melted off and away, taking the heat with them), pioneered in tandem with RAND, Huntsville, and General Electric's George Sutton, safely brought atomic warheads to their targets. The same method of fat-bottom blunt body, dissolving heat shield, and reefed parachutes would be used for the film canisters of spy satellites and the astronauts of NASA.

In Huntsville, the von Braun team was now overseen by Maj. Gen. John Bruce Medaris and maintained with a minimal budget, more as a competitive edge against the air force than out of any particular army interest in rockets. The team's signal effort at that time was a three-stage ICBM that would combine Huntsville's V-2 progeny with the army's other rocket in R&D, the Jet Propulsion Laboratory's Sergeant missile, fueled by liquid hydrogen. While the original V-2's range was 190 miles, on September 20, 1956, the multistage combination of Redstone and Sergeant known as Jupiter soared 3,400 miles, a record that held for many years. Regardless, Defense Secretary Charles Wilson decided to give the navy and air force monies for developing long-range missiles, while restricting the army to battlefield artillery weapons of 200 miles or less. Von Braun became so distraught that, in a speech before the National Association of Food Chains, he said: "I suspect we will have to pass Russian customs when we finally reach the moon." Huntsville was forced to turn its forty-five remaining Jupiters over to the air force, which based them in Turkey and Italy, from where they could be ordered by NATO to strike Leningrad, Stalingrad, and Moscow, and where their menacing presence would trigger the most dangerous moment of the Cold War.

As the Space and Missile Race now evolved from theory into practice, the Americans would discover that battling the Soviets was like facing an army of ghosts. Until the late 1970s, long after the race to the Moon had been won, the United States essentially knew nothing about the Russian manned spaceflight program. A few months before Korolyov launched his orbiting transmitter, Russia's *Radio and Astronomer's Circular* described how ham operators could listen to its transmissions and how it could be watched with telescopes, but this was about the limit of Soviet preflight publicity. While Korolyov and his fellow designers could read a translation of U.S. newspaper accounts of the latest tests of the Atlas and the successful recovery of the Jupiter nose cone, von Braun was not even aware of Korolyov's existence— only, in time, that the Soviets had a "chief designer" largely responsible for their heroic Space Race achievements. "Our 'intelligence' on the Russian space program was pretty hot stuff: notebooks with newspaper and trade journal clips pasted in them," flight director Chris Kraft admitted. "The military apparently didn't feel that the civilians in NASA had to know whatever it was they knew."

In 1958, American military intelligence believed the USSR would soon have four hundred Bisons and three hundred Bear bombers capable of striking the American heartland. Their evidence was the high serial number of a Bison that had flown at a May Day parade in Moscow. In fact, the Soviets knew the Americans were watching, and intentionally inflated that number. This kind of gross overestimation would mark Pentagon and CIA intelligence misfires for most of the Cold War: "You'll never get court-martialed for saying [the Soviets] have a new type of weapon and it turns out that they don't," CIA Soviet military analyst Victor Marchetti said. "But you'll lose your ass if you say that they don't have it and it turns out that they do."

The CIA National Intelligence Estimates from the 1950s and '60s vary wildly in accuracy, and at times contradict themselves, so that trying to make an assessment with even a full history of documents can only lead to error and frustration. In 1953, the CIA fretted that the Soviets could launch nuclear missiles from submarines (they couldn't), and in 1958 predicted the Soviets would have one hundred ICBMs by the following year and five hundred by 1962 (instead, there were four, one at Tyuratam, and three at Plesetsk, in the Arkhangelsk Oblast), and would achieve a lunar landing by 1965. In 1959, using information from Corona spy satellites, the CIA reversed the USAF's prediction on rapid ICBM deployment, with September 21, 1961's National Intelligence Estimate projecting a total of between ten and twenty-five Soviet IRBM/ICBM stations in all. "Our only solid Central Intelligence Agency information came from the Corona satellite photography," NASA deputy Bob Seamans said. "We observed the construction of the Soviets' large launch vehicle assembly building with rails leading to an underdeveloped area. Later, a launch area materialized, and on one fortuitous occasion, a large vehicle was spotted ready for the launchpad. . . . But we had no direct evidence of their lunar program until the visit of three MIT faculty to their Moscow Aviation Institute space laboratory in December 1989." Unfortunately, however, even Corona had its limits. In 1961, American intelligence predicted a Russian lunar orbit by 1966 and a lunar landing no earlier than 1970; in 1962, they estimated Russians on the Moon by 1967 or 1968; in 1965, they believed the Russians would have a space station aloft in two years, but would not beat Apollo to the lunar surface.

America's ignorance about her stirring and nervous foe, especially in the first years of the Cold War, only ratcheted up Washington's own fears. In 1947, *Foreign Affairs* published George Kennan's "The Sources of Soviet Conduct," which held that Russia was so driven by political ideology that it would not countenance a thriving, coexisting capitalist West, and that it would send its theories to "every nook and cranny available to it in the basin of world power." Communism, however, held within itself "the seeds of its own decay," and there was only one answer to keeping it in check: "containment." In 1954, Eisenhower then introduced the foreign policy theory that would hold sway over the White House for the next twenty years: "You have a row of dominoes set up, you knock over the first one, and . . . the last one . . . will go over very quickly. So you could have . . . a disintegration that would have the most profound influences." Key political support for the inauguration of NASA would in part rise from the belief that its shining example to developing nations would help keep them from being infected by communism.

These collective worries also triggered a reformulation of thinking among America's leaders. "I have come to the conclusion that some of our traditional ideas of international sportsmanship are scarcely applicable in the morass in which the world now flounders," Eisenhower said. "Truth, honor, justice, consideration for others, liberty for all—the problem is how to preserve them . . . when we are opposed by people who scorn . . . these values."

"We are facing an implacable enemy whose avowed objective is world domination," announced the 1954 Doolittle Report, a classified overview of American covert operations. "There are no rules in such a game. Hitherto acceptable norms of human conduct do not apply." Americans must "learn to be able not to be good," and the American public "must be made acquainted with, understand and support this fundamentally repugnant philosophy."

In time, the Soviet insistence on absolute secrecy (and their great power in achieving it) made American presidents, senators, and the Pentagon crave information, with the government unleashing vast resources on spycraft. On October 4, 1950, the RAND Corporation issued "The Satellite Rocket Vehicle," which outlined how satellites would be used for communications, weather, security, and reconnaissance, and how their successes would greatly expand American prestige overseas. However, since the Soviet Union might not take kindly to foreign satellites flying aggressively through its airspace, the first American efforts must clearly be civilian and scientific. Could the United States establish a precedent in international law of space overflight? Not only would the first American satellite look as civilian as possible, but as far as the president was concerned, the Soviets could go first, proving that a spaceship did not signify hostile intent as it flew high above other nations.

After calculating the speed of the planet's rotation at the launch site, the equatorial plane's inclination, and the thrust needed to settle into orbit (without falling back to Earth, or vanishing into deep space), Korolyov and his team at the Cosmodrome launched a two-stage R-7 on the one hundredth anniversary of Konstantin Tsiolkovsky's birth—October 4, 1957—at 10:26 p.m. They had picked a nighttime liftoff to foil American spy planes, in case their experiment failed. Instead, their achievement would transform a onetime concentration-camp inmate into one of the seminal figures of engineering history.

To guarantee that it would be seen from Earth, Korolyov surfaced the rocket's one-hundred-foot casing (which followed the silver moon in orbit) with prisms that caught the light. When it reached orbital velocity of 18,000 miles per hour, the R-7 jettisoned a basketball-sized orb just under two feet wide, weighing 184 pounds, and trailing strands of insect-like antennae, which would circle the globe every ninety-six minutes. Iskustvenniy Sputnik Zemli (Спутник), Artificial Earth Satellite, carried a crucial instrument—twin radio transmitters that broadcast a steady *meep . . . meep . . . meep*, which anyone with a ham radio unit could pick up around the world. The satellite (and more visibly, its trailing, prismed casing) shone like a star in the night, a shooting star that was under Soviet dominion. To the American public, "meep . . . meep . . . meep" meant one thing: *I am here.*

"At the appointed hour, our neighbors came to our yard to help me watch it," astronaut training manager Homer Hickam remembered. "My father said, 'Well, they can all go home because President Eisenhower will never allow anything Russian to fly over Coalwood!' But at the appointed moment, Sputnik flew over Coalwood. If it had been God in his chariot that had flown over, I could not have been more impressed. It was awe-inspiring. Sputnik looked like a bright star that moved with such utter purpose that nothing could stop it; and I, in that moment, realized I wanted to be part of the movement into space."

"Thanks to Comrade Korolyov and his associates, we now [had] a rocket that could carry a nuclear warhead," Khrushchev helpfully suggested. "His invention also had many peacetime uses." The Eisenhower White House was pleased that the CIA, with the precedent of Sputnik, could orbit spy satellites peaceably across Russian airspace, but Ike's political instincts on the issue failed him miserably on the domestic front. At a press conference on October 9, he was asked, "Mr. President, Russia has launched an Earth satellite. They also claimed a successful firing of an intercontinental ballistic missile, none of which this country has. I ask you, sir, what are you going to do about it?" (This was two months before the USAF successfully launched its first Atlas ICBM on December 17, 1957, with a range of 11,500 miles, a warhead of 3.8 megatons, and a load and launch of fifteen minutes.) Eisenhower replied: "The Russians, under a dictatorial society, where they have some of the finest scientists in the world, who have for many years been working on it, apparently from what they say they have put one small ball in the air."

"One small ball" was hardly how anyone else in America viewed the newest star in the sky, a Russian one, at that. Sputnik whipped American journalists and then the American public into a torrent of fear over Soviets flying anywhere they pleased, taking pictures of anything they liked (or listening in with microphones), and dropping bombs with galactic impunity. While *Pravda* had announced Korolyov's successful launch with a barely noticeable slug, both the London and *New York Times* featured oversized, alarmed headlines on their front pages. Eventually *The New York Times* would insist, "Now the Communists have established a foothold in outer space, [America is in] a race for survival." While Senator Henry "Scoop" Jackson of Washington called for "a National Week of Shame and Danger," broadcast journalist Eric Sevareid editorialized, "If the intercontinental missile is, indeed, the ultimate, the final weapon of warfare, that at the present rate, Russia will soon come to a period during which she can stand astride the world, its military master." To Eisenhower's anemic response, *Time* magazine replied with the headline: "A Crisis of Leadership," and the Dow fell 10 percent.

As an event, Sputnik, like 9/11 and Pearl Harbor, gave rise to a key fear in the minds of the American people: How could so direct and devastating an assault be launched against the United States of America? In each case, the attack was so startling, and so unanticipated, that it became mythic in its force, and even more terrifying. Added to this was the primal shock of realizing that their country, which had not been invaded since 1814, thanks to its protection by both its Atlantic and Pacific frontiers, could now be attacked at will. In the five years after Sputnik, Americans would increasingly wonder: Why were Russian space triumphs repeatedly so much more successful than the United States' homegrown efforts? Wasn't the United States smart enough to beat the Russians in outer space? It was inconceivable. Was communism superior to American culture? Was a totalitarian educational system superior to democratic efforts when it came to science, technology, and engineering? The general attitude was later summed up by Neil Armstrong: "It was disappointing that a country who was the 'evil empire' in our minds at that time would be beating us in technology, where we thought we were preeminent. . . . It did change our world. It absolutely changed our country's view of what was happening, the potential of space."

Texas senator Lyndon Johnson heard of Sputnik while he was at his Hill Country ranch: "In the open West, you learn to live closely with the sky. It is part of your life. But now somehow, in some new way, the sky seemed almost alien. [I felt] the profound shock of realizing that it might be possible for another nation to achieve technological superiority over this great country of ours." Very shortly LBJ would come to see this moment as both a national emergency and a remarkable political opportunity. A memo from Democratic Policy Committee staff director George Reedy to the senator noted that "The issue [of Sputnik] is one which, if properly handled, would blast the Republicans out of the water, unify the Democratic party, and elect you President. . . . the important thing is that the Russians have left the earth and the race for control of the universe has started." The future great champion of NASA later told colleagues, "I'll be damned if I sleep by the light of a Red Moon . . . soon they will be dropping bombs on us from space like kids dropping rocks on the cars from freeway overpasses."

The Russians could not help but be gratified by the American response. "Khrushchev was astonished by the reaction," Sputnik historian Matthew Brzezinski said. "It was as if, overnight, his nation had been vaulted to a preeminent position atop the global hierarchy. The Soviet Union, in the eyes of the world, had suddenly become a genuine superpower, not just a backward and brutish empire to be feared because of its sheer size, territorial ambitions, and aggressive ideology, but a true and equal rival of the United States, a beacon of progress that deserved respect for its technological prowess and forward thinking. Moscow, for once, held the high moral ground in this new phase of the contest, because Sputnik, as opposed to Hiroshima, could be touted as a purely peaceful and scientific achievement."

Sputnik II, launched on November 3, 1957, caught even Eisenhower's attention. Thrusting 1,120 pounds—including a husky-terrier stray named Laika (Лайка, or "Barker")—to an altitude of 1,031 miles, it proved the Soviets could launch big payloads (like hydrogen bombs) as well as the first living creature into the cosmos; *Time* helpfully explained its potential as "enough weight allowance to put a powerful atomic bomb on the moon." The Soviets at the time were able to keep the world from learning, however, that they hadn't yet solved their reentry problems, and so were planning to poison Laika before allowing her capsule to incinerate on its return. Instead, between five and seven hours after launch, Sputnik II's nose cone's insulation pulled loose during staging, and its heat-control system failed, killing its canine passenger.

If Eisenhower and his administration repeatedly minimized the import of the first Sputnik, at the highest echelons of American government, there was plenty of concern. Before Korolyov's launch, the USAF and the Oval Office's Gaither Report had insisted that the Soviet Union was dramatically ahead of the United States in nuclear missile artillery, an assessment the CIA had countered, with evidence from its U-2 program, saying there was no such "missile gap." After the second Sputnik, however, the intelligence service dramatically reversed course, with its National Intelligence Estimate of August 19, 1958, predicting one hundred Communist ICBMs ready to fire by 1959, five hundred by 1960, and the Russians landing on the Moon in 1965 (which raised another issue, less remembered today: that if a nation could engineer a method for leaving the Earth, and more important, reaching another heavenly body, then certain of its citizens could survive a nuclear holocaust). At the National Security Council meeting on October 10, the president's lead speechwriter, Arthur Larson, suggested, "I wonder if our plans for the next great breakthrough are adequate. If we lose repeatedly to the Russians as we have lost with the Earth satellite, the accumulated damage will be tremendous. We should accordingly plan, ourselves, to achieve the next big breakthrough first, a manned satellite, or getting to the moon."

"Let us not pretend that Sputnik is anything but a defeat for the United States," *Life* magazine had solemnly noted in the wake of the rising of the first Russian moon. "[It is] really and truly the shot heard round the world." After the second Sputnik was in orbit, the massively popular magazine would

publish an article colored by a whole new level of fear: "Arguing the Case for
Being Panicky." Given that an October 29, 1957, CIA report had already concluded that the Soviets would have an advanced missile program ready at any
moment and that the United States was in a period of "grave national emergency," Senator Lyndon Johnson announced on November 4 that the Senate
Preparedness Subcommittee would hold hearings on the matter. The White
House–commissioned Gaither Report of November 7 reported, meanwhile,
that the United States would be imminently surpassed in both atomic weaponry and civil defense and recommended the government immediately spend
$25 billion on fallout shelters and educational programs instructing the
American public on what to do in the event of an attack by atomic warheads.
Washington's survival brochures, which recommended shelters with one-
foot-deep concrete walls, were sent only to suburban Americans, since it was
assumed that city dwellers could do nothing to save themselves. During
three notable periods, American backyards would see an upswing in shelter
excavation: in the aftermath of Sputnik, upon the news of the building of the
Berlin Wall, and at the Cuban Missile Crisis, all of which would be milestones in the Space and Missile Race.

   With the start of what would become a public relations campaign of executive brilliance, Khrushchev did not take this moment as an opportunity
to soothe the souls of Washington. Instead, he told *The New York Times* that
his country would be making R-7 missiles "like sausages," and informed
Pravda, "The fact that the Soviet Union was the first to launch an artificial
earth satellite, which within a month was followed by another, says a lot. If
necessary, tomorrow we can launch ten, twenty satellites. All that is required
for this is to replace the warhead of an intercontinental ballistic rocket with
the necessary instruments. There is a satellite for you."

On December 6, 1957, it was finally America's turn to show the world what
it could do, as the United States Navy readied to launch its Vanguard satellite from Cape Canaveral. While the Soviets had been so secretive that their
*Sputnik*s were not publicly unveiled until they were overhead and in orbit,
from the very beginning, the United States invited the public to witness its
liftoffs, an offer of transparency that the press turned into the latest salvos in
a global battle, with (in time) NASA's (more or less) express cooperation. No
one could remember a government R&D effort that had been held so publicly before. It was a sign either of great bravado . . . or of desperation.

   As the countdown was announced, clouds of flame and smoke exploded from the booster's tail. America's first major rocket lifted on streaks of fire, and began its journey to the sky. Then, suddenly, it stuttered, sank back onto its pad, defeated, and listed sideways, slowly falling to the ground and exploding. Its satellite—a midget of 3¼ pounds—was propelled by the blast off into the brush, where its signal was picked up by the Cape's radios: "Meep . . . meep . . . meep." Reporter Dorothy Kilgallen asked, "Why doesn't somebody go out there, find it, and kill it?" At the United Nations, Soviet delegates consoled their U.S. counterparts by noting that America might now qualify for technical assistance from the USSR's program to aid developing nations.

After losing his big rockets to the air force and NATO, Bruce Medaris had been keeping his German team together by testing nose cone reentry technology. Immediately following the navy's Vanguard embarrassment, von Braun told Neil McElroy, America's next secretary of defense, "When you get back to Washington you'll find that all hell has broken loose. I wish you would keep one thought in mind through all the noise and confusion: We can fire a satellite into orbit sixty days from the moment you give us the green light."

Huntsville and Pasadena finally got their go-ahead. On January 31, 1958, Explorer I stood ready on its Canaveral pad, its Jupiter booster mating a first-stage extended Huntsville Redstone with second- and third-stage JPL Sergeants of eleven and three solid-fueled rocket motors each. The satellite had its own motor and a geiger counter designed by astrophysicist James Van Allen, and and it weighed all of eighteen pounds. Bruce Medaris: "Almost every reference to Army-developed hardware was stricken from the [launch's press materials] in a rather dishonest attempt to make our first space triumph look like a civilian effort." This feint would become a cornerstone for both nations in the Space and Missile Race.

Thirteen minutes before launch, the upper stages' outer shells began to spin, reaching 550 rpm and thus becoming their own stabilizing gyroscopes. At 10:55 p.m. EST, the stage one booster fired for 157 seconds, until it reached an altitude of sixty miles. Explosive bolts then released it to fall away into the Atlantic, while stages two and three burned, respectively, for 247 seconds, and then 6.5 seconds. They were released, stage four was ignited, and if the von Braun/JPL team was successful, the satellite would now enter orbit at Newton's escape velocity: 18,000 miles per hour.

Secretary of the Army Wilber Brucker, the Jet Propulsion Laboratory's William Pickering, and Huntsville's von Braun were connected in phone conference, with Pickering also on another line with Goldstone, the San Diego receiving station, due to be the first to capture Explorer I's signal. It was now February 1, 1958, 12:41 a.m.:

Pickering: Do you hear her?
San Diego: No sir.
Pickering [a few minutes later]: Do you hear her now?
San Diego: No sir.
Pickering: Well, why the hell don't you hear anything?
Brucker: Wernher, what happened?
Unidentified: What happened?
Pickering: Hooray! They hear her, Wernher, they hear her! Goldstone has the bird!
Von Braun: She is eight minutes late. Interesting.

Ultimately, America's Explorer I and its fifty-four successors would turn out to be far more scientifically useful than the Russian Sputniks, especially in mapping the radiation belts that surrounded the Earth's magnetic field (which would be named for Van Allen), and in proving that the Earth is shaped like a pear. But they never carried a living creature, or were of sufficient power to be used as ICBMs. Since Explorer I weighed only 3¼ pounds, compared to Laika's 1,120-pound capsule, Khrushchev returned Eisenhower's slur of "little ball" by calling it "a grapefruit."

Nonetheless, Americans were so excited about their first successful liftoff, that *The American Weekly* ran von Braun's autobiography in three installments on July 20, 27, and August 3, 1958—"Space Man—The Story of My Life"—the tale of a boy who dreamed of travels through the cosmos, was co-opted by the terror state that was Nazism but escaped, and now pursued his vision as an American citizen contributing to the security of his adopted homeland. This storyline was eventually developed by Columbia Pictures into a 1960 biopic, *I Aim at the Stars*, which in turn inspired a famous line by Kennedy speechwriter Mort Sahl, who quipped that during World War II, von Braun "aimed at the stars, but often hit London." Despite the concerted effort to rehabilitate his image, America's most beloved rocket scientist remained sufficiently controversial that Tom Lehrer wrote a well-remembered ditty about him:

Don't say that he's hypocritical
Say rather that he's apolitical.
Once the rockets are up, who cares where they come down?
"That's not my department," says Wernher von Braun.

While the public as a whole celebrated von Braun, however, America's missile R&D was still an embarrassment. Huntsville and JPL's simple Explorers were far from any missions involving living creatures, much less manned spaceflight. The navy had launched eleven Vanguards, eight of them disasters. The air force was developing successors to the X-15, supersonic rocket jets, but these weren't powerful enough to reach escape velocity. It seemed that America had defaulted its position as the world's leader in science and engineering. The need for national renewal was felt across the entire country, especially in Congress and at the Department of Defense.

Who, however, should own the responsibility of leading America in exploring this new frontier? Navy admirals tended to view spaceships as another kind of submarine (and missile development as crucial to their fleet), army generals saw rockets as an outstanding form of artillery (with theirs already the most successful rocket team in America), while the men of the USAF saw missiles as soaring bombs, either remote-controlled or, if that service had its way, commandeered by pilots. The air force's lobbyists would even invent the term "aerospace" to demonstrate that the two were intimately linked, and thus a natural extension of that branch's theater.

Even the little National Advisory Committee for Aeronautics entered this policy fray. Various chieftains of American industry and defense had originally tried to get Congress to create the NACA as early as 1915, but it was voted down. Encouraged by the navy's assistant secretary, Franklin Roosevelt, however, a ticker authorizing the agency was included in that year's navy appropriations bill, which sailed through. NACA became a team of research engineers working with airplane manufacturers and the military in an attempt to keep American aviation technology competitive with Europe's—an endeavor that would be especially appreciated two years later, when America entered the First World War against parts of Europe. After the war ended, NACA established an agency tradition by receiving special dispensation to import German engineer Max Munk to design what became a celebrated wind tunnel. In the 1920s and '30s, while the navy developed its aircraft carriers, NACA worked out various design and engineering solutions for small and light but powerful planes that could launch and then "controlled crash" on the heaving decks of floating airports, a relationship that would last for decades. Relations were not so congenial with the army, however, especially during World War II, when NACA essentially fell into a technological slumber, and Army Air Forces chief Hap Arnold was dumbfounded to learn that Britain had already produced a turbojet. In September 1941, he assigned an officer to travel with the jet's plans handcuffed to his wrist from London to

Massachusetts so that General Electric could begin making American jet engines, all of which was kept secret from the government employees of NACA.

Except for global public relations, then, everything that will be found in the signature amorphous nature of NASA is already in its forefather—a wavering position between defense corporations and the military; a close relationship with the navy; a Pentagon official actively withholding information; military research "trickling down" into civilian and corporate applications; and a history of German engineers who need their passports freshened.

In time, NACA became a haven for engineering geniuses who needed little more than three hots, a cot, and room to think. Perhaps humiliated by Arnold's example, after World War II the little agency turned itself around to a remarkable extent. Working on an engineering design with Buffalo, New York's Bell Aircraft and the USAF for a plane that could soar 35,000 feet at 800 miles per hour, NACA engineers noted that, since there was so little ballistic design information available for such a craft, they needed another model—maybe, a .50-caliber bullet?—and so the shape of the X-1 was born. On October 14, 1947, an air force officer was flown on a B-29 to 20,000 feet, slipped into that orange bullet, and shot his rocket engines to 42,000 feet. There was the sound of an immense boom that day across the Mojave, for Chuck Yeager had broken the sound barrier.

NACA next approached the air force and the navy with a plane that could reach fifty miles and 5,000 miles per hour (Mach 7). As its craft began flying beyond the upper reaches of the atmosphere, the agency was effectively forced into space research. Inspired by the concept of combining a Redstone with a plane, one design's rocket would burn LOX and anhydrous ammonia for power, and use hydrogen peroxide thrusters to control attitude in the upper atmosphere's lack of air. The air force took this proposal to North American Aviation, and eight years later, the X-15 became the greatest flying machine in the world. While the X-1 had looked like a pumpkin bullet, the X-15 was carbon black, with the menace of a dagger. (When North American Aviation CEO "Dutch" Kindelberger complained that this experimental plane wouldn't lead to any commercial product, his development vice president, "Stormy" Storms, countered, "Finally, we've got a chance to build something that doesn't have any guns on it.")

The X-15's federal champion was the dapper and nominally self-effacing Hugh Dryden, who arrived at the agency just as Yeager was creating sonic booms, and who had previously overseen development of America's first successful combat missile, the air-to-surface "Bat," which had no rocket booster, but used radar homing, a tail elevator, autopilot servomotors, gyro-stabilizers, and wind generators to destroy its target. By 1958, Dryden had become NACA's director, and in the wake of America's profound self-doubt and deep-seated terror over Russian space supremacy, he insisted that a serious American space effort needed to be ignited—with NACA at its core. On January 14, 1958, he published "A National Research Program for Space Technology," which explained why the future of the universe should be in his hands: "Both from consideration of our prestige as a nation as well as military necessity that this challenge [Sputnik] be met by an energetic program of research and development for the conquest of space . . . It is accordingly proposed that the scientific research be the responsibility of a national civilian agency working in close cooperation with the applied research and development groups required for weapon systems development by the military. The pattern to be followed is that already developed by the NACA and the military services. . . ."

Dryden then reallocated one-fifth of his agency's budget to space research, and created a Special Committee on Space Technology to organize federal research between the Pentagon, NACA, the aerospace industry, and various universities. It was the dawn of a new world, for at this committee's meetings, von Braun, creator of Nazi rockets, sat across the table from Hendrik Wade Bode, creator of Britain's automatic artillery robot, which brought down those very same rockets.

On February 4, 1958, Eisenhower invited NACA chairman General Jimmy Doolittle, Polaroid's Edwin H. "Din" Land, the Pentagon's Advanced Research Projects Agency (ARPA) chief Herb York, and Nobel physicist Ed Purcell to formulate an American plan of action for outer space. The group decided that a publicly open civilian organization should lead the way in exploring the cosmos for reasons of national prestige, science, and technology, with the military's space needs reserved for the Pentagon. Two days later, the Senate created its Special Committee on Science and Aeronautics, headed by Lyndon Johnson, to investigate the same questions. Then, in a bipartisan rush of legislation and following the precedent of the Atomic Energy Commission, Eisenhower and Johnson's Congress on July 29, 1958, merged the eight thousand employees of NACA's Langley, Virginia, headquarters; Ames Aeronautical Propulsion and High-Speed Flight at Edwards in California; Lewis Flight in Ohio; Wallops Island launch facility in Maryland; a handful of DoD ICBM executives; the navy's Research Laboratory (which would be renamed for Goddard, and would focus on satellites); the army and air force's lunar probe divisions; and, in time, the Explorer group (Pasadena's Jet Propulsion Laboratory and the Army Ballistic Missile Agency at Huntsville), to create the National Aeronautics and Space Administration—NASA.

The army was enraged. Though it had long ago been ordered to focus on artillery-level missiles, the service had in fact been planning since April of 1957 for Huntsville to develop a booster powerful enough to launch soldiers into orbit. NASA had to fight to get von Braun and his team, but they had the directive of the president. At least the army chiefs were consoled that it was losing its rocket scientists to a civilian entity, and not to the air force.

. NACA's Hugh Dryden had won the federal space showdown, with the explicit support of White House science adviser James Killian, who had convinced both Eisenhower and recalcitrant congressmen of the agency's unique role and history in American government (at the same time, Killian helped establish ARPA, the military's R&D arm, which in turn begat the foundations of the Internet). Historians would laud the onetime Supreme Commander of Allied Forces for deliberately removing American oversight of celestial exploration from the Pentagon, but this was a less than clear-cut decision. Both Congress and the Oval Office worried over how to balance this novel outer space power. Many believed there was a pressing need to dilute interservice rivalry, while the more PR-oriented specifically wanted to contrast Russian secrecy and belligerence with a transparent "for all mankind" stance. There were also international sensitivities to consider, such as the advantages of showing the world that the Western system was superior to the Soviet, and of maintaining a scientific façade, so that the overflights and privacy invasion of espionage satellites would not cause an overseas uproar. Jim McDivitt: "We had tracking stations all around the world in foreign countries. Every one of them had a different political situation, and if it looked like these countries were supporting an American military operation, then there could be a lot of problems, either domestically or with some of their foreign relations. So, this was strictly a civilian project."

Not even a presidential decree, however, could immunize NASA from its military DNA. Beyond its NACA origins, the new civilian agency used air force launchpads for liftoff, the Army Corps of Engineers to build its facilities, and the navy to recover crew and capsules at splashdown. NASA fliers were required to have a turn in the service in order to become combat test pilots, even if they had little interest in navy, marine, or air force careers (as would be true of both Collins and Armstrong). Great strata of NASA management were detailed from the air force's Minuteman program, while its corporate subcontractors were Pentagon mainstays; one reason for establishing NASA was a fear that aerospace defense contractors needed government support, as they were falling into a postwar recession. The science was the same—until von Braun's Saturn, all of NASA's missions were flown using military ballistic missiles, with Project Gemini opting for a modified-by-$20-million USAF Titan II instead of anything produced at Huntsville, while the same blunt-body and ablative heat shield capsule designs were used to safeguard the dangerous atmospheric reentry of nuclear warheads, reconnaissance photography, and astronauts. There was constant industry and government employee cross-pollination, with onetime-NACA employees who did not become engineers and executives for the new space agency joining such American defense corporations as North American, Boeing, and Douglas, where they designed and built the various modules and rocket parts that would become the missions of Mercury, Gemini, and Apollo.

NASA's determined civilian posture did, however, have a direct effect on the USSR, as historian Paul Lucas summarized: "The history of the Soviet space program is also the history of the militarization of space. Peaceful exploration was never considered a viable option in the early days, and later they pursued scientific and propaganda agendas only to match the Americans' efforts in that arena."

Even after the establishment of NASA, however, the Pentagon, and most notably the air force, continued with its own space programs, as did the federal intelligence services with their Corona and Keyhole eyes-in-the-skies, early-warnings, SIGINT and ELINT (signal and electronic intelligence, such as phone calls and later e-mails) POPPY satellites, and ASATS (antisatellite missiles that could destroy enemy spacecraft). One prominent piece of intercepted intelligence for both the American and Soviet services would be the telemetry beamed back to ground controllers by each nation's spaceships. When the shah of Iran fell, NASA, the Pentagon, and the CIA lost their best tower for eavesdropping on signals from the Soviets' Baikonur Cosmodrome.

Even at NASA's birth, its civilian role was hazy; as historian Walter McDougall flatly stated, "The space program was a paramilitary operation in the Cold War, no matter who ran it." Eisenhower's 1959 State of the Union Address tried to educate the American people on the need for such civilian/military vagaries: "What makes the Soviet threat unique in history is its all-inclusiveness. Every human activity is pressed into service as a weapon of expansion. Trade, economic development, military power, arts, science, education, the whole world of ideas—all are harnessed to this same chariot of expansion. The Soviets are, in short, waging total cold war. The only answer to a regime that wages total cold war is to wage total peace. This means bringing to bear every asset of our personal and national lives upon the task of building the conditions in which security and peace can grow." His administration would launch or expand one "total cold war" initiative after the next—from the United States Information Agency, Voice of America, and Radio Free Europe to Atoms for Peace and People-to-People—but none would have the tremendous impact, at home and abroad, that would be seized by his successor and culminate in Apollo 11.

Regardless of its military heritage and its continuing Pentagon relations, NASA's creation as an ostensibly civilian entity was a turning point in the Cold War. There were any number of military strategies that a frightened, threatened, and self-doubting America could have pursued to regain its self-confidence in the wake of Soviet space triumphs, strategies that might have triggered a series of ever more aggressive military responses by the USSR. George Reedy realized this while crafting legislation for his senator: "Without [Lyndon] Johnson, the launching of Sputnik could very well have led to an increased missile program. I think what really happened here is that Johnson pushed it into the area of outer space exploration. And thank God."

The same year—1959—that NASA was founded, the brilliant Sergei Koro-lyov was already on his way to the Moon. His first three lunar probes were failures, but Luna 2 was a success, and Luna 3 was a technological miracle. After putting itself into orbit on October 4, 1959, and using thrusters to steady itself, the Russian satellite spent forty minutes photographing the Moon's far side for the first time in history, less than two years after the first Sputnik. During this same period, NASA tried twenty-eight times to get a satellite into orbit, with only eight successes. Its first two Mercury-Redstone launches were out-and-out failures.

Both the United States and the Soviet Union, however, continued to struggle in the Missile Race. Since the Soviet SS-6 had a range of under 3,000 miles, it was based in the Arctic in order to attack the United States over the North Pole, but this locale froze the rocket's pumps and lubricants. The USAF's Atlas-D, meanwhile, had a range of 7,500 miles, but its primi-tive cryogenics meant that it needed an hour of fueling before it could be launched, and so could easily be targeted, not to mention that the fueling procedure itself seemed to regularly cause the rocket to explode before it could even get off the pad. At the same time, the Atlas-D's radio control was easily overrun, meaning that the air force had to wait five minutes between each firing. In time, the Soviets would have rockets that were bigger and faster and moved sooner than those of the United States, but what the Pentagon did not know was that Korolyov still hadn't solved his reentry problem: his rockets' warhead dummies usually incinerated in the atmosphere, far from their targets.

There was at this time one remarkable American success when, on July 1960, the USS *George Washington* submarine succeeded in test-firing two Polaris missiles while underwater. By November, it was on active patrol, armed with sixteen warheads. Then in August 1960, the Soviets launched, orbited, and safely brought home the pair of dogs Strelka and Belka (along with plants, insects, mice, rats, and sheets of human skin to test the effects of radiation), while America's only successful launch was Echo, a balloon. NASA's first Mercury-Atlas unmanned test, meanwhile, on July 29, ended in 58.5 seconds, when the craft struck Max-Q—the outermost combination of gravity fighting speed—and exploded.

At that time, the Soviet space program was going so well that Khrushchev was making it a habit of carrying around small models of the nation's ships to give out as gifts. Eisenhower was treated to a Luna, and the UN's Dag Hammarskjöld a Mars I Martian probe in 1960. (In 1962, Kennedy would receive one of cosmonaut dog Strelka's puppies.) But then came a rash of disasters. Three Soviet rockets failed back-to-back and, on December 1, 1960, Korabl-Sputnik 3, carrying two dogs, Muchka and Psholka, failed at reentry, and the module burned into ash. That night, Korolyov was stricken by a heart attack, and doctors learned he had the common gulag ailment of failing kidneys. He was ordered to reduce his workload, but the chief designer knew that, if he did not succeed in becoming the world's leader in outer space achievements, the Kremlin would end his funding. Released from the hospital, he returned to his labs and his cosmodrome to work harder than ever.

Korolyov refined the R-7 into a solid performer for launching men and satellites. However, like the USAF's Atlas-D, its cryogenic fuels made it a poor ICBM, so the R-16, designed by another Korolyov rival, Mikhail Yangel, was simultaneously being developed. Yangel rocket used the hypergolic formula of UDMH-nitric acid, a mixture so potent and toxic that it was known as "Devil's Venom."

On Monday, October 24, 1960, a prototype R-16 stood poised on the launchpad at Baikonur, awaiting a test that was considered so important to the future of Soviet defense that it was being personally overseen by Korolyov's boss, Strategic Rocket Forces marshal Mitrofan Nedelin. An error was noted, the countdown was halted, and the crew went in to make repairs. At some point, someone had left a current distributor in the wrong position. An engineer reset it, unaware that the rocket's batteries were live.

At 6:45 p.m., these two errors combined to trigger the stage two engines to ignite, throwing a cascade of flames down the sides of the rocket and across the pad, engulfing the pad workers, immediately igniting the stage one fuel tanks, and triggering a cataclysm. One hundred and fifteen people were killed, including Nedelin. The disaster so gravely delayed the progress and manufacture of Russian ICBMs that Khrushchev felt the need to erect launchpads and install Soviet IRBMs in a location adjacent to the United States. These would be ready and enabled in one year's time.

10

The Bluff at Nobleman's Grave

In 1960, American intelligence analysts came to believe that, by the following year, the Union of Soviet Socialist Republics would have five hundred ICBMs ready to launch against the United States. To verify that this projection was true, on May 1, 1960, an American pilot took off from Pakistan and, for the first time in U-2 history, flew over the whole of the Soviet Union in a mission called Operation Grand Slam. Francis Gary Powers began by tracking a rail line in hopes of photographing Soviet ICBM missile bases at Plesetsk, military installations at Sverdlovsk, and the Baikonur Cosmodrome at Tyuratam. As the CIA was at the time beginning its transition from U-2s to satellites, this May Day flight was most likely the historic skunk plane's last mission over Russian skies.

Tracking him on radar near Sverdlovsk, the Soviets sent three of their new SA-2 Guideline surface-to-air missiles in pursuit. These rockets, designed by another of Korolyov's great rivals, V. N. Chelomei (who curried favor with Khrushchev by hiring his rocket-scientist son, Sergei), could, for the first time, accurately target at an altitude of seventy thousand feet. The first SA-2 missed, and the second brought down a pursuing MiG. The third was a direct strike.

Powers's bailout was so difficult that he was unable to reach his ship's destruct button, and he did not use the poison pin hidden inside an American silver dollar. Instead, he stayed with his dying craft from seventy thousand to thirty thousand feet, bailed out, free fell for fifteen thousand feet before releasing his chute, landed safely, was intercepted, and taken prisoner.

Four days later, the American government took a U-2, painted it to look like a NACA/NASA weather plane, and casually made public mention that one like it had been lost near Turkey. Khrushchev retorted that, in fact, a CIA spy plane had been brought down. The Eisenhower administration insisted that there had to have been a mistake, since "there was absolutely no deliberate attempt to violate Soviet airspace and never has been."

On May 7, Khrushchev announced: "I must tell you a secret. When I made my first report I deliberately did not say that the pilot was alive and well . . . and now just look how many silly things [the Americans] have said." The Soviets had not only Powers, but had assembled from the wreckage a more-or-less intact U-2, including its camera, surveillance photos, 7,500 emergency rubles, and a cadre of jewels. The then-underway U.S.-USSR Paris summit collapsed, postponing détente. Privately Ike admitted that, if American airspace had been invaded to similar effect by the Soviets, he would have immediately gone to Congress for a declaration of war.

As they had been by Sputnik and the other Korolyov triumphs, Eisenhower and the United States were once again internationally humiliated by the Russians. The capture of Francis Gary Powers, the revelation that the United States had been illicitly invading the airspace of a sovereign nation, and the inept Oval Office handling of the entire imbroglio was one more in a chain of public embarrassments for America that would extend from the last years of Eisenhower's tenure through the very rough beginnings of the Kennedy administration.

Even before the Powers fiasco, however, it had become clear that the U-2 program needed to be upgraded. Americans fearful of Soviet rockets (even though the projection of five hundred ICBMs by 1961 proved to be wildly inaccurate, as the Russians in fact had only four) demanded a far more reliable and far more powerful espionage weapon, one that could again fly freely over enemy skies. By April 1962, the U-2 would be replaced by Lockheed's titanium-skinned A-12/SR-71 Blackbird, which flew three times the speed of sound. (Since the Soviet Union was at the time the world's best source of titanium, a shell corporation was created to buy the metal to clothe the planes that spied on the metal's source.) But in time, both the U-2 and Blackbird would be upstaged by an even more powerful tool. "The Soviet headstart in space was so disturbing that Eisenhower considered revealing the capabilities and achievements of the still-secret U-2 program to allay the fears of the nation," the CIA's covert action chief Dick Bissell said. "Instead, he became determined to intensify America's satellite program. In January 1958 he issued a National Security Action Memorandum establishing the development of a working reconnaissance satellite . . . [and] having been influenced by the success of the U-2, he assigned the responsibility to the CIA."

This was an idea Bissell had been developing since the summer of 1957 with Polaroid's Din Land, the White House's James Killian, and the USAF's missile chief, General Bernard Schriever. Publicly known as the Discoverer scientific research satellite, sometimes called "Keyhole" by those with the necessary security clearance, and first successfully launched by a two-stage USAF Thor from Vandenberg AFB in California on August 18, 1960, Corona—Bissell named it after his typewriter—was thrown into polar orbit, carried over a half mile of film, and was able to focus on items the size of a truck. After shooting its pictures, it would use stabilizer jets to strike just the right angle for reentry, a blunt bottom to slow its descent, a melting ablative heat shield to protect its cargo, and a drogue system of parachutes to settle gently back to Earth, to be caught by a USAF plane near Hawaii or pulled from the Pacific by navy helicopters, and rushed back to CIA headquarters for analysis. Corona's early years, though, sound very familiar to any student of rocket history:

January 1959: launch aborted due to technical errors.
February 1959: launched, but stabilizing system failed, throwing satellite out of orbit and crashing to earth.
Third attempt: successfully reached orbit, but its ejected film capsule was never found.
Fourth and fifth: did not reach orbit.
Sixth: ejected capsule never found.
Seventh: stabilizer failed and fell from orbit.
Eighth and ninth: failed on the pad.
February 1960: the tenth attempt failed to reach orbit.
Eleventh: lost by tracking stations.
Twelfth: collapsed on the pad.
August 10, 1960: Corona 13 a success, but its capsule was lost at sea.

Finally, with Corona 14, came the great triumph. "The very day that Francis Gary Powers was standing in the dock in Moscow taking his sentence [August 18, 1960] was the very first successful day the satellite was flying over Moscow taking pictures from above," CIA National Photographic Interpretation Center founder Art Lundahl noted. On August 24, 1960, at 8:15 a.m., Killian, Land, and NSA chief Gordon Gray unspooled Corona 14's reel for Eisenhower in the Oval Office. The little satellite had covered a total of 1.5 million square miles of the Soviet Union, snapping clear photographs of sixty-four airfields, twenty-six SAM launchers, and even the rocket pads at Plesetsk and Baikonur. The bounty from this single mission was greater than that from every U-2 flight combined, and its success was so astonishing that almost immediately, every branch of the federal government wanted Corona for its own. Ike, however, was so irritated with continuing air force mismanagement of its military space programs—after decades of effort and billions of dollars, its only clear success was the infrared missile-detector Midas—that he insisted the launches be run through a new department that would report directly to the civilian air force secretary: the National Reconnaissance Office.

The CIA and the NRO would fly 121 Coronas (among other spy satellites) until May of 1972, with the program becoming so well known that Cosmodrome workers wrote giant obscenities in the snow to be picked up by Bissell's eyes in the sky. Beginning in 1961, the Russians gained parity by launching over five hundred of their own Zenit spy satellites (public name: Kosmos), creating the third leg of superpower competition in the Space and Missile Race. Using the same Korolyov capsule as the cosmonaut-carrying Vostok to parachute reusable cameras and film from space to military strategic analysts, Zenit/Kosmos was not unlike Corona and all other rockets in all other nations in its historical progression, with failure after failure until finally on July 28, 1962, Kosmos 7 was launched successfully and returned with a full set of surveillance photography covering the entire United States. The Soviets would continue firing Zenit variations until 1994, matching the Americans in technology point by point.

During the third week of May 1960, America's Midas 1 satellite began patrolling Soviet airspace using infrared sensors that could detect the fire of a launching missile. Within four years, the USAF would have an array of four satellites, the "Vela Hotel," which used X-ray, gamma-ray, infrared, and neutron detectors to identify the flame of rockets. Unlike their approach with most spycraft, America made the nature of Midas and Vela's mission very clear to the Russians, and the Russians returned the favor when they launched their own nuclear monitoring system. Eventually Eisenhower's "Open Skies" proposal would be adopted by default. "During the Cold War's space race, some aspects of the military competition were held back—and for good reason," historians Michael Krepon and Michael Katz-Hyman discovered. "Satellites were—and remain—linked to the nuclear deterrents of major powers. To mess with satellites would invite nuclear danger. Washington and Moscow separately decided that the Cold War competition was hot enough without adding A-Sats [antisatellite weapons] to a volatile mix. To symbolize this understanding, the superpowers agreed formally not to interfere with satellites that monitored treaty compliance."

The only problem with eyes in the sky was that the vast majority of their pictures were of clouds. The Pentagon wanted to correct this by canceling Dyna-Soar and launching instead a Manned Orbiting Laboratory—MOL— using Gemini equipment to house two officers for thirty days of spying in orbit with a camera the size of a Chevy. Initiated under the Johnson Administration on August 25, 1965, MOL cost $1.3 billion and graduated a series of military astronauts out of ARPS—the USAF's Aerospace Research Pilots School—run by Yeager from Edwards. Like Apollo, MOL was terminated by President Nixon; a similar fate awaited the Soviet version, Almaz ("uncut diamond"), which had its last flight in February 1977, even though that mission's commander was amazed to report that, "We could see human beings on the street."

"I wouldn't want to be quoted on this," Lyndon Johnson was later quoted as telling government officials, "but we've spent thirty-five or forty billion dollars on the space program. And if nothing else had come out of it but the knowledge we've gained from space photography, it would be worth ten times what the whole program has cost. Because tonight we know how many missiles the enemy has and, it turned out, our guesses were way off. We were building things we didn't need to build. We were harboring fears we didn't need to harbor."

The same year—1960—of Powers's capture and Corona's triumph, a White House aide told Eisenhower that sending a man to the Moon would be as historic as Isabella sponsoring Columbus to go to the New World. After receiving budget estimates of $26 to $38 billion for such a program, the president replied that he "was not about to hock his jewels," and a White House staffer commented, "If we let scientists explore the Moon, then before you know it, they'll want funds to explore the planets." The Oval Office had a speech drafted announcing the end of U.S. human spaceflight with the termination of Project Mercury, but Ike never delivered it. In his final budget message to Capitol Hill, he noted that a decision needed to be made "to establish whether there are any valid scientific reasons for extending manned space flight beyond the Mercury program."

By the time Eisenhower left office in January 1961, the United States, in response to the Soviets' rocketry, had stationed in Europe 160 Atlas ICBMs supplemented by almost 100 IRBMs. The Soviets, meanwhile, had four R-7s. That same year, with four successful reconnaissance satellites operating, the CIA reduced its estimate of Soviet missile counts, from 120 to 50 (at the most) and then offered a most-likely estimate of 14. The United States had 233. The USSR would not catch up until 1969, when, at about the same time that Apollo 11 appeared to win the Space Race, the Missile Race would be tied.

"Empowered by U-2 intelligence, which proved there was neither a bomber gap nor a missile gap, Eisenhower repeatedly frustrated the attempts of America's military-industrial complex to gain financing for larger and more expensive weapon systems. An economically strong America, unhobbled by huge budget deficits, was one of Eisenhower's greatest legacies," Richard Bissell proudly concluded. But what would foreign nations think if they knew his reconnaissance satellites were so sophisticated? While millions around the world watched NASA astronauts parading through ticker tape, National Security Adviser McGeorge Bundy insisted on near-absolute secrecy for military and intelligence space missions, most of which remain classified to this day.

What Operation Aquatone had failed to reveal, which Corona unveiled in full, was the fact that the Kremlin had been, for many years, bluffing. Still recovering from the devastation of World War II, the Politburo, notably Khrushchev, was fearful of LeMay's bombers and NATO's missiles—by 1955, the United States had stockpiled 2,280 thermonuclear weapons, twenty times the Soviets' supply (a number which rose to 23,000 by 1963), with bases stretching from Greenland and Alaska to Japan, Korea, Pakistan, Turkey, Morocco, Germany, and Italy, which enabled it to launch strikes anywhere across the whole of the USSR.

Instead of merely serving as public examples of the fantastic Soviet rebound from the U.S. Army's Operation Paperclip, space triumphs for Khrushchev became immensely powerful evidentiary symbols. Using Korolyov's triumphs as a bully stick of intimidation, Khrushchev regularly implied there was an equivalent technological lead in Russian military power with ICBMs. During the first years of the Space and Missile Race, all of America's public failures were repeatedly compared with Soviet triumphs by a now-globe-trotting premier, who loved to publicly extol this disparity as hard proof of the superiority of Soviet culture, politics, economics, and philosophy.

Emboldened by the magnificent achievements of his cosmodrome at Nobleman's Grave, the first secretary blustered again and again, to everyone and everywhere, until the myth of Soviet military power took hold. When Senator Hubert Humphrey visited the Kremlin, for one example, Khrushchev pointed to a map of the world and asked where he was from. The premier then drew a blue circle around Minneapolis, explaining, "That's so I don't forget to order them to spare the city when the rockets fly." By 1960, polls found that Europeans believed in Soviet military superiority over America: 59 to 15 percent in the United Kingdom, 37 to 16 percent in France, and 47 to 22 percent in West Germany.

After he'd been driven from power, Khrushchev would admit the reason he never agreed to Eisenhower's "open skies" policy or Kennedy's many proposals for joint U.S.-USSR space missions: "Our missiles were still imperfect in performance and insignificant in number. Taken by themselves, they didn't represent much of a threat to the United States. . . . [So] we couldn't allow the U.S. and its allies to send their inspectors criss-crossing around the Soviet Union. They would have discovered that we were in a relatively weak position, and that realization might have encouraged them to attack us." Though Khrushchev may not be judged by history as one of its more admired leaders—the Cuban Missile Crisis and the Berlin Wall were not marks of honor—his use of the Space Race, for both global public relations and confidence-building within the Kremlin, was propaganda, statesmanship, and executive management at its finest. Without the glamour and prestige of the Space Race, it seems likely that at some point in the Cold War, a Russian military chief might have chosen another method to prove that the USSR was the equal of any global superpower.

In 1960, Khrushchev's bravado supplied a remarkable political chip to Senator John F. Kennedy on the campaign trail for the Oval Office against Eisenhower's vice president, Richard Nixon. Aided by a ghosted article published in *Missiles and Rockets* magazine under his name—"We are in a strategic space race with the Russians, and we are losing. If a man orbits earth this year, his name will be Ivan. If the Soviets control space they can control the earth, as in past centuries the nation that controlled the seas has dominated the continents. We cannot afford to run second in this vital race. To ensure peace and freedom we must be first. Space is our great New Frontier"—JFK turned the Soviet space, missile, and bomber gap into an indictment of Republican leadership. Senate leader Lyndon Johnson, meanwhile, drew upon the same issue to appeal to other emotions: "Control of space means control of the world. From space the masters of infinity would have the power to control the Earth's weather, to cause droughts and floods, to change the tides and raise the levels of the sea, divert the Gulf Stream and change the temperature climates to frigid. Our national goal and the goal of all free men must be to hold that position."

Kennedy returned to the Democrats' theme on September 7, 1960, capturing the national emotion that had led to the birth of NASA: "The people of the world respect achievement. For most of the twentieth century they admired American science and American education, which was second to none. But they are not at all certain about which way the future lies. The first vehicle in outer space was called Sputnik, not Vanguard. The first country to place its national emblem on the moon was the Soviet Union, not the United States. The first canine passengers in space who safely returned were named Strelka and Belka, not Rover or Fido, or even Checkers [the name of Nixon's dog]." Ironically, Kennedy's constant reference to a missile gap during his campaign would lead Soviets to regard him as an aggressive, warmongering hawk, a judgment in their minds that would be borne out by the Bay of Pigs, and would lead to the escalation at the heart of the Cold War, as described by historian John Lewis Gaddis: "One state acts to make itself safer, but in doing so diminishes the security of one or more other states, which in turn try to repair the damage through measures that diminish the security of the first state, [resulting in] an ever-deepening whirlpool of distrust from which even the best-intentioned and most far-sighted leaders find it difficult to extricate themselves."

Just as their triumphs in outer space eventually came to symbolize so many things to the Soviets, they likewise radiated a significance beyond their literal achievements to Kennedy, Johnson, NASA executives, and the American public. Starting with the goal of proving the United States technologically equal to the USSR, America's own Space Race milestones became profound sources of national pride, and evidence that the United States was still the greatest country in the world. Its astronauts became the ultimate example of American manhood (daring, courageous, and with a strong moral character), with NASA a factory of miracles producing spaceships and flying robots that photographed the solar system. Before NASA, after all, a hallmark of American classrooms was an array of painted wooden planets on wire posts that wobbled around a phosphorus sun. Today, it is NASA's Hubble photograph of clouds that are nurseries for galactic stars waiting to be born. In time, going to the Moon would become, for many Americans, just the kind of thing their country did as a matter of natural character, like saving Europe with World War II and the Marshall Plan.

In his "Farewell Address to the Nation," of January 17, 1961, Dwight D.

Eisenhower gave one of the most prescient speeches of any president in the twentieth century:

> We annually spend on military security more than the net income of all United States corporations. This conjunction of an immense military establishment and a large arms industry is new in the American experience. The total influence—economic, political, even spiritual—is felt in every city, every Statehouse, every office of the Federal government. We recognize the imperative need for this development. Yet we must not fail to comprehend its grave implications. Our toil, resources and livelihood are all involved; so is the very structure of our society.
>
> In the councils of government, we must guard against the acquisition of unwarranted influence, whether sought or unsought, by the military-industrial complex. The potential for the disastrous rise of misplaced power exists and will persist. We must never let the weight of this combination endanger our liberties or democratic processes. We should take nothing for granted. Only an alert and knowledgeable citizenry can compel the proper meshing of the huge industrial and military machinery of defense with our peaceful methods and goals, so that security and liberty may prosper together.
>
> Akin to, and largely responsible for the sweeping changes in our industrial-military posture, has been the technological revolution during recent decades. In this revolution, research has become central, it also becomes more formalized, complex, and costly. A steadily increasing share is conducted for, by, or at the direction of, the Federal government. . . . The prospect of domination of the nation's scholars by Federal employment, project allocations, and the power of money is ever present—and is gravely to be regarded. Yet, in holding scientific research and discovery in respect, as we should, we must also be alert to the equal and opposite danger that public policy could itself become the captive of a scientific-technological elite.

With Corona and Zenit, the potentially aggressive act of using a nation's territorial airspace to spy on it, which Eisenhower had admitted could well lead to a declaration of war, was now a constant occurrence. By ceaselessly monitoring each other, though, the two superpowers kept from falling into the kind of panic that is the result of ignorance, and leads to contemplating reaching for nuclear triggers. It was a panic that would only take hold once in the whole of the Cold War, and it would be triggered by the von Braun missiles that the U.S. Air Force had stationed in Turkey so many years earlier, and by the fact that the Soviet Union did not have anywhere near the missile power Khrushchev constantly implied that it had. Richard Bissell's story, meanwhile, would end in tragedy, for the father of one of the Cold War's great successes, aerial reconnaissance, is today known primarily for his role in inadvertently engineering an international catastrophe that so damaged the reputation of America and its new president that the nation was forced to redeem itself by launching the century's greatest adventure: sending a man to the Moon.

11

The Fluid Front

If there is a universally acknowledged truth about the history of NASA, it is that America's youngest president launched the greatest challenge of the Space Race by inspiring the citizens of his nation to work together in the epic task of reaching for the Moon. The truth about Apollo 11's founding father, however, is much more complicated, for at the same time that JFK set into motion the events that would culminate with the triumph of Armstrong, Aldrin, and Collins, he repeatedly called on the Kremlin to end the Space Race altogether and work jointly with the United States to explore the cosmos. As early as his inaugural address of January 20, 1961, Kennedy reached out to the Politburo, saying, "Let both sides seek to invoke the wonders of science instead of its terrors. Together let us explore the stars, conquer the deserts, eradicate disease, tap the ocean depths, and encourage the arts and commerce." Ten days later, during his January 30, 1961, State of the Union, he would repeat this offer in another form: "Specifically, I now invite all nations—including the Soviet Union—to join with us in developing a weather prediction program, in a new communications satellite program and in preparation for probing the distant planets of Mars and Venus. . . ."

Immediately upon taking office, Kennedy appointed Lyndon Johnson to oversee the federal government's space programs, partly because the vice president was the most knowledgeable Democrat on the subject from his years in the Senate laboring at the birth of NASA, and partly to keep the politically adept Johnson well occupied. As something of a counterpoint, JFK retained Eisenhower's final science adviser, MIT physicist Jerome Wiesner, who fervently believed that, when it came to rockets, America should continue its efforts in Pentagon defense and CIA reconnaissance—the Missile Race—instead of supporting NASA's manned flights. When the president then started looking for a NASA administrator, he found there was a serious problem: Between its ongoing competition with the Russians on the one hand and the Pentagon on the other, the agency did not appear to have much of a future. Additionally, it seemed more than likely that its administrator would eventually face the public relations nightmare of the death of an astronaut, and his boss would be LBJ—a notorious handful. Seventeen candidates turned Kennedy down.

He finally settled on a protégé of Oklahoma's Bob Kerr, Johnson's replacement on the Senate Aeronautical and Space Sciences Committee. The man who made the agency what it became during the Apollo era was a native of Tally Ho, North Carolina, who served with the Marine Corps' first aviation squad, became an expert on radar in World War II, worked as Truman's budget director during the Marshall Plan, and then became undersecretary of state, a job his predecessor called "the equivalent of trying to remove a man's appendix while he is carrying a heavy trunk up three flights of stairs." If there are many underrecognized heroes in the Apollo 11 story, chief among them is the bureaucrat who devoted his life to getting men to the Moon: James Webb. A bulldog in the shape of a Washington lifer, with gray-blue eyes and a chin that wouldn't take no for an answer, Webb was a Beltway insider before there was a Beltway, and talking to him was like "trying to get information out of a fire hydrant," as one colleague described it. A friend once had lunch with Webb and, on a bet, said nothing but "hello" at the start of the meal and "good-bye" at the end. Webb declared as they parted, "That was one of the best lunches we've had together. I learned so much." Alan Shepard: "NACA was a group of engineers. They didn't have a political-type administrator. But when Webb came along, I mean, what a fresh breath he was! He knew all the ins and outs of Washington. He knew which chords to play. And not that he was a lobbyist by any sense of the imagination; he didn't have to be. He had a great package: men in space. And he played it well, he really did."

Installed as NASA administrator on February 14, 1961, Webb used his extensive management expertise to establish an agency leadership triad, with Hugh Dryden now overseeing science and foreign affairs; another radar man, Bob Seamans, as general manager; and Webb himself on politics: managing Congress, the Oval Office, and the American people. For the manned spaceflight division, the Washington chief was Brainerd Holmes, who oversaw the three field directorates—Huntsville's von Braun, Houston's Robert Gilruth, and Florida's Kurt Debus. Backed by these men, Jim Webb would lead the most extraordinary engineering project in modern history to greatness.

But first, NASA engineers had a great deal to learn. "Most of us came in from aircraft flight test," Gene Kranz said. "And we knew nothing about rocketry, we knew nothing about spacecraft, we knew nothing about orbits. So it was a question of learning to drink from a fire hose. We had to learn all about trajectories and I had never heard the term *retrofire*—coming on down from orbit, and getting the spacecraft home. We had to virtually invent or adapt every tool that we used."

"We were going to launch a pig, and we put him in the [Mercury] cradle and started monitoring him, and the pig died," aerospace technician Alan Kehlet remembered of those very early days. "One of our secretaries was a farm girl, and she said, 'If you'd asked me before you had the pig in there, I would have told you that you never put a pig on his back, that the belly fat on there will suffocate the pig.' And that's exactly what happened. So, we went from pigs to monkeys."

One Apollo manufacturer, however, continued to use so many pigs in its testing that it built an on-site sty. "One day [at McDonnell in St. Louis], I walked through the hangars, where they were building the spacecraft, and I got a strong odor of like a pigpen somewhere," Guenter Wendt recalled. "So I said, 'Now that couldn't happen.' But lo and behold! In one corner of the building, there was a pigpen. There were several pigs in it. And then I found a project engineer who explained what pigs were for. See, we had to worry about that if Mercury spacecraft would land on land, how much of an impact would the astronaut experience . . . ? So in order to do that, we had to do some drop tests, because we put crushable honeycomb material underneath the couch. So then the powers-to-be found out that the intestinals and the inside makeup of pigs was very similar to that of a human being; like kidney, liver, lungs, and so on. What they would do is make a couch for pigs, then hoist them up on a rather tall facility, and drop them and see what happened to the crushable material. Some of the tests went very well. Some of them didn't go too well. But then, whatever was left over went to the orphanage in St. Louis."

In its salad years NASA had astronaut pigs, and female computers: "Computing at that time consisted of row after row of women . . . who sat and did line after line of calculations on desktop calculators," NASA research mathematician Sandra Jansen remembered. "I was a math major, so I could take the equations and translate them into the various sheets they needed to do their job. They didn't have to do the math; all they had to do was follow the instructions. . . . The girls who worked there were called computers."

For its spaceport, the new NASA decided to build adjacent to the air force's Cape Canaveral, even though the site was plagued by bad weather and corrosive salt air. It was attractive in being accessible by year-round waterways to barge in enormous rocket and module parts, and in being remote from any civilian population that might be imperiled by exploding or misguided rockets and their parts. For its Manned Spacecraft Center, the agency searched for a site that would offer a quick commute to the pads in Florida, to North American and Rocketdyne in California, and to MIT in Massachusetts. There was such a place, and remarkably enough it happened to be in the home state of NASA's House Appropriations Committee chairman, Albert Thomas, as well as the agency's great champion, Lyndon Johnson. Kennedy was at that very moment having trouble getting some of his legislation passed, and that trouble was Albert Thomas. The president would get his bills; Jim Webb would get a crucial ally on the Hill; and Albert Thomas would bring a big slice of NASA pork to Texas.

Like many such maneuverings of Lyndon Johnson and his cohorts in that era, it was both good politics and a good deal. Thomas's associates at Humble Oil (now ExxonMobil) had donated a thousand acres of mineral-leased pasture twenty-two miles southeast of Houston to Thomas's alma mater, Rice University, which in turn had donated it to the federal government. "The site had been selected and announced about September 6, 1961, I believe, several days after Hurricane Carla," public affairs officer Col. John A. "Shorty" Powers said. "We went to where the center now is, but there was nothing but a pasture. There is a famous picture that shows a windmill, watering trough, and there were cows out there. . . . [After the hurricane] the road was strewn with debris, there were telephone trucks by the hundreds on the roads attempting to get all the telephone lines repaired and serviced. . . . There were boats up on the road, pieces of houses and buildings lay around—it was very depressing first look. . . . There were stories about hundreds of snakes crawling around the streets." When the Army Corps of Engineers began its surveys to prepare what was known as Clear Lake Ranch for NASA's Houston campus, they encountered empty fields, lonesome hand-built homesteading shanties, windmills, and a cowboy skinning a wolf.

At first, the men of NACA in Langley, Virginia, were horrified to learn that they were emigrating to Texas, but in time, one would admit, "Moving to Houston was magnificent. We got away from the old fogies. We got down there and created a whole new thing with a bunch of twenty- and thirty-year-olds. Holy smokes, that move to Houston is what made the program!" The Floridians, meanwhile, had to trade palm trees for pin oak and *ropa vieja* for burritos, but otherwise were very much at home in the tropical marshland, with plenty of room to grow and an endlessly vast, big-blue sky. While NASA children were taught to recognize poisonous snakes, since all of North America's could be found in Houston backyards—rattlers and cottonmouths, copperheads and corals—their parents noted that the petrochemical exhaust that made it difficult for the naked eye to see celestial bodies from the Gulf Coast at least made for dazzling sunsets, Technicolor explosions of crimson, deep purple, and warning-lamp orange.

The oil company that had originally donated the land for the agency's campus kept the neighboring thirty thousand acres, and when the NASA town of Clear Lake City grew from 7,000 in 1961 to 30,000 by 1964, to 175,000 by 2004, ExxonMobil found that its charity had paid off handsomely. In time, the agency would serve as an engine of southern reconstruction, its facilities a scimitar across the Gulf Coast states, with rocket engineering in Louisiana, testing in Mississippi, and oversight in Alabama, connecting Houston in the west with Canaveral to the east. Corporate welfare joined with the New Frontier and, in time, the Great Society, to create an institution and an adventure that only a great nation could dream of making come true.

January 31, 1961: Mercury-Redstone 2 took off from Cape Canaveral's LC-5 carrying three-year-old Cameroonian pilot Ham (aka "#61"), who had been trained by Holloman Air Force Base at White Sands with carrots (banana pellets) and sticks (electric shocks to the soles of the feet). For those who believed NASA was ready to launch human beings, this mission upended that hope. First the training system in the capsule went haywire, adminstering to Ham repeated electric shocks, even while he was perfectly executing his chores. The capsule was supposed to travel at 1,970 miles per second; instead, it raced along at 2,298. An abort call was made, which yanked the retro rockets, but Mission Control could not slow the capsule for reentry. Then a snorkel valve lost its pin, and the cabin lost its pressure—but since Ham was in his own spacesuit, he was unharmed. He also seemed unharmed by being subjected to just under 15 g's, instead of the 11 that was expected. On splashdown, the heat shield punctured the capsule, and between the holes it made and the broken valve, by the time the navy hauled the Mercury out of the sea, it had taken on eight hundred pounds of water and was sinking fast.

After recovery, Ham got an apple and an orange for surviving his mission, but tried to bite anyone who dared draw near; as the mission log noted, "Sometime later, when he was shown the spacecraft, it was visually apparent that he had no further interest in cooperating with the space flight program." This near-disaster convinced agency chiefs that they could not yet dare to send up a human being in a Mercury-Redstone like Ham's, a decision that enraged the Original Seven. "Some of the medical members [of the committee under Jerome Wiesner] caucused . . . and turned in a very negative report . . . that we could not possibly take the risk of flying a man in a weightless flight without many additional tests of monkeys," MSC deputy director George Low said. "In fact, they wanted us to fly fifty monkeys. We found out there weren't this many trained monkeys available in the country." Dr. Gilruth suggested that, to surmount the lack of American chimps, the program could be transferred to Africa. Pad leader Guenter Wendt:

> We were in Hangar S at the time where the Air Force was controlling the chimps. And here comes what I thought of as a little bit obnoxious congressman and he said, "I want to see the apes."
>
> So I got down then and I talked to the handlers, an Air Force captain . . . and he said, "Oh, you know, Enos just came back from a training session and he is pretty much ticked off. And Ham just went into it. So I'm not sure." And he says, "You know what Enos does." I said, "Ja, I know what he does. But I got somebody who wants to see him." So he says, "I would rather not."
>
> Then I go out and I says, "Congressman," I said, "We only got one chimp right now that came out of training and he is—his disposition isn't that great. I'd rather have not that you go in." He says, "Oh, you are telling me that I can't see the apes?" I said, "No, sir, I'm not telling you you can't see the apes. But sometimes it might be better not to do that." He says, "I'm a congressman. If I want to do that, I'd like to do that." I said, "Okay, go ahead, sir."
>
> He went in and I kind of had an indication of what was going to happen. Enos saw us coming and he thought he was going to be put back in training, which he didn't like. So he hunched down and deposited something in his hand. And he hit that guy from fifteen feet square on the chest! And [it] was dripping down on this nice white shirt and tie, and he looked at me and says, "Oh. I guess I know now why you didn't want me to go in." I said, "Sir, I didn't know that was going to happen, but you know these are unpredictable animals."

March 22, 1961: Administrator Webb asked for a dramatic increase in NASA's budget, warning Kennedy "that the Russians will, for the next five to ten years, beat us to every spectacular exploratory flight," and proposing a lunar touchdown by the end of the decade. Instead, the White House approved a modest additional sum to begin development of the Saturn booster. This lack of enthusiasm may have been a judgment on NASA's early track record. "In the late fifties, early sixties, we had sometimes twenty, thirty launches a week here, and the average was, three out of the five would blow up, or didn't make it," Guenter Wendt recalled. "'Hey, look, there goes another nose cone' . . . But then you said, 'Oh, wait a minute, how can we launch people? Hmm.'"

"We did a lot of things that had never been done before and, as a result, we had a lot of failures," senior operations director Rob Roy Tillett said. "I saw a couple of the spectacular ones. I was close enough to one to just about want to run around and dive under the car. . . . One of the pieces . . . hit right by the pad and in the [monitor footage] you see this guy hoofing it just as fast as he can down the road getting the hell out of there. He had hid out to have a really good view of the launch and then it blew up right over his head. . . . What was happening was the aerodynamics weren't quite right and [the rocket] was picking up fringes of the rocket exhaust and recirculating them back up into the engine compartment and burning everything up, over- heating the 10,000 psi helium spheres. Boy, when one of those goes off, it makes a bang, let me tell you."

"As a matter of fact, at one time, the seven astronauts were here watching an Atlas launch, and that thing blew, too," Wendt said. "Shepard said, 'I hope they fix that problem before they launch us.'" One comedian called NASA's game plan "First the chimp, then the chump."

April 12, 1961, 9:07 a.m. Moscow Time: The first of what Korolyov called his "little eagles," twenty-seven-year-old air force major Yuri Alek- seyevich Gagarin, radioed "Here we go!" as he lifted off to become both the first man in space and the first to orbit the Earth. Gagarin had been anointed above the other candidates for the glory of symbolism; his mother and fa- ther were peasants on a collective farm, and he was born in a log cabin. Save for a few minutes when telemetry was lost, the flight was pitch-perfect, and its passenger was filled with joy: "The earth was gay with a lavish pal- ette of colors . . . it seemed as though my hands and legs and my whole body did not belong to me. They did not weigh anything. You neither sit nor lie, but just keep floating in the cabin. All the loose objects likewise float in the air and you watch them as in a dream."

His feat was known only to a small circle of Soviet leaders until 12:25 p.m., when Radio Moscow announced, "At 10:55, Cosmonaut Gagarin safely returned to the sacred soil of our Motherland."

"You have made yourself immortal because you are the first man to pene- trate space," Khrushchev told him on the broadcast.

"Now, let the other countries try to catch us," Gagarin answered.

"That's right," Khrushchev said, "let the capitalists try to catch up with our country, which has blazed the trail into space and launched the world's first cosmonaut."

Immediately after the program ended, at 5:30 a.m. Eastern Time, a reporter called NASA, waking up PR chief Shorty Powers, who was sleeping on a cot in his office. When asked how NASA and the United States would respond to this Soviet triumph, a groggy Powers said, "What is this? We're all asleep down here!" The next day's headline: SOVIETS PUT MAN IN SPACE. SPOKESMAN SAYS U.S. ASLEEP.

The shock of Gagarin's achievement convinced any American who wasn't already concerned about the implications of Sputnik, Luna, and Strelka to now join the chorus of complaint against NASA and a drowsy federal leadership. For Americans fearful of massed Soviet nuclear attack aimed directly at their cities in the heat of the Cold War, the fantasy of aggressive Communist soldiers (and their dogs) floating free as they pleased over United States territory was little short of horrifying. For American leaders, the most dramatic demonstration to date of the gap between the scientific and technological abilities of the United States and the USSR was both frightening and even more humiliating than previous Russian successes. *Life* editorialized that the United States had become a nation of decayed Romans, while the Russians were the twentieth century's energetic Visigoths; Speaker of the House John McCormack predicted that America was on the verge of "national extinction," while Wernher von Braun warned Congress that orbiting Soviet bombs would imminently appear. Even Lyndon B. Johnson offered a chilling history lesson: "The Roman Empire controlled the world because it could build roads. Later—when it moved to the sea—the British Empire was dominant because it had ships. In the air age, we were powerful because we had airplanes. Now the communists have established a foothold in outer space. . . ."

At one hearing, House Space Committee member James Fulton (R-PA) asked deputy NASA administrator Hugh Dryden if the Soviets could "create a red dust and turn the whole moon red," and Dryden admitted that, yes, they might be able to do just that. Fulton then declared, "Tell me how much money you need, and this committee will authorize all you need," while Rep. Victor Anfuso (D-NY) announced: "I want to see our country mobilized to a wartime basis, because we are at war." Meanwhile, at a White House press conference, an embarrassed Kennedy conceded that "while no one is more tired than I am" of America's trailing Russia when it came to achievements in outer space, "we are, I hope, going to go in other areas where we can be first and which will bring perhaps more long-range benefits to mankind. But here we are behind." The national conversation started framing the Space Race as a war of opposing philosophies, as described by *Life*: "If this is a race, do we want to catch up? What's wrong with Mercury and our other lagging programs? Should we change our established priorities in favor of some great leapfrog effort? At what cost? Must we really (as one scientist testified last week) get up $25–$40 billion to speed an American to the moon? What is our national space objective: Survival, knowledge, or prestige? [Then there] is the question of the role of science in the organized life of mankind.

"For the Soviets, this question does not exist. They answered it long ago by making science their god. Despite the wholly unscientific character of their political dogmas, they look to science as the chief tool and ultimate vindication of their system. It is a science-oriented system. Thus there is an eerie element of truth in Khrushchev's wild boast that Gagarin's flight 'contains a new triumph of Lenin's ideas, a confirmation of the correctness of Marxist-Leninist teaching.' Gagarin, too, gave the credit to 'our own Communist party,' the vanguard on the 'great road to penetrating the secrets of nature.'"

John Kennedy had entered office on a tide of joyous hope for the future, a promise of new energy and new beginnings. In the months after the inauguration, though, both the country's and Kennedy's optimism was put to the test by an endless parade of Russian space triumphs and overseas disasters. On the campaign trail, JFK had relentlessly criticized the Eisenhower administration's laggard reaction to Soviet space and missile advances, with special counselor to the president Ted Sorensen observing that, to the senator, this failing "symbolized the nation's lack of initiative, ingenuity and vitality under Republican rule." Yet, in his own first months running the Oval Office, Kennedy had given no indication of addressing the problem. And the worst was still to come.

April 14, 1961: Two days after Gagarin's triumph, at a White House cabinet meeting with NASA administrator Jim Webb, deputy Hugh Dryden, special counselor Ted Sorensen, scientific adviser Jerome Wiesner, and budget adviser David Bell, the president asked, "Is there anyplace we can catch them? What can we do? Can we go around the Moon before them? Can we put a man on the Moon before them? . . . When we know more, I can decide if it's worth it or not. If someone can just tell me how to catch up . . . There is nothing more important."

Dryden told the president that, if the United States was willing to designate putting a man on the Moon as the equivalent of a Manhattan Project, at a cost of $40 billion, there was a fifty-fifty chance of success. Sorensen later said of this meeting that Kennedy "immediately sensed that the possibility of putting a man on the moon could galvanize public support for the exploration of space as one of the great human adventures of the twentieth century. . . . The lunar landing program, by accelerating and organizing our efforts to avert the Soviet militarization of space, and by increasing the possibility of international scientific cooperation, including work with the Soviet Union, helped his efforts for peace. Part of the motivation [was] to assure the world that neither superpower would dominate this new ocean through hostile, military means."

April 17–21: After Fidel Castro had aligned with the USSR in March 1960, Eisenhower had broken off diplomatic relations, begun an economic embargo, and signed a directive for the CIA to remove the new leader of Cuba from office. One of the agency's projects, Operation Zapata, outfitted and trained 1,511 anti-Castro Cuban refugees in Guatemala to invade and reconquer their island. Since the CIA had already successfully overthrown other governments (such as Iran's in 1953 and Guatemala's in 1954), there was great expectation that Zapata would be similarly successful.

The plan was to have the rebels invade on April 17 at Bahía de Cochinos' Playa Girón, a Caribbean tourist beach of emerald seas lapping against a strip of sand and palm. The counterrevolutionaries expected that fellow Cubans would immediately join their cause, even though the American government's own December 1960 intelligence reports found Castro "firmly in control" and the country's political opposition "generally ineffective." Indeed, ten thousand potential sympathizers had been mass-arrested in the days before the invasion, as Castro had had advance warning via the KGB, Cuban security agents, and loose lips in Florida. Radio Moscow had even broadcast, in English, four days earlier, that "a plot hatched by the CIA" would strike Cuba within a week.

The CIA knew that the Soviet Union had had advance word of the attack, but did not inform Kennedy; the CIA had predicted an inept Cuban military response, but Castro had extensive Soviet armament and military advisers in place. Zapata had provided for American bomber strikes to destroy the Cuban air force, but after the Cuban military brought down fourteen B-26s, the rest of the aerial support was cancelled.

The failed Bay of Pigs invasion left the Kennedy administration appearing grossly inept and the CIA incompetent; both its chief, Allen Dulles, and its black-ops manager, Richard Bissell, were forced to resign. "There are a number of paradoxes in the story of the Bay of Pigs," Ted Sorensen said. "The largest covert operation in CIA history was too large to remain covert and too small to be successful, with too few men on the ground, too few pilots in the air, too few ships in the water, and too little ammunition in reserve."

Even so, the Bay of Pigs did have some beneficial consequences. The terrible advice that the Pentagon's hawks offered the president during this incident would give him leave to ignore their equally inept counsel during the Cuban Missile Crisis. The fiasco would also inspire Kennedy to take a great leap in an attempt to restore his and his nation's pride, as he later explained to John Kenneth Galbraith: "There are limits to the number of defeats I can defend in one twelve-month period. I've had the Bay of Pigs, pulling out of Laos [in the face of Communist insurgency from the Pathet Lao], and I can't accept a third." Presidential science adviser Jerome Wiesner noted, following those foreign embarrassments (and, just before them in fact, a third, in Congo), "I don't think anyone can measure it, but I'm sure [the Bay of Pigs] had an impact. I think the President felt some pressure to get something else in the foreground. . . . We talked a lot about do we have to do this. He said to me, 'Well, it's your fault. If you had a scientific spectacular on this earth that would be more useful—say, desalting the ocean—or something that is just as dramatic and convincing as space, then we would do it.' . . . If Kennedy could have opted out of a big space program without hurting the country in his judgment, he would have. . . . I think he became convinced that space was the symbol of the twentieth century. It was a decision he made cold-bloodedly. He thought it was right for the country."

April 20: JFK sent a memo to Vice President Johnson:

1. Do we have a chance of beating the Soviets by putting a laboratory in space, or by a trip around the moon, or by a rocket to land on the moon, or by a rocket to go to the moon and back with a man. Is there any other space program which promises dramatic results in which we could win?
2. How much additional would it cost?
3. Are we working twenty-four hours a day on existing programs? If not, why not? If not, will you make recommendations to me as to how work can be speeded up.
4. In building large boosters should we put our emphasis on nuclear, chemical, or liquid fuel, or a combination of these three?
5. Are we making maximum effort? Are we achieving necessary results?

April 22: Lyndon Johnson met with NASA executives, who predicted that the Soviets would be first to orbit a space station, but that America might take the lead in landing on the Moon "if a determined national effort is made." To begin that effort, they estimated that NASA's 1961–1970 budget of $12 million would need to be increased by 2,000 percent to $22 to $34 billion. Two days later, LBJ met with navy R&D chief Admiral John Hayward, the air force's Systems Command head General Bernard Schriever, and NASA's Wernher von Braun to get their perspectives. Schriever stated, "We need a major national space program for prestige purposes," and pointed out that such a direction would help the aerospace industry at a time when many in the USAF believed that the ICBM missile race would be leveling off. Such corporate munificence had in fact been one of the original reasons for Eisenhower and Congress to launch NASA, as the American aerospace industry had suffered a post–World War II collapse, with nine of the twelve major firms losing $35 million in 1946 and $115 million in 1947.

At his meeting with the vice president, von Braun concurred that "We do not have a good chance of beating the Soviets to a manned 'laboratory in space.' . . . we have a sporting chance of sending a three-man crew around the moon ahead of the Soviets . . . we have an excellent chance of beating the Soviets to the first landing of a crew on the moon . . . in the space race we are competing with a determined opponent whose peacetime economy is on a wartime footing. . . . I do not believe that we can win this race unless we take at least some measures which thus far have been considered acceptable only in times of a national emergency."

LBJ's own view of the matter was plain. Though he actively investigated a range of opinions, whenever he met with criticism about a dramatically expanded space program, he would say, "Would you rather have us be a second-rate nation, or should we spend a little money?" After the costs of World War II, the Manhattan Project, and the Marshall Plan, after all, the price of Apollo did not seem particularly daunting. It was an era when a great country did great things.

April 25: The CIA released an intelligence report that predicted a Russian lunar orbit by 1966, and a Soviet lunar landing no earlier than 1970—that is after "this decade is out."

May 2: Jim Webb asked Jerome Wiesner for his support in keeping the political and public relations demands of beating the Russians in various "firsts" from eclipsing NASA's scientific and technological work, including satellites for communications, weather monitoring, and unmanned probes to survey the Moon, as well as the science to be conducted by Project Apollo astronauts. Wiesner would not be particularly cooperative on this front, or indeed, on any other NASA matter, while Webb's attempts to make Apollo more than just a Space Race and engineering triumph would bring him into conflict often with his ultimate boss, Kennedy. In this aspect of his mission as NASA chief, Webb would fail, and after he left the agency, NASA would be subjected to decades of criticism for not doing more of exactly what Webb had hoped to achieve.

May 5: Forty-five million Americans watched as Alan Shepard—selected to be first into space since he was considered the smartest of the Original Seven—got fed up with the endless delays at launch (during which his heart rate hit 200) and demanded that von Braun "light this candle." On television sets across America, out of the flat blue-green plain of Florida, Freedom 7 rose, an explosion on the ground, a line of flame, a trail of smoke—an immense bomb exploding against itself. From Edwards, Chuck Yeager reminded Shepard to "brush the monkey shit" off his seat before buckling in.

Shepard's public triumph and personal heroism in becoming America's first astronaut, riding a ballistic missile into space and back, triggered an outburst of patriotism in the American public, which inundated Congress with thrilled letters and phone calls. "The mission only lasted twenty minutes, but it was the purest, happiest twenty minutes of our lives," flight director Gene Kranz exulted. "Our first man in space. Total joy."

Overnight, President Kennedy seemed just as enraptured, as Shepard tells it: "There's a picture of me sitting on the sofa [at the White House], Jack is in the rocking chair, and I'm telling him how I was flying the spacecraft, and he's leaning forward listening intently to this thing. We talked about the details of the flight, specifically how man had responded and reacted to being able to work in a space environment. And toward the end of the conversation he said to the NASA people, 'What are we doing next? What are our plans?' And they said, 'There were a couple of guys over in a corner talking about maybe going to the Moon.' He said, 'I want a briefing.' Just three weeks after that mission, fifteen minutes in space, Kennedy made his announcement: 'Folks, we are going to the Moon, and we're going to do it within this decade.' After fifteen minutes of space time! Now, you don't think he was excited? You don't think he was a space cadet? Absolutely, absolutely!

"When Glenn finished his mission, Glenn, Grissom, and I flew with Jack back from West Palm to Washington for Glenn's ceremony. The four of us sat in his cabin and we talked about what Gus had done, we talked about what John had done, we talked about what I had done. All the way back. People would come in with papers to be signed and he'd say, 'Don't worry, we'll get to those when we get back to Washington.' The entire flight. I tell you, he was really, really a space cadet. And it's too bad he could not have lived to see his promise."

"The president was impressed with the world's reaction to the Shepard flight," Bob Gilruth remembered, "and wanted to know more about what we were going to do." At a meeting immediately following Shepard's great success (perhaps they were the "couple of guys over in a corner"), Gilruth and Low described the rest of Project Mercury. They then went on to detail a study that had recently been completed, which would send a man to orbit the Moon. Kennedy immediately wanted to know, "Why aren't you considering landing men on the moon? If we're going to beat the USSR, don't we need to do something more than just flying around the moon?"

This notion caught the men from NASA off-guard. "I didn't want to sound negative," Gilruth said later, "so I told him that landing on the moon was probably an order-of-magnitude bigger challenge than a circumlunar flight. But he didn't let go."

"What do you need?" Kennedy asked.

"Sufficient time, presidential support, and a congressional mandate," Gilruth said, off-the-cuff.

"How much time?"

Gilruth and Low discussed what might be needed, and gave him an estimate: "Ten years."

May 6: Jim Webb and Robert Seamans went to the Pentagon to meet with Secretary of Defense Robert McNamara and prepare a joint memo for the White House for what would become Project Apollo. After Webb explained that NASA was thinking of landing on the Moon by the end of the 1960s, McNamara argued that this was short-sighted. America, he believed, should forget about the Moon and go directly to Mars. An aghast Webb and Seamans "found his suggestion horrifying and pointed out that we had neither the technology nor the physiological understanding to proceed with such a mission," as Seamans recalled. Since the Pentagon was going to cut dramatically back on a number of defense contracts to pare the military's budget, however, a Moon mission would give the U.S. aviation industry a corporate safety net, just as Schriever had suggested. Wiesner said that when McNamara explained to Kennedy what Apollo would do for the economics of the nation's aerospace industry, "this took away all argument against the space program."

May 7, 1961: Johnson (who would now become one of NASA's many unsung heroes), sent JFK a memo entitled "Recommendations for Our National Space Program: Changes, Policies, Goals," which summarized his findings and echoed Eisenhower's "total cold war" thinking: "It is man, not merely machines, in space that captures the imagination of the world. . . . Dramatic achievements in space . . . symbolize the technological power and organizing capacity of a nation. . . . Major successes, such as orbiting a man as the Soviets have just done, lend national prestige even though the scientific, commercial or military value of the undertaking may, by ordinary standards, be marginal or economically unjustified. . . . Our attainments are a major element in the international competition between the Soviet system and our own. The non-military, non-commercial, non-scientific but 'civilian' projects such as lunar and planetary exploration are, in this sense, part of the battle along the fluid front of the cold war."

A few days later, the president told NASA executives: "All over the world we're judged by how well we do in space. Therefore, we've got to be first. That's all there is to it. . . . I want you to start on the moon program. I'm going to ask Congress for the money. I'm going to tell them you're going to put a man on the moon by 1970."

May 17: In the midst of preparing to ask Congress and America for the unprecedented federal commitment that would become Project Apollo, however, Kennedy tried to circumvent the Space Race by having Secretary of State Dean Rusk ask Soviet foreign minister Andrei Gromyko if he would convey to his country the American president's enthusiasm for exploring the universe together. Gromyko was not optimistic.

May 21: Robert Kennedy reiterated his brother's overtures when he met with his backdoor Soviet counterpart, Soviet military intelligence officer Georgi Bolshakov, explaining that the president wanted to develop an agreement with Khrushchev on cooperative ventures in outer space at the upcoming Vienna summit.

May 25: JFK decided to counter the many stumbles of his first months in office by delivering what was in effect a second State of the Union address before a joint session of Congress on the topic of "Urgent National Needs." In addition to presenting requests for a greatly expanded army, marines, United States Information Agency, aid to developing nations, and a nationwide program of civil defense to survive Soviet nuclear attack, President Kennedy declared: "Now is the time to take longer strides, time for a great new American enterprise, time for this nation to take a clearly leading role in space achievement, which in many ways may hold the key to our future on earth. . . . For while we cannot guarantee that we shall one day be first, we can guarantee that failure to make this effort will make us last. We take an additional risk by making it in full view of the world; but as shown by the feat of Astronaut Shepard, this very trial enhances our stature when we are successful.

"I believe that this nation should commit itself to achieving the goal, before this decade is out, of landing a man on the moon and returning him safely to earth. No single space project in this period will be more impressive to mankind or more important for the long-range exploration of space, and none will be so difficult or expensive to accomplish. . . . In a very real sense, it will not be one man going to the moon—if we make this judgment affirmatively, it will be an entire nation. For all of us must work to put him there."

The combined military and civilian space effort, Kennedy explained, might cost as much as $40 billion, but "a great nation was one that undertook great adventures."

In another speech, JFK urged his countrymen to sacrifice: "We choose to go to the moon in this decade and do these other things, not because they are easy, but because they are hard. Because that goal will serve to organize and measure the best of our energies and skills. . . . only if the United States occupies a position of preeminence can we decide whether this new ocean will be a sea of peace or a new, terrifying theater of war." Ted Sorensen believed that sending a man to the Moon "embodied everything [Kennedy] had said for a year and longer about striving to get this country moving again, about joining the Russians in peaceful space exploration, about crossing 'new frontiers.'"

The at times pugnacious Kennedy had ultimately arrived at his decision on Apollo by realizing that there were three avenues of competition possible between Russia and America: war, business, and technology. The first meant the unacceptable course of nuclear war, while the second would involve engaging in a long contest before the winner would be known. With the very public and deeply symbolic Space Race, however, the entire world would immediately know who had triumphed. He told Sorensen that shifting "our efforts in space from low to high gear . . . [was] one of the most important decisions he would make as President." If the great theme of the Kennedy administration was New Frontiers, then space would be the biggest New Frontier of them all.

"There's been much conjecture about President Kennedy's motivation when he addressed Congress and recommended a lunar landing and safe return within the decade," Bob Seamans remembered. "Was he a true space cadet fantasizing about a lunar mission from Earth? Or was he impressed with the scientific importance of learning more about our universe, particularly our own solar system? Some have suggested that he felt the need for a major effort so that the Soviets would agree to negotiate a joint program. My meetings with the President at the White House on 21 November 1962, and during his visit to Cape Canaveral on 16 November 1963, showed me that he had one straightforward goal, and it wasn't any of the above. He wanted the United States to conduct a major, readily discernible mission in space prior to an equivalent Soviet Union achievement. The Soviets, thanks to Khrushchev's opportunism and Korolyov's mastery of Soviet technology, had embarrassed the United States again and again with cleverly devised and well-conceived forays into space. Were we, as they claimed, a degenerating civilization and they the wave of the future? Kennedy wanted to prove it wasn't so."

Still, privately, Kennedy would continue to ask, "Can't you fellows invent some other race here on earth that will do some good?" and commented about getting a man on the Moon, "The cost, that's what gets me." It was a huge decision—what responsible human being wouldn't have second thoughts? At a 1962 cabinet meeting, Kennedy argued, "This is, whether we like it or not, in a sense a race. . . . Why are we spending seven million dollars on getting fresh water from saltwater, when we're spending seven billion dollars to find out about space? Obviously, you wouldn't put it on that priority except for the defense implications. . . . the Soviet Union has made this a test of the system. So that's why we're doing it. . . . we've been telling everybody we're preeminent in space for five years and nobody believes it! . . . [Apollo is] one of the two things—except for defense—the top priority of the United States government . . . otherwise we shouldn't be spending this kind of money, because I'm not that interested in space."

Whatever his personal ambivalence, in his public rhetoric JFK inspired Americans at every level with the glory of the space agenda. "It's that challenge that best explains the emotional hold of the Kennedys. . . . that we could step beyond our narrow personal concerns to achieve great things, that we could do better, be better, if only we had the strength and courage to work harder and dream bigger," journalist Bob Herbert remembered. "['Ask not what your country can do'] was a call to civic engagement, to national commitment and sacrifice. [Jack and Bobby] never failed to implore voters to make the effort to touch the best in ourselves. The Kennedy brothers helped bolster our capacity to believe."

So magnificent an undertaking did meet with considerable skepticism at the time, however, and many of the most dubious would turn out to be charged with making this dream come true. When the Senate Appropriations Committee interviewed NASA's Hugh Dryden and asked what purpose was served by going to the Moon, Dryden replied bluntly: "It certainly does not make sense to me." The head of America's Space Task Group remembered being "aghast" by Kennedy's proposal, while President Eisenhower emerged from retirement to publicly call the program "just nuts." "I always noticed that when we became NASA, President Kennedy said to go to the Moon by the end of the decade, and all our badges that were issued to us would expire on December 1969," aerospace engineer Milton Silveira remembered. "Well, that's a message for you. Either you do it or you're not employed anymore."

The man who would be responsible for building a spacecraft to achieve Kennedy's vision was Bob Gilruth, whose NACA Space Task Group in Virginia had become the Manned Spacecraft Center in Houston. After the president's May 25 speech, he went into shock: "I could hardly believe my ears. I was literally aghast at the size of the project being undertaken. At that time, there was not detailed plans and no studies in depth, on how the landing could be done." At that point Shepard's flight—all of fifteen minutes—represented the sum total of NASA's experience in manned spaceflight.

June 4: Nine days after announcing that "I believe that this nation should commit itself to achieving the goal, before this decade is out, of landing a man on the moon," Kennedy once again directly asked Khrushchev at their Vienna summit to end the Space Race and explore the universe together. The premier immediately rejected the idea, but then wondered aloud, "Why not?" He then categorically insisted that the two superpowers had to pursue military disarmament before joint civilian space missions could be considered.

August 6: Khrushchev insisted that Korolyov rush his next mission—seventeen orbits in twenty-four hours—with predictable results. Air force major Gherman Titov became horribly nauseated, and then mentally confused, suffering from a life-threatening attack of space sickness—another historic Russian first. After seven hours of flight, he announced, "Now I'm going to lie down and sleep. You can think what you like, but I'm going to sleep." He was saved by ground controllers and automatic reentry systems, and able to make an appearance with Khrushchev the next day in Red Square.

Five days later, a barbed wire and prefab-concrete fence was erected across the middle of Germany—the Berlin Wall. Khrushchev had a perfectly reasonable rationale for the move, explaining that "more than 30,000 people, in fact the best and most qualified people from [East Germany] left the country in July. . . . The economy would have collapsed if we hadn't done something soon against the mass flight." His decision concerned the West sufficiently that even more Americans were encouraged to excavate their backyards and install nuclear fallout shelters, including the Kennedys at Hyannis Port, and the von Brauns in Huntsville.

The more the American public learned such news as the need for fallout shelters, the building of Berlin Walls, and the dazzling achievements of cosmonauts, the more they came to idolize their nation's modern-day gladiators. When the Original Seven became instant national heroes, though, everyone at NASA was taken aback. Suddenly a tidal wave of glamour had engulfed the agency's pilots and engineers, a glamour similar to Kennedy's, that warm combination of good looks, vigor, youth, a drive to the future, and a conquest of New Frontiers. Before the astronauts, America's best-known military heroes tended to look like Eisenhower, or Patton, just as before Kennedy, American presidents tended to look like Truman, or FDR. Now, they looked like movie stars, and what started as a mass public crush turned into a form of hysteria. "After we had the initial press conference in April 1959, and the astronauts' pictures (and in the next week their seven wives) appeared on the cover of *Life* magazine, everybody in the world knew who they were and nothing was sacred," public affairs officer (and "voice of Mercury") Shorty Powers recalled. "You could be in the head and someone would come up and start talking to you. The astronauts' homes were invaded, their kids were followed to school and back, the kids' teachers were questioned about how the kids were doing, if they were smart as their daddies, et cetera. . . .

"Gus Grissom's house has no windows on the side facing the street and it was built that way purposely. He simply did not want people peering in his windows. John Glenn's house in Arlington, Virginia, at the time of his flight was overrun. We finally got the county and state police to guard the house. I think it was the Junior Chamber of Commerce or Lions Club or another organization that came in after the flight and resodded John's lawn for him because the public had stomped it right into the mud."

As their celebrity grew, the astronauts were besieged not only by press requests but by business opportunities. The NASA PR office turned to tax lawyer Leo De Orsey for help. "I insist on only two conditions," he told the Original Seven. "One, I will accept no fee. Two, I will not be reimbursed for expenses." It was an era when the Cape Canaveral technicians were better paid than NASA's fliers, since the astronauts lived on military pay scales (with the exception of civilian Neil Armstrong, the agency's highest-paid Apollo candidate). America's new heroes made a base salary of $5,500 to $8,000, with an additional $2,000 in housing allowance and $1,500 in flight pay—a compelling reason why the astronauts insisted on flying themselves between Houston, Canaveral, and the various subcontractor facilities. This was not a terrible salary in the 1960s, but it was minuscule if you were a famous public figure who needed employees to handle everything from fan mail to security, constant changes of phone numbers, and to buy a few moments of privacy for yourself and your family.

Adding to the astronauts' precarious financial status was the fact that they could not even obtain life insurance. Though not one astronaut would die on a mission until *Challenger* in 1986, the insurance companies were well aware of the mortality rates of test pilots—in 1948 alone at Edwards, thirteen fliers were killed, including Captain Glen Edwards, for whom the base was named. In fact, when *The New York Times* asked flight director Chris Kraft what the odds were on John Glenn's having a successful launch the following day, Kraft replied, "If I thought about the odds at all, we'd never get to the pad."

To solve this particular problem, De Orsey approached every insurance company he could find, and the only offer came from Lloyd's, which quoted a premium of $16,000 for John Glenn's 4.5-hour flight. De Orsey ended up writing Annie Glenn a $100,000 check out of his own account for her security if her husband did not return. Houston congressman Albert Thomas then tried to get a special insurance program set up for the fliers, but that idea was shot down, since it wasn't fair to all the boys drafted to fight in Vietnam— they couldn't get life insurance, either.

Ultimately, De Orsey auctioned the astronauts' "exclusive personal stories" to *Life* for $500,000, with the additional provision that each pilot be insured for $50,000. This half-million-dollar payment would be shared by all astronauts, so as annually their ranks grew, the individual payouts declined. It was a contract that gave *Life* exclusivity on "the personal background of each man, his family, his children and their attitudes, his church relationship, his childhood remembrances, and his by-lined personalized 'human experience' view of the mission outside of normal scientific and project reporting," as John Glenn described it. "In other words, these were the interesting sidelines to the real historical accomplishments of each flight. To know these details and feelings of each man and his family, it was obviously necessary that we open our lives to reporters to whom we would otherwise deny access. . . . I was certainly willing to do that if I could be recompensed for the trouble enough to know that my children could be guaranteed an education I could not otherwise afford or we could enjoy some of the things we did not have on the basis of straight military pay scales."

When the contract reached agency head Jim Webb for approval and he opposed it, John Glenn went over his head. While sailing with Kennedy on *Honey-Fitz,* the astronaut told the president why the *Life* proposal was a good idea. On August 24, at an Oval Office meeting of Pierre Salinger, Bundy, Sorensen, and Webb, JFK repeated Glenn's arguments, including the fact that the new heroes "were burdened with expenses they would not incur were they not in the public eye." In the end the astronauts would have their payments, their insurance, and their bylines in *Life*; Kennedy was unhappy, however, with a house-flipping scheme the astronauts were involved with in the Houston suburb of Sharpstown, and insisted they back out of that contract. When a public outcry then erupted over astronaut ownership of the Cape Colony Inn, a luxury hotel near Canaveral, De Orsey recommended they sell their shares in that, as well; each made about $6,000 on the deal.

The *Life* contract would remain controversial with both journalists and NASA executives until it expired shortly after Apollo 11. On August 27, 1962, *New York Times* managing editor Turner Catledge complained to Webb that "whatever property rights there may be in the stories of the astronauts . . . these property rights belong to the American people and not to individual citizens." Others were dismayed when *Life* began issuing its bland articles, each censored in turn by NASA, the relevant military service, and the astronauts (or their wives). With so many editorial hands involved, *Life* ended up publishing stories that portrayed the astronauts and other agency employees the way that NASA, the Pentagon, the air force, the navy, the marines, and their wives wanted them to be seen, instead of anything remotely resembling human beings. In hindsight, however, Webb came to see the value of the deal, reasoning that, "If a society editor called up and said, 'I want to see Annie Glenn,' we couldn't have said, 'No, you can't see her,' but since Annie Glenn signed a *Life* contract, she could say no." But public affairs chief Julian Scheer sided with the *Times*: "Before I came to NASA, when I was on the other side of the fence, I was against the astronauts' contracts on principle. I still feel the same way."

For the pilots and their families, the *Life* agreement was a simple matter of economics. Jim McDivitt: "As soon as we got selected, we were told that, 'No more uniforms.' And we all of a sudden, you know, here we are, 95 percent of your wardrobe is extinct! And so, now you have to re-equip. And we didn't have a heck of a lot of money to do that. We had to have tuxedos and suits and all that other stuff. We were the representatives of our country to the foreign countries and to all the people who were paying for the program. So, we really scrambled around. And then to top all that off, we weren't living on base housing. Fortunately, things were not too expensive in Houston in those days, and so we could afford to buy a house and things. And I must say

that we ended up with a contract, ultimately, with *Life* and World Book for our
exclusive stories, which provided a little extra money. Very little extra money!
But at least it gave us enough money so we could afford a wardrobe."

The wives eventually found that this bargain cut two ways. For some, the
magazine's journalists became so close to the astronauts' families that "*Life*
was the one thing I trusted, because you couldn't trust NASA," as Rene Carpenter said. But when Joan Aldrin was given a job hosting a local radio talk
show, magazine executives decided it was a violation of their agreement, even
though it wasn't specifically mentioned in the contract, and she was forced to
turn down the offer.

While its astronauts were becoming universally admired symbols of American youth, bravery, and vigor, the agency itself came to be seen as a shining
example of top-notch American can-do ingenuity, even though the first years
of NASA's manned program were hardly an unqualified success. Although
Mercury was considered a triumph, the specific details of its missions raised
a chorus of doubts.

"All of the Mercury flights had trouble, Shepard had leaks in his thrusters,
the hatch on Liberty Bell 7 blew before the capsule had been secured, Grissom almost drowned, John Glenn had endless delays in lift-off, weeks went
by, in flight the automatic steering did not work too well, the left thruster
failed, then the right, the gyroscope indicators were at odds with what Glenn
could see for himself, ground control received signals that the lock holding
the heat shield in place for reentry had opened, if that were true then Glenn
would be dead on return," as Norman Mailer enumerated. "Carpenter had
trouble with his control system; his deployment switch failed to release the
landing parachute; he had to throw the switch by hand. Schirra had a flight
suit which overheated, and the launch vehicle did a clockwise roll after it left
the pad, the booster engines were misaligned. Cooper had a host of malfunctions, the suit, the gravity light, the carbon dioxide level, the automatic pilot
system failed; he had to fly the ship into reentry himself."

Astronaut nurse, First Lieutenant Dee O'Hara:

> In November of '59, I was working in the labor and delivery room at the
> hospital there at Patrick [Air Force Base in Cape Canaveral, Florida], working
> nights. I had a message the next morning that the "old man," meaning the
> commander of the hospital, wanted to see me when I got off duty the next
> morning. Well, naturally, I was terrified, because I'd only been there six

months and I knew that when you went to see Colonel Knauf, it was for two reasons: one, you were in trouble; or, two, it was for a promotion. Well, I knew it was not for a promotion because I'd only been there six months. So I kept thinking, oh, boy, what have I done? I didn't remember harming anybody or harming a baby.

I gave morning report the next morning and went to his office, and here sat his exec officer, the chief nurse, and all these people. I really was terrified because I didn't know why exactly I was there. I literally sat on the edge of the seat. Anyway, he started talking about Mercury, and I thought, well, there's a planet Mercury and there's mercury in a thermometer, and then he mentioned astronauts. That, of course, didn't mean anything to me. I didn't know what they were. He mentioned NASA, and I thought he was saying Nassau, because of the island of Nassau. I had just been there, and I thought, "How the heck did he know I was down there?" Anyway, I was quite confused.

He turned and said to me, "Well, do you want the job?" I kind of turned around, because I didn't think he was talking to me. He said, "Well, you haven't heard a word I've said, have you?" I said, "No, sir." And he said, "Well, do you want the job or not?"

I didn't know what else to say, so I said, "Well, I guess so," absolutely not knowing at all what I had committed myself to. Of course, the chief nurse, who was there, was furious with me afterwards, because she was losing me out of the hospital. Also, she thought NASA was crazy because they were going to be putting a man on the top of a rocket.

That's how I got involved. Colonel Knauf decided that he wanted a nurse, and NASA said, "Well, we don't want a nurse." He said, "Well, you're going to have a nurse." NASA said, "Well, we didn't want one, anyway."

He wanted someone that would get to know the astronauts so well that she would certainly know if they were ill, because, as we all know, pilots, let alone astronauts, are not about to tell a flight surgeon when they're sick, and that's understandable because, as you know, pilots are so afraid of being grounded, and the flight surgeon's the only one that has that authority over them. So they're not usually very friendly with their flight surgeons. Anyway, his whole idea and concept was to put someone out there at the Cape to be with them all the time and just to get to know them so well that she would certainly know if they were ill or not. . . .

[Before Alan Shepard's flight] was probably the most emotional, excruciating time for me. Since we had never put a man on top of a rocket or launched one, so to speak, it was very nerve-wracking, and I think it was nerve-wracking for everyone simply because we didn't know quite what was going to happen. As you know, they were always testing missiles, and they'd go up and explode, or they'd nose-dive into the ocean, and now we were going to put some guy on top of one of these rockets. So it was very scary.

After Alan's flight, as soon as he was launched, I don't think I breathed for quite some time, and even then, his flight was, what, sixteen minutes. But those were very, very long minutes until we got the word that he was down-range and he had landed exactly where he was supposed to.

It never really got easier. I think we all had just a tetch more self-confidence, but not much, because, again, it was still all very new and very experimental. Gus got into trouble upon landing. As you know, the hatch blew, and the valve on his suit opened, and water started pouring in, and the helicopter that was hovering over him, here Gus was frantically waving at them and trying to tell them that he was in trouble, and they thought he was just waving and being friendly, and so they started waving back at him.

I remember John Glenn's flight. He was being launched on an Atlas, and we had watched those blow up right and left, and there was tremendous anxiety, at least, again, on my part.

I'd made an agreement with the astronauts, actually each group that came in, a long time ago that they could come to me with anything. It didn't matter what it is, and that I would never betray them. But there is one condition. I said that "You have to understand that if, in my opinion, it would jeopardize you or the mission, then ethically I will go to a flight surgeon with this. So don't come to me with anything you don't want them to know," and that was the understanding, and that's the way it worked all these years.

Spaceflight is certainly not for sissies. It's like getting old is not for sissies either. Spaceflight is dangerous, and I think we lose sight of that because we've been so very successful.

Given NASA's mixed record of success, it must have been very difficult for most at the agency to believe in the feasibility of the Apollo program. The key Saturn booster component, the F-1 engine, had begun as an ARPA/von Braun project on August 15, 1958, and even after five years in development, it appeared to be beyond the abilities of American engineering talent. NASA general manager Robert Seamans: "Its 1.5 million pounds of thrust was an order of magnitude greater than the thrust of the existing engines used in ballistic missiles. The fuel and oxidizer pumps were driven by 55,000-horsepower engines, and the engine itself produced 160 million horsepower at ignition. The F-1 was a brute, and it was nearly impossible to tame. Blowups occurred, apparently at random, during its 3.5-minute burn. Its development was more an art than a science."

The essence of rocketry is combining fuel, flame, and oxidizer in a combustion chamber to create a continuously exploding bomb, the force of which is directed through a throat, and then a nozzle. If the liquid fuel pumped into the chamber isn't immediately ignited, however, it builds up, causing varying degrees of explosion—known as combustion oscillations—some of which can be more powerful than the chamber is designed to hold. The result is that the motor regularly blows itself up, which is precisely what happened in the case of the F-1. The solution was to find the perfect design for the engine's injector, a showerhead that controls the flow of propellant. Again and again, Rocketdyne ran the engine through bombing tests, detonating explosives inside the engine to simulate erratic firing. Time and again, the engine was destroyed.

The work was frustrating and didn't lend itself to mathematical analysis. Holes in the injection plate for the oxidizer and fuel could be rearranged and retested. Baffles of various dimensions were introduced to determine their effectiveness. Finally, stability was achieved by moving the burn closer to the mouth of the nozzle, resulting in a loss in efficiency of only a few percentage points. It took nearly seven years for the F-1 to pass its flight-rating test, on March 8, 1965.

The liquid hydrogen for the Saturn's upper-stage J-2 engines also turned out to be problematic, since it would boil at −423 degrees F, and its tiny molecules leaked out of piping that was perfectly suitable for more robust fuels. Propellant testing became a serious concern, since bubbles and vapor pockets could cause such vibrations that they would damage the missile's plumbing, and there would be a significant variance in fluid flow with the loss of gravity. How would frozen gas behave in a zero-g tank? It was a problem that could not really be tested by Earth-bound engineers. But there was an answer: to have little rockets fire to force the gas down into the bottom of its tanks for pumping—a mechanism called the ullage engine.

Given the danger and problems inherent in this process, why were enormous explosions even used as a method of interplanetary transportation? Because there is nothing else—all other methods require something to push against. A wing curved in air. The rubber meets the road. Rockets move from the force of exhaust pushing against their own bodies, and so can function in the great nothingness of space—exactly where *The New York Times* erred in its slur against Robert Goddard.

On September 1, 1963, Manned Spacecraft Center director Brainerd Holmes was replaced by George Mueller (pronounced *Miller*), a TRW man with a Ph.D. in physics as well as experience in Minuteman performance, budgeting, and scheduling. Mueller brought along with him twenty-one air force colleagues, notably Brigadier General Samuel C. Phillips (the architect of the

USAF's Minuteman program), to run the greatly expanded Office of Manned Space Flight for Apollo, overseeing design, procurement, manufacturing, testing, logistics, and crew training and support, as well as North American Aviation's Command Module overseer, Joe Shea. Mueller was a workaholic who expected everyone beneath him to toil as hard as he did; he was extremely opinionated, not much liked, and a key force in getting Apollo 11 to the Moon before 1970. Engineering executive John Disher said that Mueller was "always the epitome of politeness, but you know down deep he's just as hard as steel."

As soon as Mueller got to NASA, he had a study prepared by Disher and rocket propulsion director Adelbert Tischler to assess the probability of Apollo's reaching the Moon by 1970—Kennedy's deadline and a winning date for the Space Race if Corona satellite coverage (and resulting CIA analysis) of the Soviet program was accurate. Their estimate? That NASA had a 10 percent chance of reaching the Moon by 1970, a finding that could so damage the agency in congressional eyes that Seamans and Mueller had the authors destroy the report and its supporting documents after they had presented their findings. Mueller then led a wholesale revamping of NASA, centering on one key managerial innovation.

Under von Braun's and Debus's direction—and a tradition carried over from the V-2 era—Huntsville and Canaveral did remarkably thorough, careful, and methodical piece-by-piece testing of each subcontractor-delivered component individually, and then another round of tests when the components were assembled. Each individual stage was fired and flight-tested, then two stages were stacked and tested, then three, then the full rocket—requiring six separate launches. When problems arose with the Saturn second stage from North American, for example, the Huntsville team disassembled the entire missile and examined it, piece by piece, looking for defects. Del Tischler: "Some of the people at headquarters referred to Marshall [Huntsville] as the Chicago Bridge and Ironworks. All of their vehicles were very conservatively designed with safety factors that were probably excessive by today's standards at least. However, there's one thing that has to be acknowledged: they worked."

After circulating Disher and Tischler's damning report, Mueller had a solution—"all-up" testing, in which the entire ship would be mated and flight-tested as quickly as possible. Since Apollo would end up involving thirty-three flights (with twenty-two being dress rehearsals for the eleven manned flights), this was a crucial strategy for meeting Kennedy's target.

George Mueller: "By the time you had to do all of the work necessary to fly a single stage by itself, you hadn't really done the work you needed to fly the whole stack. If you lost a vehicle, you were likely to lose it at any stage so you might as well go as far as you can and find out where the problems are. It turns out that Kurt Debus was a strong supporter of the idea as soon as he thought about it, because he'd seen the same things I had, things blowing up all over the place, and indiscriminately stage-wise."

Von Braun: "To the conservative breed of old rocketeers who had learned the hard way that it never seemed to pay to introduce more than one major change between flight tests, George's ideas had an unrealistic ring. Instead of beginning with a ballasted first-stage flight as in the Saturn I program, adding a live second stage only after the first stage had proven its flight worthiness, his 'all-up' concept was startling. It meant nothing less than that the very first flight would be conducted with all three live stages of the giant Saturn V. Moreover, in order to maximize the payoff of that first flight, George said it should carry a live Apollo command and service module as payload. The entire flight should be carried through a sophisticated trajectory that would permit the command module to reenter the atmosphere under conditions simulating a return from the Moon.

"It sounded reckless, but George Mueller's reasoning was impeccable. . . . the arguments went on until George in the end prevailed. In retrospect it is clear that without all-up testing the first manned lunar landing could not have taken place as early as 1969."

While Mueller was reengineering the engineers at Huntsville, Khrushchev was beginning to believe that Washington would try to atone for the Bay of Pigs by attempting another invasion of Cuba. In fact, both the Pentagon and the CIA did want another chance at victory, with the intelligence agency planning any number of operations.

On November 1, Operation Mongoose (an attempt to inspire Cubans to counterrevolt against the new regime) began, with four CIA teams inserted into the island in April of 1962. On August 11, 15, and September 27, three of those groups were captured by Cuban authorities, along with their radios, weapons, and explosives. During this period, meanwhile, NBC News broadcast film of Americans training anti-Castro Cubans in Florida.

As reports of these various schemes reached Castro, the Cuban leader asked Khrushchev for tanks and SAMs (surface-to-air missiles) to defend against the United States. Instead, he received Soviet aid of a very different sort, for the leader of the Soviet Union wanted, as he described it, to "throw a hedgehog at Uncle Sam's pants."

"The fate of Cuba and the maintenance of Soviet prestige in that part of the world preoccupied me," Khrushchev later explained. "We had to think up some way of confronting America with more than words. We had to establish a tangible and effective deterrent to American interference in the Caribbean. But what exactly? The logical answer was missiles." Citing NATO's forty-five von Braun Jupiter rockets aimed at the Soviet Union from Turkey, the premier began a nuclear quid pro quo, believing that Americans would then learn "just what it feels like to have enemy missiles pointing at you; we'd be doing nothing more than giving them a little of their own medicine." Khrushchev accordingly sent to Havana an array of ballistic missiles, both short-range as a defense against the next American invasion, and IRBMs for attacking American cities.

October 14, 1962: After a U-2 on reconnaissance—used instead of Corona, since the United States mistakenly believed there were no SAMs in the region sophisticated enough to take it down—returned with photos showing missile trucks, cranes, and launchpads in Cuba, Pentagon and CIA analysts identified the rockets as Soviet SS-4s, their intermediate-range weapon that could travel 1,100 nautical miles and carry three-megaton warheads, which meant that from their Cuban base they could bomb Washington in ten minutes. Defense Secretary Robert McNamara had one reaction: "Total shock. Why in God's name would the Soviets do such a thing? They must've known we'd react. . . . There was a fear that if we did not force the missiles out, the Soviets would move aggressively elsewhere in the world against Western interests, and was a very deep-seated fear."

October 18: Additional photo reconnaissance revealed a group of sites containing SS-5 IRBMs, missiles that American intelligence analysts believed could travel 2,200 miles with a five-megaton warhead, meaning they could reach any target in the continental United States, save northern California, Oregon, and the state of Washington.

The Joint Chiefs at the Department of Defense proposed an annihilation bombing run by the Eighty-second Airborne that would "mop up Cuba in seventy-two hours with a loss of only ten thousand Americans, more or less," as Ted Sorensen remembered. While one group at Kennedy's Cuban Missile Crisis committee, ExComm, pursued the strategy of blockade and quarantine, Curtis LeMay called this notion "almost as bad as the appeasement at Munich" and demanded "direct military intervention right now," while the

Marine Corps commandant insisted that "You'll have to invade . . . as quick as possible." Vice President Johnson advised, "All I know is that when you were walking along a Texas road and a rattlesnake rose up ready to strike, there was only one thing to do. Take a long stick and knock its head off."

Dean Acheson believed that, if the Americans bombed the Cuban missile sites, the Soviets would retaliate by bombing missile sites in Turkey. The United States would then be forced by NATO treaties to bomb missile sites within the Soviet Union. What the Americans did not know was that in case of an American invasion, the Soviet general in Havana was preauthorized to launch nine of his nuclear warheads against U.S. targets. "If the American president gives in to pressure from his armed forces and mounts an attack on Cuba, then an invasion of 250,000 American troops could be stopped with one atomic bomb," Sergei Khrushchev later revealed. "The Americans didn't know anything about those tactical nuclear weapons on Cuba. If they had attacked, I believe the commander of our forces there would have used those weapons. Then, the Americans would be dead, their ships sunk . . . what would Kennedy do then? He would deploy his own nuclear weapons. But then what?"

October 22: In a television broadcast, Kennedy told the nation that he had ordered a military quarantine of Cuba, while seeking support from the Organization of American States and the United Nations. It was clear to anyone watching that the world was on the brink of nuclear war. Two hundred and fifty thousand troops were mobilized for an invasion, nonessential personnel were evacuated from America's Guantánamo base, and the whole of Florida was placed under a state of emergency. On Cuba, 43,000 Soviet troops were in wait to assist local forces.

October 24: Twenty Soviet cargo ships approached America's sixty-ship Cuban naval blockade and came to a halt. One slipped through; nineteen reversed course. One of those returning to the USSR, it was later revealed, contained twenty nuclear warheads to supplement the twenty already on the island. Secretary of State Dean Rusk: "We're eyeball to eyeball, and I think the other fellow just blinked."

October 26: Kennedy received a cable from Khrushchev, saying that if the Americans promised to never again attempt to invade Cuba, the missiles would be removed. A second letter from the Soviet premier then arrived, demanding that the Jupiter missiles stationed in Italy and Turkey be removed as well. RFK advised his brother to accept the terms of the first letter, and ignore the second, which, officially, he did. Kennedy told Khrushchev, however, that the United States would invade Cuba if an agreement was not reached within twenty-four hours.

October 28: Khrushchev and Kennedy came to terms, and the crisis was resolved. Cold War historian John Lewis Gaddis believes that this defining moment of the Space and Missile Race "persuaded everyone who was involved in it—with the possible exception of Castro, who claimed, even years afterward, to have been willing to die in a nuclear conflagration—that the weapons each side had developed during the Cold War posed a greater threat to both sides than the United States and the Soviet Union did to one another. This improbable series of events, universally regarded now as the closest the world came, during the second half of the twentieth century, to a third world war, provided a glimpse of a future no one wanted: Of a conflict projected beyond restraint, reason, and the likelihood of survival." Within a year, the two nations had signed the Limited Test Ban Treaty, ending atomic tests in the atmosphere, and would continue this progress in 1968 and 1972 with a Nuclear Non-Proliferation Treaty and a Strategic Arms Limitation Interim Agreement. They also installed a private and direct phone circuit, a hotline, between Washington and Moscow, to be used in the event of future crises.

Date unknown: At a meeting with Soviet ambassador Anatoly Dobrynin, Bobby Kennedy pledged to remove the von Braun–NATO missiles from Turkey, by April 1, but only if the USSR promised to keep this element of the agreement a secret. The gesture meant little real reduction in the area's American military power, however, as U.S. submarines armed with Polarises now patrolled the Mediterranean. Later that year, America's solid-fueled and silo-based Minuteman ICBMs, which solved all of the key technical problems of their predecessors, became operational. It would take the Soviets six years to achieve parity in the Missile Race (in 1969, the year of Apollo 11), which effectively meant America's long reign of global nuclear dominance had begun, as McNamara admitted in 1965: "No national government would initiate a strike against us because our superiority insures that we will survive an attack with sufficient power to respond and literally destroy the attacking nation."

After the most terrifying confrontation in the history of the Missile Race had been resolved, Kennedy tried once again to end the Space Race as well. In the fall of 1963, he told Webb that he was going to propose to the Soviets a joint mission to the Moon, and wanted to know, "Are you in sufficient control to prevent my being undercut by NASA if I do that?" Webb assured him that he was. On September 18, 1963, National Security Adviser McGeorge Bundy wrote the president:

Webb called me yesterday to comment on three interconnected aspects of the space problem. . . .

The Soviets. He reports more forthcoming noises about cooperation from Blagonravov [Soviet representative to the United Nations' Committee on the Peaceful Uses of Outer Space], and I am trying to run down a report in today's *Times* (attached) that we have rebuffed the Soviets on this. Webb himself is quite open to an exploration of possible cooperation with the Soviets and thinks that they might wish to use our big rocket, and offer in exchange the advanced technology which they are likely to get in the immediate future. . . . The obvious choice is whether to press for cooperation or to continue to use the Soviet space effort as a spur to our own. The *Times* story suggests that there is already low-level disagreement on exactly this point.

The Military Role. Webb reports that the discontent of the military with their limited role in space damaged the bill on the hill this year, with no corresponding advantage to the military. He thinks this point can and should be made to the Air Force, and he believes that the thing to do is to offer the military an increased role somehow. . . .

If we compete, we should do everything we can to unify all agencies of the United States Government in a combined space program which comes as near to our existing pledges as possible.

If we cooperate, the pressure comes off, and we can easily argue that it was our crash effort in '61 and '62 which made the Soviets ready to cooperate.

I am for cooperation if it is possible, and I think we need to make a really major effort inside and outside the government to find out whether in fact it can be done. . . .

Two days later, in a September 20, 1963, United Nations address, Kennedy made public his offer of a joint mission. The man who had initiated the most significant leg of the Space Race now said, "In a field where the United States and the Soviet Union have a special capacity—in the field of space—there is room for new cooperation, for further joint efforts in the regulation and exploration of space. I include among these possibilities a joint expedition to the moon. Space offers no problems of sovereignty. . . . Why, therefore, should man's first flight to the moon be a matter of national competition? Why should the United States and the Soviet Union, in preparing for such expeditions, become involved in immense duplications of research, construction, and expenditure? Surely we should explore whether the scientists and astronauts of our two countries—indeed of all the world—cannot work together in the conquest of space, sending someday in this decade to the moon not the representatives of a single nation, but the representatives of all of our countries."

"It was with great attention that we studied President Kennedy's proposal for a joint moon project," Khrushchev responded. "What could be better than to send a Russian and an American together, or, better still, a Russian man and an American woman. . . . If we could all agree on further easing of tension, not just moral but in concrete terms such as disarmament, this would give greater resources, namely international resources for the development of science."

Three weeks after JFK's speech, the U.S. House of Representatives voted 125 to 110 to reject any expenditure of federal funds for "participating in a manned lunar landing to be carried out jointly by the United States and any Communist-controlled, or Communist-dominated country," with Appropriations Committee chair Clarence Cannon (D-MO) calling Kennedy's notion "a moondoggle . . . Scientists say there is absolutely nothing we could learn [by going to the moon] except the origin of the solar system. Gentlemen, I consider the origin of the solar system is of inferior importance when we need jobs in this country."

But the president would not be deterred, writing Webb on November 12: "I would like you to assume personally the initiative and central responsibility within the Government for the development of a program of substantive cooperation with the Soviet Union in the field of outer space, including the development of specific technical proposals . . . including cooperation in lunar landing programs. . . . In addition to developing substantive proposals, I expect that you will assist the Secretary of State in exploring problems of procedure and timing connected with holding discussions with the Soviet Union and in proposing for my consideration the channels which would be most desirable from our point of view. . . . I would like an interim report on the progress of our planning by December 15."

In 1997, Khrushchev's son, Sergei, revealed, "My father decided that maybe he should accept [Kennedy's] offer, given the state of the space program of the two countries. He thought that if Americans wanted to get our technology and create defenses against it, they would do that anyway. Maybe we could get [technology] in the bargain that would be better [for] us. I think if Kennedy had lived, we would be living in a completely different world." One intriguing "what if" question suggested by this history is: If Kennedy had indeed served out his terms as president, would he have ultimately succeeded in ending the Space Race, which after his death would be run, in great measure, in his name, and on behalf of his legacy?

On Saturday, November 16, 1963, the president flew to Cape Canaveral's Launch Operations Center. After von Braun demonstrated mock-ups of the Saturn that would take Americans to the Moon, the president accompanied Gus Grissom, Gordon Cooper, and Canaveral's new director, von Braun's colleague Kurt Debus, on a helicopter tour of the spaceport. He asked Webb, if we beat the Russians in the race to the Moon, what other purpose would Apollo serve? "The nations of the world, seeking a basis for their own futures, continually pass judgment on our ability as a nation to make decisions, to concentrate effort, to manage vast and complex technological programs in our own interest," Webb replied. "It is not too much to say that in many ways the viability of representative government and of the free enterprise system in a period of revolutionary changes based on science and technology is being tested in space. . . . [Society has] reached a point where its progress and even its survival increasingly depend upon our ability to organize the complex and do the unusual. We cannot do these things except through large aggregations of resources and power. [It is] revolution from above."

In a San Antonio speech on November 21, Kennedy praised NASA, and that night he attended a dinner in Houston in honor of Albert Thomas, one of the key figures in bringing the Manned Spacecraft Center to Texas. The following day, November 22, the president was assassinated in Dallas. Within a week, on November 28, NASA renamed its Canaveral operations the John F. Kennedy Space Center.

In their grief, many at the time believed that Kennedy's death would mean the end of NASA. But administrator Jim Webb, who'd used his relationship with the president to bully anyone who got in the way of the agency's mission, now used Kennedy's death to even better effect. Any attempt to cut NASA's budget, he would imply to a mere congressman, would destroy the cherished dream of our great martyred leader, cut down in his prime and unable to guide his legacy to fruition. Working in tandem with Webb was President Johnson, who sincerely considered the Moon shot a Kennedy legacy he was duty-bound to enact. There was additionally a fundamental difference in the two presidents' sensibilities, which accrued to NASA's benefit, as described by Dean Rusk: "You could never get President Kennedy to think beyond what he had to do at nine o'clock tomorrow morning, whereas with Johnson it's always, 'Well, where are we going to be ten years from now?'" Apollo also remained well supported by voters at the time; in October 1964, 77 percent of Americans agreed that the race to the Moon should either continue as is or be ramped up, and 62 percent wanted the same or more money spent on NASA and its dreams. After Johnson's reelection, his OMB director pointed out that they could save a great deal of money by not trying to land on the Moon by 1970. Johnson replied that NASA's budget could not be cut so drastically, since he owed Kennedy a Russian-beating lunar landing.

During the JFK era, Project Mercury had become a very popular signature program of the forward-thinking president, his country, and the nation's New Frontier. Under Johnson, however, Project Gemini—the stop-gap interim between Mercury and Apollo that would develop docking and rendezvous— became lost in the teeming crowd of projects that comprised the Great Society, including civil rights, Medicare, Medicaid, consumer and environmental protection, the Job Corps, VISTA, Upward Bound, food stamps, Head Start, the Public Broadcasting Service, National Public Radio, the National Endowment for the Arts, the Kennedy Center for the Performing Arts, new federal agencies for housing, transportation, and urban development, and the first African American justice on the Supreme Court. "We're the richest country in the world, the most powerful," President Johnson said. "We can do it all." At the same time, NASA itself failed to educate the public on how the achievements of Gemini would help land an American on the Moon, or just how tightly contested the Space Race continued to be.

Under these conditions, not to mention the escalating human and financial costs of the war in Vietnam, public enthusiasm dramatically waned. A July 1965 Gallup Poll found twice as many Americans believing that the budget for space exploration should be decreased as increased. Something of the same divided attitude was rising even within the agency. "Gemini was an unsung hero in terms of the readiness of the American space program to go do Apollo in many, many ways," Gene Kranz said. "We had learned the new technologies of space, we had learned to work with computers, we had learned to navigate, we had learned to dock. Perhaps the most important way was to create the team of people, the band of brothers, that were ready to go do Apollo when the time came."

Others had a very different opinion. Aerospace technician Alan Kehlet: "There was a very bitter feeling among a lot of us that Gemini should not have been done. . . . that it was a sapping of effort that could have gone into the Apollo program and we could have gotten the Apollo program along a lot quicker if we hadn't had a divided house of a strong effort on Gemini and a strong effort on Apollo concurrently. . . . we always seem to wait till the last minute to decide what to do next. And that's the place we were in Mercury. You could see the end of Mercury and something had to be done and Apollo looked like it was going to be a lot more than just a couple of years. So there is going to be a big gap between Mercury and Apollo. So Gemini was stuffed in there."

On October 13, 1964, the Soviet Central Committee informed Nikita Khrushchev that he was rude, arrogant, and incompetent; that he had embarrassed the tenets of Marxist-Leninist thought by erecting the Berlin Wall; that he had destroyed Soviet farming and almost started a nuclear war; and that he was now deposed.

Soon after President Kennedy's assassination, his widow, Jacqueline, sat down with Theodore White for an interview that remained unpublished until 1995, a year after her death. In it, she commented on the various memorials planned for her late husband: "I've got everything I want; I have that flame in Arlington National Cemetery and I have the Cape. I don't care what people say. I want that flame, and I wanted his name on just that one booster, the one that would put us ahead of the Russians . . . that's all I wanted."

12

The Transfiguration

From the February 1962 orbit of John Glenn to the June 1965 spacewalk of Ed White, NASA in the early 1960s became a shining legend to the American people, a Valhalla of heroic astronauts and genius engineers, agents of a series of achievements that the nation as a whole was deservedly proud of, all cohering into a national myth that was both true and exaggerated. This pride and faith would be shattered by one of the Space Age's greatest tragedies, and then would be reborn in one of its most astonishing triumphs.

For the first Apollo mission, Deke Slayton selected on March 21, forty-year-old Gus Grissom, the second American commander ever in outer space; thirty-one-year-old newcomer Roger Chaffee; and the first American to spacewalk, Armstrong's next-door neighbor and Buzz Aldrin's best friend from Germany, the thirty-six-year-old Ed White. From the very start, however, there were serious problems with Apollo 1. Gemini, which was supposed to have been the shakedown effort for the Apollo program, had achieved only limited success, its limits exacerbated by Apollo engineers' taking a "not invented here" attitude, and in many ways ignoring Gemini's achievements. There was, across the whole of the agency, an extravagant amount of reinventing the wheel for an enormous project on a deadline that loomed ever closer.

"I remember one of the first times I went out [to North American Aviation] and flew the Apollo simulator," Frank Borman remembered. "And I pulled the hand controller back, and the nose went down; and I reversed it, and the nose went up [i.e., the opposite of every stick in the history of flight]. I called the engineer over and I said, 'You got the polarity reversed on this hand controller.' And he said, 'Oh no, that's the way we're going to use it. That's the way we're going to fly it because it makes rendezvousing easy. It makes docking easier because, when you pull back on the stick, your nose goes down but the target goes up. You see,' and 'That's the way we were going to do it.' But this is another example of NASA. I said, 'Well, look, that may be the way you're going to do it, sitting here on your ass as an engineer, but that's not the way we're going to do it.' And I called back to the Apollo Program Office, and I got it changed right there."

The backward stick design was only one example of the idiosyncrasies in the new Apollo Command Module, which had between two and three million parts; no one at its manufacturer, North American Aviation, knew for sure. They did know that it was the most complicated machine in human history. About his CM, Commander Gus Grissom told Astronaut Office chief Alan Shepard: "There are a lot of things wrong with this spacecraft. It's not as good as the [Mercury and Gemini] ones we flew earlier. There's something different about this thing." He became so vexed by Apollo 1's flaws that he nailed a lemon to its skin.

"The first flight in a program is very intense, and it's the first big milestone that kicks off the program," flight control director Gene Kranz said. "So there's a lot of personal contact between astronauts, flight directors, flight controllers, and we're all pooling our knowledge, trying to get to the point where we all feel that we're ready. And this feeling just never quite seemed to get there [with Apollo 1]. It always seemed that every time we'd turn a corner there were things that were left undone or answers that we didn't have or we were moving down a wrong path, but we had the confidence that we'd been through this before. We'd been through it in Mercury, we'd been through it in Gemini, so we had the confidence that by the time we got to launch date, all the pieces would fit together."

Late in their training, the Apollo 1 crew posed for a photograph with their heads bowed, and their brows furrowed in prayer. They gave a print to North American's space division head Stormy Storms, and another to Apollo program manager Joe Shea, which was inscribed: "It isn't that we don't trust you, Joe, but this time we've decided to go over your head."

On January 27, 1967, at Launch Complex 34 of the Kennedy Space Center, four weeks before their mission was to begin, the Apollo 1 crew and ground team were running a countdown simulation. The three astronauts sat in their capsule atop a Saturn 1B, twenty stories up in the sky. There was endless trouble with the radio, so much so that Grissom lost his temper: "How can we get to the moon if we can't talk between three buildings? I can't hear a thing you're saying."

There was also a strange odor coming from somewhere, like milk gone bad.

As they would in space, Gus Grissom, Roger Chaffee, and Ed White were breathing pure oxygen. (To put this into perspective, the air we breathe consists of about 78 percent nitrogen, 21 percent oxygen, 1 percent argon, one-third percent carbon dioxide, and traces of various other gases.) On the X-15, North American had used a mixture of nitrogen and oxygen as a deterrent against cockpit fires, but NASA had insisted on oxygen alone for Apollo, because having one system was simpler and less trouble-prone, and because it had been used successfully with both Gemini and Mercury. Gemini and Mercury, however, had not run ground tests with pure oxygen in ships filled with all sorts of flammable material, notably Velcro. Additionally, the Apollo cabin had been pressurized for five and a half hours, to test the equipment, and, partially as a result of Grissom's troubles with a faulty outward-opening hatch on his Mercury Liberty Bell 7 splashdown, North American had designed a new set of double hatches for Apollo 1, one of which opened inward.

The problems with the radio transmission continued. At 6:31 p.m., Deke Slayton was listening in on the squawk box when he thought he heard someone say "fire." North American engineer John Tribe asked another man, "Did he say 'fire'? What the hell are they talking about?" Slayton looked at the CCTV screen. The capsule window, in black-and-white, was glowing. A technician watching another camera saw Ed White's gloved hand reach through to the hatch bolts. Slayton then heard Chaffee say, "We've got a fire in the cockpit." Spacecraft test conductor Skip Chauvin screamed for the technicians running electrical to shut everything down. Then Chaffee's voice was heard again: "We've got a bad fire . . . Let's get out . . . We're burning up!"

From the first word, "fire," to the final explosion took all of twelve seconds. North American technician Jim Gleaves was running toward the craft when there came a rush of wind, a flash, and a blast that threw Gleaves against the door and bombarded him with flame, smoke, and debris. "It was like when you were a kid and you put a firecracker in a tin can, and it blew the whole side out of the tin can with the flames shooting out," he recalled. Apollo systems engineer Jesse Owens: "It all happened much faster than I can tell it. I turned to my left and a sheet of flame went out." In eighteen seconds, monitor readouts showed the capsule interior's temperature spiking to 2,500 degrees F.

"The technicians, once they knew they could not put the fire out in the capsule and open the door, because it was sealed from the inside, concentrated on getting firebottles and spraying the coolant, not at the fire, but at the nozzles under the [launch escape] rocket engines to keep it from igniting," spacecraft manager Ernie Reyes remembered. "Had it ignited, [the launch escape tower] would have gone up . . . and it would have settled down . . .

then on impact, it would have collapsed the service module . . . then it would have hit the S-IVB, then . . . it would have hit the first stage of the S-1C. That would have gone high order and we'd have killed everybody on the pad."

After being driven back twice by exploding walls of smoke and flame, pad technicians tried repeatedly to open the module's double hatches, but the heat and smoke were overwhelming. Finally, the capsule doors were released. The pad director was heard on the radio: "I'd better not describe what I see." All three astronauts were dead.

Flight director Gene Kranz:

I've never seen a facility or a group of people, a group of men, so shaken in their entire lives. John Hodge and myself had grown up in aircraft flights tests, so we were familiar with the fact that people die in the business that we were conducting, so we had maintained maybe a little bit more poise relative to the others, but the majority of the controllers were kids fresh out of college in their early twenties, and everyone had gone through this agony of listening to this crew over the sixteen seconds while they—at first we thought they had burned to death, but actually they suffocated, but it was very fresh, very real, and there were many of the controllers who just couldn't seem to cope with this disaster that had occurred. . . .

Kraft had declared a total freeze on operations to protect the data, terminating phone calls and directing the controllers to write down every event, any and every recollection of what they had seen and heard. With any ground or flight accident, it was essential to the investigation to bring everything to a dead stop while memories and data were still fresh and uncontaminated by the inevitable aftershock, confusion, and second-guessing. . . .

You keep playing—and I think this is the characteristic in Mission Control—you'll play the data back. You're looking for something so you can try to find a cause or an answer, and you're actually doing what it amounts to as meaningless work. You're basically trying to kill the time. And we couldn't find any answers. The next morning, we came back out to work again, trying to see if there was any answers, because in that kind of an environment you're trying to find answers, you're trying to find out why, what happened, etc., and there were no answers. We worked through the Sunday time frame, again just sitting in offices almost just paralyzed, we were so stung.

I tend to be maybe one of the more emotional of the controllers. I believed that that's part of a leader's responsibility, to get his people pumped up, and I gave what my controllers came to know as the "tough and competent" speech, and concluded the talk identifying that the problem throughout all of our preparation for Apollo 1 was the fact that we were not tough enough; we were avoiding our responsibilities, we had not assumed the accountability we should have for what was going on during that day's test. We had the opportunity to call it all off, to say, "This isn't right. Let's shut it down," and none of us did.

We had become very complacent about working in a pure oxygen environment. We all knew this was dangerous. Many of us who flew aircraft knew it was extremely dangerous, but we had sort of stopped learning. We had just really taken it for granted that this was the environment, and since we had flown the Mercury and Gemini Program at this 100 percent oxygen environment, everything was okay. And it wasn't.

I had each member of the control team on the blackboards in their offices write "tough and competent" at the top of that blackboard, and that could never be erased until we had gotten a man on the Moon.

NASA administrator Jim Webb immediately created a review board to oversee an investigation. One document revealed that, during an August 1966 meeting between NASA and North American engineers, the very issue that would kill the three men was repeatedly discussed. "I would like to observe that we have some materials in spacecraft 012 [Apollo 1] which, when we considered spacecraft 008 safety, we did have removed," said NASA's William M. Bland at that meeting with North American. "These are made of nylon and Velcro. Piece by piece and individually they are not too bad, but in certain places there are large amounts measured in feet and probably in pounds where these chafing straps are gathered. I consider them hazardous."

"Regarding the Velcro chafing guards, our position is that our criteria are satisfactory," North American Apollo program manager Dale D. Myers replied. "There are other materials that could be used but we do not see the necessity of changing these particular ones."

Bland did not give up: "The only thing I can say is that when you have a broken conductor ignited it burns like mad and sputters all over, as tests have shown. . . . In the case of a flight spacecraft, I think it would be a little more difficult, but I imagine a fiberglass cover to protect these same wire bundles could easily be made. Fiberglass being much more preferred than a nylon-Velcro cover."

After the discussion grew rancorous, Joe Shea tried to sum up the group's concerns: "I think we are trying to take a rational position, which says that the real concern is to get these flammable materials in a position where they cannot be ignited. The only place where we can see these getting ignited, since we are not going to carry matches on board, is the possibility of a wire breaking and some kind of shorting. . . . I think we should stop bitching at each other and go clear the thing up. . . . It's like we're standing on a matter of principle instead of fixing something that is relatively easily fixed."

"You will 'kill' this before it leaves here," Gus Grissom told the North American team. "You are not going to fool around with wire bundles after we test it down at the Cape."

North American did change those particular Velcro parts, but the Apollo 1 test capsule that arrived at Kennedy had an additional nineteen pounds of combustible materials that no one had reconsidered in light of that conversation. Astronaut Frank Borman, who served on the Apollo 1 investigatory committee, later concluded, "We think that what happened, there was probably an electrical short down at the lower equipment bay near Gus's left foot that created a spark. With 100 percent oxygen and a PSI of around twenty-one pounds, that spark propagated rapidly and became an explosion." Others knowledgeable about the tragedy came to theorize that the Environmental Control Unit's metal door, repeatedly opening and closing, scraped against a cable running under it just enough to expose the wires and cause a spark, which ignited the fumes of a nearby pipe carrying glycol coolant. "Hindsight is wonderful," Max Faget said. "We had the same atmosphere in Mercury and Gemini as we had in Apollo. They never had any fires. But, you see, after I started thinking about it, kicking myself for being so stupid, I realized that the difference between Mercury and Apollo was that one Apollo experience was probably equivalent to maybe twenty or thirty Mercurys, simply because there's so much more volume in Apollo and there's so much more stuff in Apollo, so that it's going to burn just as badly."

Everyone had expected that, sooner or later, an astronaut might die on a mission. But on the pad, in a test run? Neil Armstrong: "I'd known Gus for a long time. Ed White and I bought some property together and split it. I built my house on one-half of it, and he built his house on the other. We were good friends, neighbors. Some very traumatic times. You know, I suppose you're much more likely to accept loss of a friend in flight, but it really hurt to lose them in a ground test. That was an indictment of ourselves. I mean, we didn't do the right thing somehow. That's doubly, doubly traumatic. It just hurts."

During the countdown for Alan Shepard's Freedom 7, Jim Webb had publicly warned, "Each flight is but one of the many milestones we must pass. Some will succeed in every respect, some partially, and some will fail." As he prepared for liftoff of Gemini III, Grissom told his wife, "If there is a serious accident in the space program, it'll probably be Gemini, and it'll probably be me," and to a group of reporters, "If we die, we want people to accept it. We're in a risky business." But outside the circle of astronauts themselves, few could accept it.

Joe Shea had given a speech at a UPI correspondents' dinner a few months earlier, in which he had said that, eventually, the space program would suffer a casualty, and that the effect might be more severe on the ground personnel than on the surviving astronauts. This is in fact what happened, and most seriously to Joe Shea himself. "You can't believe the impact that this fire and these deaths had at the Cape," Frank Borman said. "People got to drinking too much, taking uppers and downers and the damn doctors were handing out stuff so that people could go to sleep and so they could wake up. It was really sick. . . . One of the really respected civilian-contractor people had a nervous breakdown, and they had to haul him away in a straitjacket. He started drawing an organizational chart of Heaven. And I never will forget—he had Big Daddy. And then he just lost it all." Max Faget remembered that, every night, Joe Shea was getting dead-drunk. Chris Kraft:

> We met in one of our big conference rooms at the Manned Spacecraft Center, with General Sam Phillips running the meeting. He introduced Shea and gave him the floor. What happened next shocked us all.
>
> Joe Shea got up and started calmly with a report on the state of the investigation. But within a minute, he was rambling, and in another thirty seconds he was incoherent. I looked at him and saw my father, in the grip of dementia praecox. It was horrifying and fascinating at the same time. I've seen this before, I thought, and tried to look away. But I couldn't. After our early problems, Joe had become a friend. Now he was falling apart in front of me. Whatever was happening in Joe's head, it all came out in a jumble of mixed words and meaningless sentences. Sam Phillips listened only a few more moments before he stood up and put his hand on Joe's shoulder.
>
> "Thanks, Joe," he said. "Why don't you sit down while we get on with the agenda."
>
> That was the end of Joe Shea.

"It's one thing you learned as an engineer, was that pilots are going to die," flight control chief John Hodge admitted. "That was the nature of the game, especially if you're working on high-performance airplanes like I was, fighters. And, sure enough, you'd lose the odd flight-test guy every couple of years or so, and you have some responsibility for that. . . . And what you had to do was insulate yourself from that, because, I mean, it would kill you. . . . So I deliberately did not get close to the crew. I mean, close enough to be able to work professionally with them, but not in a sort of friend kind of a thing. . . .

"As an engineer in the airplane business, you have to learn to die a little bit every day, because you know that you're going to lose somebody. You know you're going to lose somebody. And if it comes to you all at once, it will kill you. Joe was never able to do that."

Americans had thought of NASA as absolutely the nation's best, brightest, and most courageous. The fire seriously damaged that faith. "How do you explain to the public at large that there's a certain risk, and you've got to accept that risk," George Mueller asked. "We haven't been able to do that in our society. People are quite willing to accept risk for race drivers, for example . . . when one of them gets killed, you don't have a congressional investigation." Many Americans felt baffled and humiliated that, when technological catastrophe struck, it had struck in the United States. In fact, the Soviet program had suffered losses of life, but those were admitted to only a very few, and one of the disasters that was kept secret for twenty years mirrored the Apollo 1 fire exactly. On March 23, 1961, the nation's youngest cosmonaut, the twenty-four-year-old Ukrainian Valentin Bondarenko, was also training in a cabin pressurized with 100 percent oxygen. After removing biosensors from his torso, he wiped his skin with a cotton ball soaked in alcohol and tossed it aside. The cotton struck a hot plate, ignited, and the flash fire leaped onto Bondarenko's suit. It took several minutes to get the pressurized hatch open, and by the time he was pulled from the raging fire, the cosmonaut was covered in third-degree burns. There was only one place on his body with skin that hadn't been destroyed—the soles of his feet, which had been protected by his boots—and it was there that the doctors inserted their IVs. Eight hours later, he was dead.

While NASA's internal review of the Apollo 1 fire continued, Congress in 1967 decided to call for its own series of investigations. During the discovery process, it was revealed that NASA had issued the Command and Service Module Request for Proposal in July 1961 and received bids from Martin, McDonnell, Convair, General Dynamics, and North American Aviation. The agency's Source Evaluation Board picked Martin as the clear winner, with Convair and North American tied for second. Webb, Dryden, and Seamans, however, overrode this decision and awarded the contract to NAA. When asked why, Webb testified that the astronauts themselves were "strongly of the view that they would prefer to have a company like North American which had made the X-15—according to their experience a very high performance manned aircraft—as against a company that had developed their experience primarily in the unmanned field." In addition, as NAA's lobbyist had noted earlier, "North American was five times as large as Martin, so when they got the contract you had five times as many happy people as you had unhappy people. That's democracy in its finest form."

"[North American] made the first good flight of an X-15 in November and they couldn't have picked a better time," engineering designer Caldwell Johnson explained. "It established the fact that the X-15 was a joint Air Force–NASA–Navy affair; it flew like a bird; it was more or less within cost, within plan, in time; it demonstrated that they were technically capable; and it demonstrated that they could work with NASA and with the Air Force to the extent of producing this thing. You might say from an overview of it, it was a good demonstration of the capability of the company. They weren't just talking about it, they had done it and there it was. . . . Alan Shepard was on a panel that I was on . . . and he said, 'This is all a waste of time. North American is going to be the one.'"

"I walked out behind Gus Grissom talking to some guys," NASA researcher Ivan Ertel remembered, "and he said, 'I don't care what happens. I'm going to do everything I can do to make sure North American gets this damn contract.'"

Senator Walter Mondale then unearthed a report overseen by Apollo program manager General Sam Phillips that found "continual failure by North American to achieve the progress required to support the objective of the Apollo program." There were "slippages in key milestone accomplishments, degradation in hardware performance and increasing costs. . . . Key performance milestones in testing, as well as end-item hardware deliveries, have slipped continuously . . . the delivery of the common bulkhead test article was rescheduled five times for a total slippage of more than a year, the All Systems firing rescheduled five times for a total slippage of more than a year, and S-II-1 and S-II-2 flight stage deliveries rescheduled several times for a slippage of more than a year . . . effective planning and control from a program standpoint does not exist. . . . The condition of the hardware shipped from the factory, with thousands of hours of work to complete, is unsatisfactory to NASA. . . . North American quality is not up to NASA required standards."

When questioned about this document, Jim Webb testified that he had never seen or heard of it, which was the truth—Seamans had not forwarded it to him. Before Congress, though, a confused and ashamed Webb inadvertently appeared less than honest, and at times, dissembling, to the extent that his agency became joked about as "NASA=Never A Straight Answer." Webb, believing that in Apollo he had created a new form of corporate and project management that could revolutionize American business and government, was devastated by the fire's revelation of agency incompetence. It was "a terrible blow to him," Bob Seamans said. "Jim was not interested in investigating the engineering. He wanted to know what individuals had failed him. He felt personally betrayed." The one who had betrayed him most grievously, Webb felt, was Bob Seamans, who soon left to become secretary of the air force.

Roger Chaffee and Gus Grissom were buried at Arlington; Ed White, at West Point. In the wake of this tragedy, many astronauts came to appreciate that the bravest of them all were the wives, who had accepted a life of great risk and no glory, and who would have to face raising their children alone on welfare-grade military allotments if their husbands were killed in the line of duty. "My wife is a wonderful, wonderful person who was a complete support system," Frank Borman said. "When we got married in the '50s, it was like a team. She had her job and I had my job. And one of her jobs (I guess she had assumed it) was to not in any way show any kind of fear or any kind of trepidation over what I was doing. And so I never really understood her concern. Particularly after the fire, because she and Pat White were close and she saw how Pat White was devastated. And, look, it's as I said: It wasn't as if she hadn't consoled widows before, because she had. But in any event, I misread that; and so I just assumed that she was stronger than she really was."

Scott Carpenter's wife, Rene, remembered a clarifying moment from the early Mercury days: "I was on the beach with Jo Schirra for the last Atlas test firing, and it blew up right in front of us! It was terrifying, but there was fatalism among the wives, a lot of gallows humor. You'd say, 'Oh, thank God the monkey wasn't in that one.'" The wives also knew that NASA was about the best option an American military pilot could hope for. "We all knew that if we weren't there, we would have been on a base and our husbands would have been flying in Vietnam," Valerie Anders said.

Betty Grissom, fed up with being ignored by federal and corporate bureaucrats, would later sue, winning $350,000 out-of-court from North American, with $300,000 split between Mrs. White and Mrs. Chaffee. Nearly twenty years later, Pat White, in the middle of organizing an Apollo wives' reunion, committed suicide. Launchpad 34 is left today in ruins as a memorial, marked by the federal version of Do Not Disturb—"Abandoned in Place"—and with the plaque:

Ad astra per aspera
A rough path leads to the stars

As for the astronauts themselves, they grieved the tragedy much like any group of men and women whose professional lives included an element of risk.

Beyond Apollo 1, four astronauts had died in fatal crashes of the T-38 commuter jet, and one in a car accident—NASA fliers were known in Texas and Florida for driving with their cars barely a foot from each other while topping out at a hundred miles an hour. Being an astronaut was perhaps the most honored profession in the nation, but of a total of sixty-six pilots accepted since the program began, eight were dead and eight had resigned. Buzz Aldrin had even gotten his slot on the last Gemini mission upon the death of the original candidate, his next-door neighbor Charlie Bassett. Bassett had been commuting with fellow astronaut Elliot See to McDonnell's plant in St. Louis to inspect their capsule. The weather suddenly turned into fog, rain, and snow, and the men crashed into the hangar where their spaceship was waiting. Buzz Aldrin: "The night I received this news, Joan and I crossed the backyard to Jeannie's house to tell her. I felt terrible, as if I had somehow robbed Charlie Bassett of an honor he deserved."

Mrs. Bassett responded in the armed services' historic and remarkable tradition of the widow who consoles her grievers. Buzz Aldrin: "Jeannie reacted graciously. 'Charlie felt you should have been on that flight all along,' she said, gripping my hand. 'I know he'd be pleased.'"

A journalist once asked a number of astronauts, "If something happens to you during spaceflight or otherwise, what thought would you most want to leave as a public legacy?"

Ed White: "Press on!"

Gus Grissom: "If we die, we want people to accept it. We hope that if anything happens to us it will not delay the program. The conquest of space is worth the risk of life."

The tragedy of Apollo 1 eventually led to an across-the-board overhaul of NASA and its subcontractors. North American designed a new hatch that opened outward in three seconds. New technologies were introduced to fireproof the cabin's fabric and paint. There would be no manned missions for the following eighteen months, and during that time the entire agency would be reengineered. NASA history would now be divided into two distinct periods: Before and After the Fire. "Apollo 1 was a greater contributor to the entire Apollo program than had it flown and been a success," Gene Cernan insisted.

"All of us, every single one of us [was] part of a group that had gone through Mercury, had gone through Gemini, man, we thought, we're leading! We're beating the Russians! We thought nothing could go wrong," Alan Shepard said. "And it led to a sense of false security. Deke and I remember talking about it. Gus would come back and he'd have a complaint about this. He said, 'This is the worst spacecraft I've ever seen.' He complained about that. And of course he was complaining to engineers as well as to Deke and to me. But Deke and I insidiously became part of the problem because we said, 'Okay, Gus, go ahead and make a list of this stuff and we'll see that it's fixed by the time you fly.' Not that, 'We'll see it's fixed before they stick you back in there for a test where you're using 100 percent oxygen.' You see, there was that sense of security, a sense of complacency that everyone had—including myself and including Deke."

While undergoing its overhaul, NASA faced a series of other problems on nearly every front. The tests of von Braun's Saturn V rocket were disturbing, and the Russians showed signs of once again besting America with an engineering triumph. On February 4, 1966, Tass reported: "The Luna-9 automatic station launched on January 3 soft landed on the moon surface in the Ocean of Storms at 21:45:30 Moscow time on February 3 . . . The TV camera on board the Luna-9 operates well . . . snapshots of the moon surface have been received and analyzed. Those snapshots will be broadcast . . . tonight. Thus the Luna-9 has brought us four major victories: 1. Soft landing. 2. Earth-moon radio communications. 3. Television transmissions from the moon to the Earth. 4. Pressurized cabin with conditions suitable for living beings in it."

At the time, no one at NASA realized, however, that the great power behind the Soviet space effort was gone. On January 14, 1966, chief designer Sergei Korolyov was in the hospital for a routine hemorrhoid operation when the surgeons discovered that he had advanced colon cancer. When they tried to cut away the tumor, an artery was severed, and the Soviets' greatest rocket engineer bled to death. He was replaced by his deputy, Vasily Mishin, but after little progress (and, out of four tries, four launch explosions) with the giant N1 (which had thirty LOX-kerosene engines compared to the Saturn V's five), Mishin was replaced in 1972 by a man who'd become a great opponent of going to the Moon and a champion of developing shuttles instead: Valentin Glushko. A Soviet technician later admitted, "We could not compete with you Americans. There was confusion, disorganization. Too many 'chiefs' and not one boss. You can't get anywhere without a big man in charge."

Four months after the Apollo 1 fire, on April 23, 1967, the forty-year-old Vladimir Komarov, whose backup, Yuri Gagarin, was a close friend, lifted off in Soyuz 1. Three cosmonauts would launch the next day in Soyuz 2, joining him in outer space to prove the Soviets could do everything that Gemini had accomplished—sophisticated navigation, rendezvous, docking, and worthwhile spacewalking. There was such a widespread sense of foreboding about this mission, however, that just before launch Komarov publicly reasoned, "If I don't make this flight, they'll send the backup pilot instead. That's Yuri, and he'll die instead of me."

Komarov's launch was excellent. His craft used solar panels instead of batteries for power, but on reaching orbit, one of its solar wings refused to deploy. Then, thruster trouble caused his capsule to begin tumbling end over end, so severely that ground controllers couldn't maintain a communications link. Komarov tried to keep the ship aligned so that its heat shield would be effective on reentry, but he could not do it. National Security Agency analyst Perry Fellwock was monitoring Soviet radio from a NATO building in Turkey: "They knew they had problems for about two hours before Komarov died, and were fighting to correct them. We taped the dialogue. The Kremlin called Komarov personally. They were crying, and they told Komarov he was a hero. The guy's wife got on, too, and they talked for a while. He told her how to handle their affairs, and what to do with the kids. It was pretty awful. Towards the last few minutes, he was falling apart." The cosmonaut fell to earth, his spinning capsule tangling its parachutes into each other, striking the ground at four hundred miles an hour.

An enraged Gagarin announced that "If I ever find out [Leonid Brezhnev] knew about the situation and still let everything happen, then I know exactly what I'm going to do." On March 27, 1968, the great Soviet hero was in training to recertify as a MiG pilot; an on-hand witness reported that another jet "without realizing it because of the terrible weather conditions, passed within ten or twenty meters of Yuri and Seregin's plane while breaking the sound barrier." The eyewitness then heard two booms. One, the breaking of the speed of sound; the other, Gagarin's plane falling into an out-of-control spin from the other jet's afterburner turbulence, and exploding into the ground.

The American public's adulation of its astronauts was slight in comparison to the Russians' adoration of Yuri Gagarin, who was idolized across the whole of the USSR as the Soviet ideal. Though a great majority of citizens thought the nation's money should be spent on more pressing issues than outer space, Yurochka was loved beyond all other Russian heroes, living or dead. In a memorial that staggered a grief-stricken nation, his cremated remains were ushered through the streets of Moscow by a phalanx of nine cosmonauts, and laid to rest in the wall of the Kremlin alongside Komarov and Korolyov. Just as Cape Canaveral became Kennedy Space Center, Star City became Gagarin Memorial Cosmonaut Training Center. Across the Soviet Union, statues and busts were erected to the man who represented the country's highest ideals and greatest aspirations, and the day of his first flight into space is still celebrated as a holiday in Russia and many other nations.

To fulfill President Kennedy's promise, America needed a lunar landing by the end of 1969, and it was becoming increasingly clear that NASA was in no respect close to ready for such an audacious leap. Jim McDivitt echoed the fears of many in 1967: "I came home so many nights thinking, 'There's no way we're ever going to do this thing.'" Joe Shea's replacement as the new chief of Houston's Apollo Program Office was George Low, a meticulous, soft-spoken, yet hard-nosed émigré from Vienna, who had created along one office wall a museum of switches, valves, lamps, and other components that had been tested for NASA missions and had failed. His daily reports from this era were a litany of woe:

8 January 1968.
Indeed a blue Monday.
Losing ground on almost every schedule.
LM-2 6 days behind
LM-3 9
LM-4 on schedule
2TV-1 lost 6 days last week because of further leaks in the glycol and oxygen system
101 15 days behind
103 is in DITMCO and is at least 2 weeks behind. . . .
Continue to find electrical problems with 020. . . .
Some corrosion inside alum. tubing of 103. Spec. it came from some of the servicing equipment. Potentially serious problem which as yet has no solution.

While North American's Command Module was in all-out disarray, not much better could be said for Grumman's Lunar Module. Delivered in June of 1967, the first LM didn't fit within its size or weight specifications, didn't have working radios, and had fuel tanks that regularly popped leaks. After seeing the Cape's test results, Rocco Petrone had a few words for Grumman: "What kind of two-bit garbage are you running up in Bethpage? What kind of so-called 'tests' do they do before sending this wreck to us? They'd better get this fixed, and fast! Your name is mud around here until they do."

In November of 1967, Huntsville's Saturn V was assembled for its first all-up test at the Cape. The three-day countdown became a three-week endurance marathon as one problem after another arose. Critical systems—telemetry, wires, batteries, computer navigation and control—failed again and again. On two occasions, pad technicians were forced to pump out the 500,000-gallon first-stage LOX and kerosene tanks and begin all over again.

On November 9 at 0700 EST, Apollo 4 launched. Two F-1 rockets abruptly quit during liftoff, at which the stack pulled a U-turn and headed screaming back at the ground. But the guidance system righted the vehicle, and the CM dummy capsule was successfully put into orbit. Apollo 5 met with about the same level of success—two of its engines died, and the stage-three rocket, which would have carried the craft to the Moon, refused to fire.

During the last week of April 1968, George Low met with Frank Borman, who had followed up his investigation of Apollo 1 by heading an oversight committee at North American's Downey, California, plant, which was then in the process of testing and assembling a reengineered Command Module. Borman assured Low that, in fact, North American had made dramatic strides forward in their operations, and that very shortly the ship would be ready to sail. After consulting with flight operations director Chris Kraft on the pending status of the mission's remaining elements, Low decided to take the giant leap: he notified his senior executives that they should launch an "all-up" manned mission that would use the Saturn V, test navigation computer software, attempt radio communications from a distance of 240,000 miles, perform a hypersonic reentry back to the Earth's atmosphere, and fire a trajectory that would include an orbit of the Moon. Chris Kraft: "In the hallway, Deke was almost bouncing with anticipation. 'I think I'll have to do some crewswapping. I gotta make some calls.' I hit my desk and made calls of my own. My head was abuzz with all the things we'd have to do. But it was one hell of a challenge. If we could pull it off, it would be absolutely pivotal to landing men on the moon."

Low's decision to launch Apollo 8 was, for many, shocking, starting with the fact that the mission would be only the third time that a Saturn V had ever been flown. Afterward, when asked about the risk, Chris Kraft said, "A better question is: 'What *couldn't* have gone wrong?'" George Mueller told *Life* reporter Bob Sherrod that the chances of success were one in ten. Additionally, he worried, "What can we accomplish in it? If we could clearly advance our knowledge, I'd be more enthusiastic." A mere three days before launch, safety chief Jerry Lederer said that the mission "would involve risks of great magnitude and probably risks that have not been foreseen. Apollo 8 has

5,600,000 parts and one and one half million systems, subsystems, and assemblies. Even if all functioned with 99.9 percent reliability, we could expect 5600 defects." Even Buzz Aldrin's father weighed in with an opinion, writing Jim Webb on September 29, 1968, "I do not favor a manned flight of Saturn V until the changes being made have been proven. What is the value of risking lives at this stage? You really need less Yes-men in the space program." Jim Webb had in fact asked Sam Phillips, flat out: "Are you out of your mind? You're putting our agency and the whole Apollo project at risk!" When Lyndon Johnson told Webb that he wouldn't be seeking another term, Webb, after the heartache of the Apollo 1 fire and the immense risk of Apollo 8, replied, "I'm going to walk out the day you do. Let's go out together." On October 7, he resigned.

Many assumed that, under the new Nixon administration, top-ranking NASA Republican George Mueller would become the new agency administrator. Then Webb sent word to Mueller that he would testify against his confirmation. Instead, NASA's new chief would be a onetime metallurgist and manager of General Electric's Center for Advanced Studies, a solid and hearty executive who would see the agency through to Apollo 11—Thomas Paine.

September 14, 1968: At 9:42 p.m. an immense Proton booster carrying a capsule containing a collection of bacteria, seeds, plants, flies, worms, and turtles lifted off from Baikonur and, two days later, Zond 5 circumnavigated the Moon. On reentry, it reached 16 g's and 23,432 degrees Fahrenheit, killing its living cargo. Still, it had beaten NASA to a lunar orbit, and its problems could probably be fixed and cosmonauts dispatched on a lunar flyby with two missions, or in about three months. The Soviets were still very much in the race, with even von Braun quoted as saying he doubted the United States would beat Russia to the Moon: "It is very important that we are there first, but in view of the spectacular performance of the Soviet spacecraft Zond 5 in late September, I am beginning to wonder. It will undoubtedly be a photo finish . . . the Soviets are spending more money than we are. . . . They are spending almost two per cent of their gross national product on space, while we spend less than one per cent. . . . I'm convinced that, unless something dramatic happens, the Russians are going to fly rings around us in space for a period of five years. . . . The trouble is that most people seem to think that in space we are No. 1, which we aren't."

October 26: Soyuz 3 performed exactly as planned, ending with a perfect landing in the snow-covered steppes to the north of Baikonur. It did not, however, dock two manned craft, an essential procedure for both countries to achieve before landing on the Moon.

The following month, Zond 6 performed a remarkable maneuver, "bouncing" against the Earth's atmosphere before initiating final reentry to minimize the heavy g-forces that had plagued Russian flights.

Ten months before Apollo 11, a series of memoranda indicated that, at the highest levels of government, American leaders were convinced that the Space Race was being run as hard as ever. On September 26, 1968, Johnson's science adviser, Donald Hornig, wrote the president to complain that "Recent press reports of the Soviet Zond-5 circumlunar flight have grossly exaggerated the importance and significance of this event. This view has been fueled by a series of unconscionable statements by Mr. Webb, Mr. Paine and other NASA officials which have unnecessarily inflated the Soviet accomplishment and were undoubtedly motivated by their budgetary problems. In these statements, it was variously suggested that the Soviets have demonstrated a 'capability that could change the basic structure and balance of power in the world,' that the U.S. was clearly second in space and that a Soviet manned lunar landing could be achieved in the next year. . . . I believe the best evidence we have available refutes these points . . . we have successfully flown the Saturn V launch vehicle twice . . . while the equivalent Soviet vehicle has yet to fly. I conclude . . . that we are at least one year ahead of the Soviets in this area—and not behind."

On October 1, Webb replied, "In defending our FY 1968 budget, I explained to the Congress the situation we faced in these terms: 'I cannot say that this budget or this program will give us preeminence in all major aspects of space and aeronautics. . . . We have no reason to believe that the Soviets have abandoned their stated goals of preeminence in space or the strong views they have expressed as to its importance in building Communist power.' . . . the work force now engaged in our program is about two-thirds the level reached in the peak year 1966. This means that a number of key design and engineering teams have already been broken up. Our rate of successful space launchings has fallen off sharply since the peak year 1966 . . . we are terminating production of both the Saturn IB and Saturn V boosters as soon as the Apollo requirements are clearly met . . . we have to limit our planetary programs. . . . In contrast, the Soviets show every indication of continuing to build upon their capabilities to demonstrate their power in aeronautics and to master space. . . . I have tried to dispel any illusions that we can with the reduced efforts prevent Soviet superiority in key areas."

Johnson eventually told Hornig: "I wanted [Webb] to succeed and it was only with great reluctance that for the past two years I have taken action to meet the overall fiscal requirements laid down by a determined group in the Congress by accepting cuts made in the House Appropriations Committee. I note with interest your statement that 'a continuing vigorous Soviet program coupled with a constrained U.S. program' could reverse the positions you believe now exist. This is very close to what I believe Webb has been saying."

For the crew of Apollo 8, Deke Slayton selected Commander Frank Borman, Command Module pilot Mike Collins, and Lunar Module pilot Bill Anders. Susan Borman:

> My husband [Frank] came home and, as best he could, he said what he had volunteered for. And I was trying to absorb what he was telling me. "This is August. You haven't tested the capsule yet? December? That's what? Three some-odd months? But usually you train for a year. . . ."
>
> We'd [tell the press] how proud we were. How confident we were. And then I'd go back in the house and kick a door in. I thought, "They're rushing it, they're leapfrogging, they're too anxious to get it going," and I just figured, "maybe you'd better face up to that and give it some thought and stop living in this cocoon, because this time it's not just another test flight. . . ."
>
> Chris Kraft came by one night, and I said, "Hey Chris, I'd really appreciate it if you'd level with me. I really, really, want to know what you think their chances are of getting home." . . . And he said, "OK, how's 50/50?" . . .
>
> I really didn't think they'd get them back. I just couldn't see how they could. Everything was for the first time. Everything.

Her fears were far from exaggerated. During crew training, a journalist overheard NASA officials discussing the question: "Just how do we tell Susan Borman, 'Frank is stranded in orbit around the moon?'" Meanwhile, Mike Collins, the astronaut corps' handball champion, started noticing something strange: "My legs felt peculiar, as if they didn't belong 100 percent to me. I'd heard prizefighters talk about their legs going and I thought, well, instant old age." Then, from time to time, he lost sensation in one leg, or both, to the point where he would fall over. Doctors found that both a neck disk and an incipient bone spur on Mike's spine were pressing against his spinal cord. He needed surgery, and could choose either an easy fix that would provide relief, or a more difficult and dangerous procedure that might be a cure, or might paralyze him. Mike chose the second course, which pulled him off the

Apollo 8 mission and out of the flight slot list, perhaps forever. His replacement was Buzz Aldrin's Gemini XII partner, Jim Lovell.

"We were very excited about [Apollo 8]," Neil Armstrong said. "We thought it was very bold, because we still had the pogo problem on the Saturn and we'd had a couple of problems with [both] Saturn V launches, so to take the next one, and without those problems being demonstrated as solved, and put men, a crew on it, not just take it into orbit, to take it to the Moon, it seemed incredibly aggressive. But we were for it. We thought that was a wonderful opportunity. If we could make it work, why, it would make us a giant jump ahead. It showed a lot of courage on the part of NASA management to make that step. One of the things that I was concerned with at the time was whether our navigation was sufficiently accurate, that we could, in fact, devise a trajectory that would get us around the Moon at the right distance without, say, hitting the Moon on the back side or something like that, and if we lost communication with Earth, for whatever reason, could we navigate by ourselves using celestial navigation. We thought we could, but these were undemonstrated skills."

A few weeks before launch, Huntsville had discovered two serious problems: repeated failures on the J-2 engines that powered stages two and three, and the vibrations of erratic firing causing longitudinal oscillations, or "pogo," which is similar to what happens to a car that desperately needs a tune-up, and goes ka-chugging down the road. In the case of a rocket, that ka-chug is so powerful that it compresses the ship's structure like a squeezebox.

The J-2 fix was fairly straightforward—repairing a flexible bellows-joint in the liquid hydrogen plumbing that kept cracking when it hit the ceiling of Earth's atmosphere. The pogo, however, required NASA-level brainstorming of four hundred technicians and 31,000 man-hours of engineering detective work. It, too, turned out to be a matter of plumbing. Von Braun: "An important Marshall facility was the Dynamic Test Tower, the only place outside the Cape where the entire Saturn V vehicle could be assembled. Electrically powered shakers induced various vibrational modes in the vehicle, so that its elastic deformations and structural damping characteristics could be determined. . . . In sync with the pogo oscillations, pressures in the fuel and oxidizer feed lines fluctuated wildly. If these fluctuations could be damped by gas-filled cavities attached to the propellant lines, which would act as shock absorbers, the unacceptable oscillation excursions should be drastically reduced. Such cavities were readily available in the liquid-oxygen prevalves, whose back sides were now filled with pressurized helium gas."

After General Electric automatic checkout equipment tested the ship with 3,173 measurements at 24,000 samples a second, Apollo 8 was finally "Go." CBS news anchor Walter Cronkite remembered that liftoff on December 21, 1948: "It was all there in our emotions as they took off . . . a combination of concern for their safety, and knowing that this was a great pioneering adventure . . . an event beyond the outer limits." Jim Lovell's son Jay, however, remembered a very different story: "We had our own paparazzi. Being kids, we went out and had fun with them. I got in a fight in the front yard with some other kid, and the next morning it was in the *Houston Chronicle*: 'Son Protects Home While Dad's in Space.' My mom dragged me into the laundry room and said, 'What is this?'" Bill Anders's daughter, Gayle, said, "I remember *Life* magazine taking our family pictures. We would all eat ice cream around the table or my dad pushed me on the swing. I remember thinking that this was a bit odd because we never really did that in real life."

On Apollo 8's first night in space, Frank Borman couldn't fall sleep, After two hours of tossing in his hammock, he took a Seconal. A few hours later, he awoke convulsed in vomiting and diarrhea. Houston's Mission Control actually has a twin of itself on the floor below, in case two flights are launched simultaneously. That room is also used when astronauts need to communicate in confidence, as Borman did when he informed ground control that he was seriously ill. NASA doctors became alarmed that a virus might have invaded the capsule, which could quickly spread to the other astronauts and incapacitate the entire crew. Frank Borman: "At the time I didn't know it was motion sickness. . . . You know the damn doctor's—what—100,000 miles away; he doesn't know what's going on. And I got over it very rapidly. And Jim and Bill both told me that they felt queasy, too, when they started moving around. And I threw up a couple of times, and it was over with. It wasn't a big deal. . . . Somebody said that Glenn puked when he got back, too."

Though not contagious, the commander's illness did have quite an effect on his crew. "The one nice thing about being on Earth, if someone gets sick, it lands on the floor," Bill Anders said. "In zero-g, it floats around. So when poor Frank became ill, fortunately the spacecraft was in good shape, we were coasting along; it wasn't a particularly critical time. . . . The stuff was floating around. I grabbed an emergency oxygen mask because the smell was really bad and clamped it on because I didn't want to get sick as well. . . . But I do remember watching this blob of vomitus drifting up out of the navigation bay; Lovell and I were up in the—flying the spacecraft. And it had a three-dimensional oscillation to it and I reverted to my scientific physics training and I was just amazed at the three-dimensional oscillation of this thing. You know, you don't see that on Earth because you can't simulate zero-g.

"And so I'm watching it and Lovell's sitting next to me and it's drifting kind of like this and then it split. It oscillated too much and in the laws of conservation and momentum, see, that the one piece that's going this way has to be balanced by the other piece going that way, and so we watched it and as it headed towards Lovell—thank God—and he was almost out of focus watching this thing splat onto his chest and spread out like a fried egg. So we spent, you know, a couple of hours going around the spacecraft with kind of paper wipes, like butterfly catchers trying to catch butterflies, mopping this stuff out of the air or off the walls. It was bad. You know it's amazing, after a while you didn't smell it. Not because the smell went away—'cause you get used to it. I guess a skunk can learn to live with himself."

Regardless of his initial gastrointestinal distress, Frank Borman proved to be one tough chief. When Anders made some religious comments during training, his commander barked, "Are you going to take communion every thirty seconds before the flight?" And when the two junior astronauts tried to enjoy their extraordinary experience, they were told: "Stop looking out the window and get back to work!"

Borman: "[On Apollo 8] I had two great concerns: I think the worst fear that I had was that somehow the crew would foul up, and that was the one thing that I did not want to happen. . . . We were killed more times in simulation than you could shake a stick at. . . . I had a great team in Anders and Lovell, and I wanted to make certain that we did and we could handle whatever was handed our way. The second thing was, I didn't want—really want the mission to get fouled up because we really weren't certain that the Russians weren't breathing down our backs. So I wanted to go on time."

Early in the flight, a test of the SPS propulsion—the service module's rocket, which brought Apollo crews home—uncovered a malfunction. Should Apollo 8 be aborted? They decided to continue. Flight director Gerry Griffin:

On Apollo 8, when we did the translunar injection burn—the burn that sent us toward the Moon—you could have heard a pin drop in that control center. I mean, there was nobody even breathing hardly; and it was almost like a religious experience. And then when cutoff—and the engine cutoff and the trajectory—we did a quick check on the trajectory and it was good, we were headed out, we all kind of looked at each other and said, "Well, we've done it now."

[Then] when we lost signal, as they went behind the Moon, they were going to do a maneuver back there to slow themselves down so they would go into lunar orbit. I never will forget how quiet that whole room was for that entire—I think the backside took about fifty minutes. And nobody—hardly nobody—hardly anyone moved that entire time. . . .

I never, except for one occasion, remember Kraft looking visibly nervous and agitated. And that was when he was waiting for the spacecraft to come out from behind the Moon after they had supposedly fired their engine to start them back to Earth. And of course, that was on his shoulders. And that was a single engine that had to work. One engine bell. A lot of redundancy in the piping and all that. But there was only one rocket chamber, and only one set of fuel tanks, and it had to work. And it did.

Bill Anders remembered that riding on Apollo 8 "was like being on the inside of a submarine." Jim Lovell said of being one of the first three humans to see the far side of the Moon: "We were like three schoolkids looking through the candy store window. I think we forgot the flight plan! We had our noses pressed against the glass, looking at those craters go by. It is really an amazing sight, an amazing sight."

Frank Borman was amazed by a different landscape: "It was the most awe-inspiring moment of the flight when we looked up and there coming over the lunar horizon was the earth. It was the only object in the universe that had any color to it, basically blue with white clouds, and everything that we held dear was back there. A long way away." Bill Anders agreed: "It was almost as if we were discovering the earth for the first time." When Anders then joked to his rigid commander, "Hey, don't take that [picture], it's not scheduled," Borman laughed, grabbed the Hasselblad, and turned it over to Anders, who in two shots caught the famous image of a blue-marble Earth rising over a desert moon that graced the cover of the *Whole Earth Catalog* and 188,380,000 U.S. postage stamps.

As would be true for all of Apollo, traveling 240,000 miles away gave the 8's crew a perspective on their own spaceship. "After all of the preparation, after the split-second planning of everything on the mission, the big surprise waiting for them at the other end was not about the Moon, it was about seeing our home planet as a very tiny, very precious oasis of life and color in the blackness of space," historian Andrew Chaikin remarked.

A few weeks before liftoff, a NASA official had mentioned to Borman that, when the spacecraft was in orbit on Christmas Eve, "More people will be listening to your voice than that of any man in history. So we want you to say something appropriate." On the night of December 24, 1968, the TV camera installed in Apollo 8's window broadcast images of a barren, lifeless, and crater-pocked Moon, while behind it rose a living planet of blue seas, brown lands, and white atmospheres. Bill Anders opened the flight manual to the back, where something had been printed up on fireproof paper. He began: "For all the people on Earth the crew of Apollo 8 has a message we would like to send you.

"In the beginning, God created the heaven and the earth. And the earth was without form, and void; and darkness was upon the face of the deep. And the Spirit of God moved upon the face of the waters. And God said, Let there be light: and there was light. And God saw the light, that it was good: and God divided the light from the darkness."

Jim Lovell: "And God called the light Day, and the darkness he called Night. And the evening and the morning were the first day. And God said, Let there be a firmament in the midst of the waters, and let it divide the waters from the waters. And God made the firmament, and divided the waters which were under the firmament from the waters which were above the firmament: and it was so. And God called the firmament Heaven. And the evening and the morning were the second day."

Frank Borman: "And God said, Let the waters under the heavens be gathered together unto one place, and let the dry land appear: and it was so. And God called the dry land Earth; and the gathering together of the waters called he Seas: and God saw that it was good. . . .

"And from the crew of Apollo 8, we close with good night, good luck, a Merry Christmas, and God bless all of you—all of you on the good Earth."

At Mission Control, nail-hard engineers found their eyes filling with tears. Gene Kranz: "I think that was probably the most magical Christmas Eve I've ever experienced in my life, to actually have participated in a mission, provided the controllers, worked in the initial design and the concept of this really gutsy move, and now to really see that we were the first to the Moon with men. . . . I mean, you can listen to Borman, Lovell, and Anders reading from the Book of Genesis today, but it's nothing like it was that Christmas. It was literally magic. It made you prickly. You could feel the hairs on your arms rising, and the emotion was just unbelievable." An estimated worldwide audience of one billion was either watching the flight on television or listening to it on the radio. The *Washington Post* reported that "At some point in the history of the world, someone may have read the first ten verses of the Book of Genesis under conditions that gave them greater meaning than they had on Christmas Eve. But it seems unlikely. . . ."

Frank Borman remembered another detail about that holiday flight: "When we opened up the dinner for Christmas and I found somebody had included brandy in there, you know, I didn't think that was funny at all. Because you and I both know, if we'd have drunk one drop of that damn brandy and the thing would have blown up on the way home, they'd have blamed the brandy on it."

Apollo 8 was judged an enormous success and a great breakthrough for NASA—*The New York Times* called it "the most fantastic voyage of all times"—but it wasn't over yet. Borman: "We hit the water with a real bang! I mean it was a big, big bang! And when we hit, we all got inundated with water. I don't know whether it came in one of the vents or whether it was just moisture that had collected on the environmental control system. . . . Here were the three of us, having just come back from the Moon, we're floating upside down in very rough seas—to me, rough seas. Of course, in consternation to Bill and Jim, I got good and seasick and threw up all over everything at that point."

Bill Anders: "By now the spacecraft was a real mess, you know, not just from him but from all of us. You can't imagine living in something that close; it's like being in an outhouse and after a while you just don't care, you know, and without getting into detail . . . messy. But we didn't smell anything. . . . And I did notice a very strange odor when I [finally] got out of the spacecraft and it turned out to be fresh air."

Apollo 8's rockets, navigation systems, various communications setups, and Command Module had all performed spectacularly. Many small details, however, were still far from acceptable. There was too much undissolved hydrogen and oxygen in the water for food hydrating, making everything taste the same—bad. The urine condoms leaked, as did the diapers. The crew's rest was constantly interrupted by radio calls and equipment noises. "The crew was well taken care of, except that they couldn't eat, drink, sleep, or relieve themselves," as administrator Tom Paine summed up.

Every NASA mission was in essence designed to prove two theories. The first was one of aeronautical engineering itself—that humans could use the laws of physics and build upon a legacy of aviation technology to create viable spaceships. And the second was that ground controllers and astronauts could advance this technology step-by-step through simulations, and then flight. Some missions inched forward, and others took great strides.

With Apollo 8, for the first time in its eleven-year history, America took the lead in the Space Race. Anyone believing that the race was actually won, however, was in for a surprise. On January 14, 1969, Soyuz 4 and 5 successfully completed the mission that was to have been Komarov's. With the surviving crewmen performing their original roles, the two ships launched a day apart, rendezvoused, docked, were spacewalked, undocked, and each came home to a perfect landing. Two manned craft successfully coupling was a feat NASA had not yet achieved, and it would have to be mastered before sending anyone to the Moon.

The Soviets had now trumped all of NASA's Gemini triumphs with a single mission. A mere seven months before Apollo 11, as far as anyone at NASA knew about the Russian space program, the two countries were once again in a dead heat.

PART III

13

The Great Black Sea

G **ET 00:01:00:** Aboard Apollo 11, the astronauts were gradually and insistently pressed down further and further by the weight of their rocket's velocity, a smothering force hitting 3 g's just as their ship disappeared from view and hurtled into the atmosphere. The onlookers craned their heads to follow the very last moment of visibility. In the sky, now, there was only a blinding incandescent white, a smoky tail, a distant roar, and a vanishing trail of smoke and exhaust.

GET 00:01:21: At 1,800 miles per hour and 43,000 feet of altitude, the Saturn hit that combination of velocity and atmospheric density known as maximum dynamic pressure (Max-Q). The craft's skin and structure had been engineered to withstand exactly this moment of extreme duress, which had now reached 4.5 g's. The pressure, in combination with the vibration of the rocket, makes the skin on each man's neck flap. In the unlikely event that any one of them was smiling, that smile was frozen hard to his face. "If I'd had to reach a switch with all that vibration going on," Jack Schmitt remembered, "I wouldn't have been sure where I was putting my hand."

Other than this force straining against them, Armstrong, Aldrin, and Collins were aware only of a great and distant roar, and the sensation of being aboard a very strong freight elevator. The speed and altitude dials on their computers, however, registered exactly the force of the power that was thrusting them forward into the future—now sixty-one kilometers above the earth, they were blasting away at 8,600 kilometers per hour.

GET 00:02:40: The S-IC, 11's 138-foot initial, bottom stage, which accounted for two-thirds of the craft's launch weight, had now burned two million kilograms (4.5 million pounds) of fuel in two minutes and forty seconds—forty-five times the amount used on a seven-hour flight from New York to Paris. At an altitude of 217,000 feet and a speed of 6,100 miles per hour, its five engines shut down. Eight retrorockets fired to separate S-IC from the rest of the ship—retros were used to ensure that discarded sections wouldn't crash into the rest of the spacecraft—and it fell away, to sink into the Atlantic.

The first stage's power was such that its thrust compressed the ship above it. With its shutdown, Apollo 11 decompressed, sending the crew in a kick-back from a maximum "eyeballs in" to a full "eyeballs out." "There's a big change, it's from 4 g's to a minus one and a half, as the whole stack unloads, to a plus one and a half, as you go on the second stage," said Jack Schmitt. "And that all happens in just slightly over a second. So that is probably the most dynamically exciting point in the mission, certainly in the launch part of it."

"[On Apollo 8,] we went from plus six to minus a tenth g, suddenly, which had the feeling, because of the fluids sloshing in your ears, of being cata-pulted through the instrument panel," Bill Anders remembered. "So, instinc-tively, I threw my hand up in front of my face, with just a third-level brain reaction. Well, about the time I got my hand up here, the second stage cut in at about, you know, a couple of g's and snapped my hand back into my hel-met. And the wrist string around my glove made a gash across the helmet face plate. And then on we went. Well, I looked at that gash and I thought, 'Oh, my gosh, I'm going to get kidded for being the rookie on the flight,' be-cause you know, I threw my hand up. Well, after we were in orbit and the rest of the crew took their space suits off and cleaned their helmets, I no-ticed that both Jim and Frank had a gash across the front of their helmet."

The external effect of staging was just as dramatic. "This big fireball comes roaring up the length of that booster and it's out in front of you, and then the second stage fires, and you fly right through the fireball, and you're on your way again," as Gene Cernan described it.

GET 00:02:41: One second after stage one fully separated, stage S-II fired its four 21,000-pounds-of-thrust ullage motors for five seconds to settle the zero-g fuel in its tanks, followed by five J-2 engines for six minutes, which would take the ship's altitude to 600,000 feet and its speed to over 15,000 fps. At each staging, there was a brief period of weightlessness as the prior rocket exhausted itself and fell away, and a renewed jolt of speed as the new one took its place, not unlike the shifting of gears in a manual transmission.

GET 00:03:17: No longer needed, the launch escape tower released, fired, and jettisoned. For the first time, the crew could see out all of their windows, but the only view was of black sky.

Armstrong: We've got skirts up.
CapCom (Capsule Communicator): Roger, we confirm. Skirts up.

GET 00:09:11: At an altitude of 610,000 feet and a velocity of 15,500 miles per hour, the S-II's engines fell silent, its retrorockets fired, and it detached and fell away. The single J-2 engine of the S-IVB third stage now ignited.

This apparently cavalier disposal of wildly expensive rocket parts might seem a peculiar method of transportation, but there are only so many ways of getting to the Moon, even when you have relatively unlimited financial resources. The bullet-with-fins, porthole window, and pop-up ladder style of rocket, so popular with children and cartoonists, would have to be immense to escape Earth's gravity, a force so strong that it takes two hundred pounds of fuel for every pound of load to achieve release. (In fact, the details of how gravity works are still about as much of a mystery today as the purpose of sleep, and the substance of dark matter.) A single-stage rocket with sufficient power to blast away from both the Earth and the Moon, to carry a crew, human support and equipment, as well as the heat shield needed for reentry into the atmosphere, would have to be a giant beast of a missile. If it had been used for Apollo 11, Armstrong and Aldrin would have needed a method to land a ninety-foot-long rocket on its tail, and viewers at home would then have had to watch "one small step" turn into two men in suits and backpacks climbing down a ninety-foot ladder to step onto the Moon.

As early as 1952 the von Braun rocket design team was pondering this conundrum. Their solution was a space station that allowed pieces of a ship to be launched separately and then assembled in orbit. While still with the army, von Braun had in fact suggested this strategy to NASA chief Keith Glennan on December 15, 1958. He had wanted to launch fifteen Juno Vs—the rocket he had at the time—to be assembled into one giant multistage booster, which could then fly to the Moon and back. This idea became known as Earth Orbit Rendezvous, but it seemed to require more time to develop than NASA thought it had available under the constraints of its race with the Soviets over who would control the cosmos.

In 1959, a team of engineers under Tom Dolan at Vought Astronautics came up with a different solution to the payload question by proposing modules, a spaceship chopped into parts that, as each was used, would simply be thrown away. Instead of bringing the whole rocket to land on the Moon, they devised a dinghy approach, in which a module would travel from the ship to the surface and back, just as a rowboat ferries sailors between a great schooner and land. This method, distinguished by a mother ship's waiting in lunar orbit for its rowboat's return, was called Lunar Orbit Rendezvous.

Almost every aeronautic engineer alive at the time passionately believed in EOR over LOR, for the simple reason that, if something went wrong with the technologically difficult rendezvous, it seemed more likely that astronauts could be brought home safely from Earth orbit than from a failed docking over the Moon . . . 240,000 miles away. "We were horrified at the lunar rendezvous approach the first time we saw it," admitted Manned Spacecraft Center director George Low. LOR, however, acquired a serious evangelist in the person of Langley engineer John Houbolt, who proselytized its benefits to every NASA chief who would listen (and to many who wouldn't). The LOR-verses-EOR debate grew so contentious that North American Aviation produced a study claiming that, if the Saturn V was given a little more power and the crew reduced from three to two men, direct ascent could be used, with no need for an LM (this approach would also make North American the Apollo spaceship's sole manufacturer).

"When we became convinced that [LOR] was the only way we could hack this, North American was bitterly opposed, then, even though at one time they hadn't thought to get the whole thing, once they got the whole job, they just fought tooth and nail any scheme that would dilute North American's influence," design engineer Caldwell Johnson remembered. "And the prospects of an LM, man, that just horrified them. To think they might have to cut up this big piece of cake half-and-half with some other contractor. So they fought this thing tooth and nail. They had a hell of a lot of influence in places, old Stormy and [Lee] Atwood would go right straightaway to Washington, knowing everybody there. So we were really fighting for the program. You would think that as the agency running the thing that we certainly wouldn't have to battle our contractor on something as fundamental as this, but it was almost a fight to the death there for a while."

When NASA refused to cut the number of the crew, NAA took the issue behind its back to the White House's science adviser, Jerome Wiesner, who thought LOR was nonsense. When Wiesner in turn tried to get the president to back him and NAA on this issue over NASA administrator Jim Webb, Kennedy had the two present their arguments, and then ceded the decision to his space agency, joking that Jim "had all the money, Wiesner had only me."

Caldwell Johnson: "[Finally] von Braun got up and he said, . . . 'I think the lunar orbit rendezvous concept is the way to go. We should not do the earth orbit.' From the time he said that, that was the end of that. North American knew that they were licked, were beat. And there was no more argument. I sure was glad to hear that. In the meantime, you see, nothing was happening on Apollo. Not a damn thing was doing. North American had all its engineers working on a competitive scheme. There they were, our contractors and getting paid to do the thing, and they were spending all our money assembling arguments against what we were trying to do."

During launch, Armstrong, Aldrin, and Collins were strapped in so tightly to their couches that they only realized they'd hit zero-g when little items gone missing during construction—a random nut, or bolt, or washer—suddenly came rising up into the air. Once they unbuckled, it required serious effort to remain wherever they wanted to be. Just reaching down to scratch an itchy leg meant being thrown ass-over-teakettle if they didn't stabilize themselves first. Zero-g could also lead to a very severe form of motion sickness, potentially lethal in outer space.

There had been little trouble with nausea and dizziness during the Mercury or Gemini programs, but a number of reports, notably from Frank Borman and Rusty Schweickart, had made every flier apprehensive about the Apollo capsule. Aldrin had spent untold sessions in the Vomit Comet to prepare himself, while Collins had learned everything he could to avoid getting sick. "Barfing anywhere is no fun, but barfing in space is different," Rusty Schweickart explained. "You've got the helmet on, you're locked in the space suit, your hands are out at the end of these gloves. You can't pull them in to do something here. So if you barf in weightlessness in a space suit, you die."

GET 00:11:11: After the white-on-black data streaming from their computers indicated that all was ready, Mission Control radioed Armstrong: "Apollo 11 is go for orbit." Ten minutes later, the commander acknowledged "Shutdown," and the three men began their 1.5 circumnavigations of the Earth at a height of 101 nautical miles. (Like all aviators, NASA uses nautical miles—one minute of latitude, or about 2,000 yards—instead of statute miles—1,760 yards—in its calculations.) On Apollo missions, NASA used earth orbit to ensure that the spacecraft was ready for its great voyage, especially looking into zero-gravity questions that couldn't be fully answered on the ground, and an inspection of damages inflicted by the violence of launch.

"Up there you go around every hour and a half, time after time after time," Rusty Schweickart said. "You go around across North Africa and out over the Indian Ocean, and Ceylon off to the side, Burma, Southeast Asia, and up across that monstrous Pacific Ocean—you've never realized how big that is before. And you finally come up across the coast of California and look for those friendly things: Los Angeles, and Phoenix, and there's Houston, there's home. You identify with Houston, and Phoenix and New Orleans. And the next thing is you're identifying with North Africa. When you go around in an hour and a half you begin to recognize that your identity is with that whole thing.

"You look down there and you can't imagine how many borders and boundaries you crossed. At the Mideast you know there are hundreds of people killing each other over some imaginary line that you can't see. From where you see it, the thing is a whole, and it's so beautiful. And you wish you could take one from each side in hand and say, 'Look at it from this perspective. Look at that. What's important?'"

At the start of orbit, in order to use the sextant, Apollo 11 flew upside down for 2.5 hours, the crew's heads facing earth and their feet pointed at the black sky above. At their speed of three hundred miles a minute, day and night arrived in blinks, the sunrise a rapid-fire dialing-up of reds and golds and blinding whites, the sunsets as fast as the snuffing of a candle. In minutes, the coal-black sky would dissolve, like fog, into a gray, which would lighten to a deep blue. Glowing bands of gold would appear, and the sun would suddenly leap forward to illuminate the blues, browns, greens, and whites of home. Gene Cernan called it "Sitting on God's front porch."

"The atmosphere on edge presents a striking sight. You see many distinct layers, all a different shade of iridescent blue. Through binoculars, I have counted six," ISS science officer Don Pettit remembered. "The most amazing aspect of this view is how thin this life-preserving blanket is when compared to the full extent of the planet. Like an orbital eggshell, our atmosphere looks so frail that it might crack and be gone in an instant, rendering Earth as barren and lifeless as any other baked hunk of rock orbiting the sun."

At one particularly clear view of the baby-blue wash that was Earth's atmosphere and its foamy peaks of cloud, the entire 11 crew was awestruck.

Collins: "Jesus Christ, look at that horizon! Goddamn, that's pretty, it's unreal."

Armstrong: "Isn't that something? Get a picture of that."

Collins: "Ooh, sure I will. I've lost a Hasselblad. Has anyone seen a Hasselblad floating by? It couldn't have gone far, a big son-of-a-gun like that . . . I see a pen floating loose down here, too. Is anybody missing a ball-point pen?"

The camera was finally found, too late to capture that particular sunrise, but at least, secured, it wouldn't be a danger during the next burn.

As they orbited and watched the Earth turning below them, the astronauts' radio signals were caught by a web of seventeen ground stations—Vanguard and Redstone tracking ships in the Caribbean, Goldstone in California, Guaymas in Mexico, Hawaii, Carnarvon and Honeysuckle in Australia, Ta-

On May 20, 1969, at 12:30 p.m., an eleven-man crew rolled out the 363-foot-high, 12.5-million pound Apollo 11-Saturn V and its umbilical tower 3.5 miles from NASA's Vehicle Assembly Building to Launch Complex 39A. As this behemoth could travel at less than one mile per hour, the task was accomplished with the spaceship set upon its launch pad seven hours later, at 7:46 p.m.

In the run-up to launch, Neil Armstrong, Mike Collins, and Buzz Aldrin had to juggle a massive training regimen with time for the press, including this casual photo shoot before their missile.

Except where noted, all photographs are courtesy of the National Aeronautics and Space Administration.

Buzz Aldrin training in the padded Boeing 707 "Vomit Comet," which flew swooping parabolic arcs to simulate the zero-g of space and the one-sixth g of the Moon for twenty seconds. Behind the NASA team is a mock-up of the lunar spacecraft's hatch.

Neil Armstrong before one of the first Lunar Module simulators at the Lunar Landing Research Facility on February 12. When one of his training flights went awry, Armstrong had to eject and parachute away from the crashing trainer with less than a second to spare.

ad leader Guenter Wendt working
ith the mission's backup crew—
m Lovell, Bill Anders, and Fred
laise—during an altitude chamber
est on March 24.

t a Houston facility around
pril 15, Armstrong transferred
simulated lunar samples box to
agle's porch with the Lunar
quipment Conveyor. On the
loon, he would call it the
Brooklyn clothesline."

At a fourteen-hour July 5 press conference, the crew answered questions from a plastic box engineered with blowers to keep out any of the audience's virii or bacteria.

Michael Collins and Flight Crew Operations director Deke Slayton arrive on July 12 at Florida's Patrick AFB in the T-38 jet which NASA fliers used to commute to its various facilities and contractor

A diagram of the Command and Service Modules and their Launch Escape tower that would yank the crew to safety in the event of a pad meltdown.

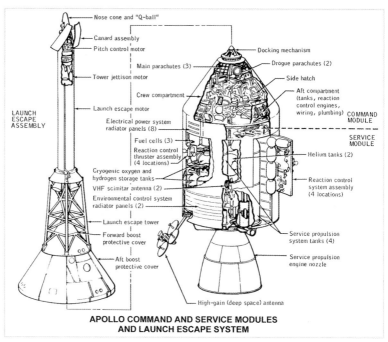

**APOLLO COMMAND AND SERVICE MODULES
AND LAUNCH ESCAPE SYSTEM**

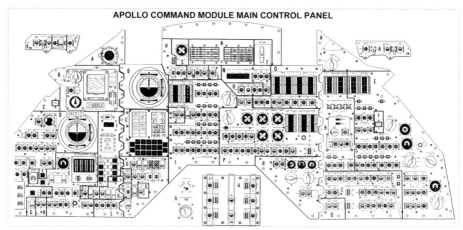

A schematic of spaceship *Columbia*'s elaborate dashboard.

The great pressures of NASA's Space and Missile Race with the Soviet Union can be seen in the last-minute changes tucked into the walls of *Columbia*'s cabin.

A photograph taken by an umbilical tower camera captures the five F-1 rocket engines of Apollo 11's first-stage booster igniting at 9:32 a.m. EDT on July 16, 1969. After each engine had reached full throttle, tiedown arms were released and the ship began its slow rise into the air, the snow and ice condensation from its liquid oxygen and hydrogen tanks crashing to the ground.

Lady Bird Johnson, former NASA Administrator Jim Webb, former President Lyndon B. Johnson, former NASA Deputy Administrator Robert Seamans, and Vice President Spiro Agnew watch the liftoff. President Richard Nixon did not attend in case anything went wrong.

The Kennedy Space Center Launch Control team watches as their rocket soared into the sky.

The American father of rocketry, Dr. Robert H. Goddard, with his liquid oxygen-fueled missile nestled in its firing frame on March 16, 1926, at Auburn, Massachusetts. Though Goddard discovered and refined all the essentials of rocket technology, his bitterness at early criticism of his work and ensuing reclusiveness (along with the Pentagon's dedication to bombers) left the United States a distant second to the Soviet Union in space technology across much of the 1950s and '60s.

Soviet Chief Designer Sergei Korolyov at the Kapustin Yar firing range in 1953. Korolyov's refinements of the V-2 rocket created by Nazi Germany would ignite the Cold War's Space and Missile Race. His genius would make the USSR the world's leader in space technology, and his death would enable the United States to leap forward with Apollos 8 and 11.

The Huntsville Times

Man Enters Space

'So Close, Yet So Far,' Sighs Cape

U.S. Had Hoped For Own Launch

Soviet Officer Orbits Globe In 5-Ton Ship

Maximum Height Reached Reported As 188 Miles

Hobbs Admits 944 Slaying

To Keep Up, U.S.A. Must Run Like Hell

Praise Is Heaped On Major Gagarin

First Man To Enter Space Is 27, Married, Father Of Two

'Worker' Stands By Story

Reds Deny Spacemen Have Died

No Astronaut Signal Received At Ft. Monmouth

Reds Win Running Lead In Race To Control Space

April 12, 1961: After launching the first satellite with *Sputnik*, Korolyov orbited the first human being around the Earth, one of his "little eagles" and the most admired man in all of Russia: Yuri Gagarin. The achievement would startle the world and shock NASA's émigré rocket team working under Wernher von Braun at Huntsville, Alabama.

On the White House lawn, John F. Kennedy awards the first American in space, Alan Shepard, with the NASA Distinguished Service Award after his May 5, 1961, ballistic launch aboard Freedom 7. To the left are Shepard's wife and his mother, with his fellow Mercury astronauts and NASA Administrator Jim Webb standing behind the astronaut and the president.

Before becoming an astronaut, Neil Armstrong worked for NASA's predecessor, the NACA, as a test pilot in the California desert flying one of the greatest planes ever built: North American Aviation's X-15, a fifty-foot-long, thirteen-foot-high, and twenty-two-foot winged missile with a skin of black Iconel X. Armstrong called this "the most fascinating time of my life." The plane seen in this picture now hangs over Apollo 11's *Columbia* at the Smithsonian's National Air and Space Museum.

NASA's rocket chief, Wernher von Braun, and his masterpiece, the Saturn V. Armstrong once commented that today, NASA staffers are shocked when the shuttle doesn't work, but in the Apollo era, they were always surprised with the Saturn V consistently launched successfully.

vo years after joining NASA's
tronaut program, Neil
rmstrong posed with a model
the Apollo Command
odule spacecraft in 1964.

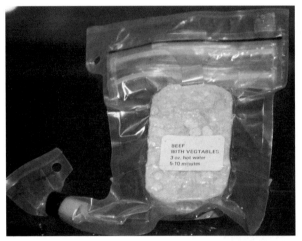

A freeze-dried and vacuum-packed Apollo dinner of beef with vegetables. A hot-water gun was inserted into the vent at lower left and given two shots; the "meal" was then kneaded until it was the consistency of toothpaste. [Photograph courtesy of the author.]

After *Eagle* had undocked from *Columbia*, Armstrong pirouetted his ship so that Collins could confirm that explosives had successfully telescoped the lander's legs into position.

A schematic of *Eagle*'s cabin. The craft was flown standing, like a trolley, with Armstrong on the left and Aldrin on the right, their shoes Velcroed to the floor and their waists tethered to the walls.

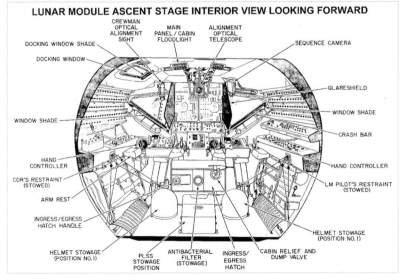

LUNAR MODULE ASCENT STAGE INTERIOR VIEW LOOKING FORWARD

CREWMAN OPTICAL ALIGNMENT SIGHT

MAIN PANEL / CABIN FLOODLIGHT

ALIGNMENT OPTICAL TELESCOPE

DOCKING WINDOW SHADE

DOCKING WINDOW

SEQUENCE CAMERA

GLARESHIELD

WINDOW SHADE

WINDOW SHADE

CRASH BAR

HAND CONTROLLER

HAND CONTROLLER

CDR'S RESTRAINT (STOWED)

LM PILOT'S RESTRAINT (STOWED)

ARM REST

INGRESS/EGRESS HATCH HANDLE

HELMET STOWAGE (POSITION NO.1)

HELMET STOWAGE (POSITION NO.1)

PLSS STOWAGE POSITION

ANTIBACTERIAL FILTER (STOWAGE)

INGRESS/ EGRESS HATCH

CABIN RELIEF AND DUMP VALVE

A panorama reveals the Sea of Tranquility from *Eagle*'s windows.

Earthrise over the first spaceship to land on the Moon.

July 20, 1969, 9:56 p.m. (Houston time): With six hundred million people—one-fifth the earth's population—watching on television or listening on radio, Armstrong's foot touched the ground of another world for the first time in human history. He said, "That's one small step for (a) man. One giant leap for mankind."

Armstrong became so engrossed in picture taking that he postponed what was to have been his first chore: Grabbing a moon rock in case the men had to immediately return to the safety of their ship. Here, Armstrong photographs Aldrin descending from *Eagle* hatch and porch onto the surface of the Moon.

In one of the very few pictures of Armstrong on the lunar surface, he stands to the left while finishing what both he and Aldrin considered the most difficult and nerve-wracking of their moonwalking chores: Planting an American flag into the lunar soil and making sure it did not fall to the ground in front of the untold millions watching on TV. This picture was taken by what NASA called its 16mm Data Acquisition Camera, mounted to *Eagle*.

Armstrong and Aldrin had so many chores to perform during their moonwalk that their gloves had to-do lists printed on the sleeves.

While Armstrong's primary duty was in photographing as much of the Moon as possible, Aldrin was in charge of the mission's basic scientific equipment, including a seismometer to measure moonquakes. This instrument had been compressed for travel aboard *Eagle* and then, after Aldrin pulled a trigger, it automatically deployed its legs and solar fuel-cells with such robotic beauty that Aldrin called it "fantastic. I wish we'd had the television set up to see that."

'hen *Eagle* rose to reunite with *Columbia*, Mike Collins captured this extraordinary oment of the Earth, the Moon, and the first spaceship to touch down on another anet returning home in what he called the mission's happiest moment.

After a July 24 splashdown at 11:49 a.m. CST 812 nautical miles southwest of Hawaii, the Apollo 11 crew was assisted by Navy frogman Clancey Hatleberg into a life raft to wait for helicopters to ferry them twelve nautical miles to recovery aircraft carrier *Hornet*.

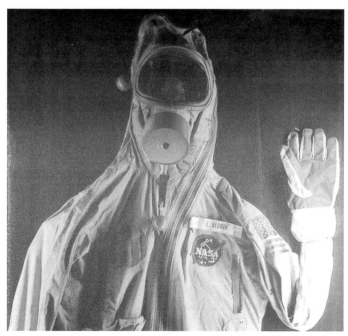

Hatleberg and the men wore NASA-designed Biological Isolation Garments doused with bleach and iodine to prevent lunar pathogens from infecting the Earth. [Photograph courtesy of the author.]

The whole of Houston's Mission Control erupts in cheers and shouts after it has been confirmed that the Apollo 11 crew is safe, sound, and aboard the Navy helicopter heading for the *Hornet*. While one MOCR screen broadcast the text of Kennedy's original speech on sending a man to the Moon, the other said: *Task Accomplished, July 24, 1969.*

Traditionally after recovery, NASA crews cut and served a cake to celebrate a successful mission. Apollo 11's ceremony on July 24 had to be altered since the men were confined to quarantine.

After dancing a jig of happiness, President Richard Nixon greeted the crew inside their Mobile Quarantine Facility trailer during what he called "the greatest week in the history of the world since the Creation."

On August 13, 1969, the three crewmen were ticker-tape celebrated by four million Americans down the Canyon of Heroes in the largest parade in New York City history. Armstrong: "Sometimes they threw a whole stack of [computer] punch cards from the eighty-seventh floor of a building, and, when they didn't come apart, it made like a brick. We had a couple of dents in our car from cards that didn't quite open." Aldrin: "We were advised not to reach out to shake hands because we could be pulled from cars and couldn't be rescued easily."

October 11, 1969: Michael and Pat Collins, Neil and Jan Armstrong, and Buzz and Joan Aldrin visit Norwegian Defense Minister Otto Greig on the outskirts of Oslo during their international goodwill tour "Giant Steps."

nanarive in the Indian Ocean, Madrid, the Canary Islands, Corpus Christi in the Bermudas, and six Boeings always aloft with transceivers in their noses, the Apollo Range Instrumented Aircraft, or ARIA.

GET 02:44 (Hours:Minutes): In the wake of continuing normatives— as-expected news—from its readouts, Mission Control ordered: "You are GO for TLI"—Trans Lunar Injection. The J-2 engine of stage three fired again for five minutes and forty-seven seconds, boosting velocity to 24,182 miles per hour—six miles per second, over ten times faster than the bullet of a Winchester .270—releasing the craft from Earth's gravity and catapulting it off to the Moon. Powered by stage three's great, final thrust, Apollo 11 would continue to fly for 200,000 miles over sixty-six hours, eventually reaching its destination in as much time as it takes an ocean liner to sail from London to New York. The ship now weighed 300,000 pounds—4 percent of its liftoff weight.

GET 03:15: As if he were piloting a souped-up airplane stick-and-pedal, Michael Collins flew the Command Module with two hands. In his left, a dashboard stick could be twisted left, right, up, down, and pushed in and out. This would signal the attitude system to fire peroxide thrusters and sail the ship. The right controller came out of the floor, and combined stick-and-pedal with autopilot to produce attitude. Pull to swivel up; push to nose down; left and right for rolling; turn the knob for yaw. Astronaut John Young compared it to "driving an aircraft carrier with an outboard motor."

Five hundred miles over the Pacific, the main capsule retrofired, detaching itself from stage three. Using a mix of manual and computer, Collins moseyed forward a bit at one-half miles per hour, and then, at two degrees per second, swiveled in a topsy-turvy U-turn. Now, while the two craft soared in parallel at 17,000 miles per hour, doors on the S-IVB stage opened up and released, like flower petals, revealing the Lunar Module, cradled inside the rocket's assembly. Drawing nigh, Collins used a rangefinder eyepiece with crosshairs to insert the CM's probe (like the ballhead on a tripod) into the LM's drogue (a dish with a hole at its center), a procedure so technologically and visually lyrical that Stanley Kubrick choreographed it as a signature waltz in his *2001: A Space Odyssey* (a movie that Aldrin and Armstrong have both said best represents the look of outer space; it may have helped that NASA's Frederick Ordway, Harry Lange, and Bob McCall were all hired as consultants on the film).

After a set of latches engaged in soft dock, Collins waited until the computer confirmed that alignment had been achieved. He then triggered a shot of nitrogen gas to yank a ring of clamps, and the two ship-modules were hard

docked together. The LM was released from the S-IVB, and that stage's engine was refired to slingshot the craft around the Moon. By flying close enough
to another heavenly body (but not so close as to be pulled into orbit), a spaceship can use the power of that body's gravity like a slingshot—a velocity
boost as good as any engine burn—now known as "gravity propulsion" or
"gravity assist," and discovered by NASA's Michael Minovitch in 1961. The
discarded S-IVB fell into a solar orbit.

With the danger of launch passed, the crew was finally allowed to get out
of their cumbersome space suits and helmets, stripping down to space-age
versions of long johns. In their tiny ship, about the size of the front seat of an
American sedan, any such maneuver resembled the mobbed-stateroom scene
from the Marx Brothers' *A Night at the Opera*.

After Jim Lovell had suggested that an eagle be used as Apollo 11's emblem, Collins went through a set of *National Geographic*s to get ideas, and
sketched out a bald eagle with an olive branch in its mouth, about to land on
the Moon. "I was going in to see Bob [Gilruth] about some of the stuff we
were doing," flight control chief John Hodge said. "Somebody had just brought
in from the Art Department—you know we have these little badges for each
mission, and this one for Apollo 11, which was going to be kind of special. Of
course, the centerpiece of Apollo 11 was a bald eagle. He says, 'Here. What
do you think of this?' That's the way Bob was with things. If somebody just
brought it in, he threw it across the table. I said, 'You know, it looks preda-
tory.' And it was, because at that time what he had was—you know the way
an eagle lands, with his talons out. There was this thing about to attack the
moon." The olive branches were moved from the eagle's mouth to its talons.

After Collins had disassembled the LM's drogue apparatus, cleaning the
passageway between the two craft, Aldrin went to inspect *Eagle*, to see if it
had suffered any damage from launch. He encountered what he felt was the
strangest sensation of the entire mission. Since the ships were connected
head-to-head, when moving between the two in zero-g, he traveled right-side-
up in one and, immediately on traversing the connection, was upside down in
the other. The sensation was, once more, something like a scene out of *2001*:
the 180-degree stewardess, ascending a staircase. Buzz Aldrin: "Now you
crawl through the tunnel as you're going towards the Moon, you crawl through
the tunnel and go up through the ceiling, you go through the hatch, now all of
a sudden, bingo you're coming down through the ceiling. And everything is
twisted sideways, now to me that was exhilarating, maybe it was notable, it
was curiously different."

The now svelte 96,000-pound Apollo 11 sailed on through the hush of space.

The capsule itself was not so quiet, though, for Armstrong had brought along a cassette tape of one of his favorite records, Samuel Hoffman's sci-fi classic of zoomy theremin compositions, *Music Out of the Moon*. Every so often, the craft would have to swivel .3 degrees per second—one revolution per twenty minutes—in what was called "barbecue" or "rotisserie" mode—to keep the sun from boiling the fuel tanks, and the shade from freezing the radiators.

As they rotisseried across space, through the windows appeared, in series, the sun, the Moon, and the Earth, each in turn shining like the days, and fading like the nights. Otherwise, there was no sense of motion, even though they soared at six miles per second . . . except that, at times, a crewman would notice that the Earth seemed a little smaller than the last time he'd looked.

Both the Command and Service Modules had a skin like Sputnik's, a foil polished until it was as reflective as a mirror, and for the same reason—to be seen. The Soviet chief designer Korolyov wanted his satellite to be seen by Earth's telescopes; NASA wanted Collins and Armstrong to be able to see each other when the two craft reunited.

On reentry, the Command Module's polished foil skin would burn away as part of the heat shield, revealing a charred, orange carcass. But for now, Apollo 11, the head-to-head mating of *Eagle* and *Columbia*, was a tiny silver ship, floating upon a great black sea.

14

The Birth of the Moon

After leaving home orbit, Aldrin and Armstrong busied themselves with preparing for their landing, while Collins attended to the tasks of a working astronaut: purging fuel cells, throwing waste water overboard, checking the chlorine in the drinking water, recharging batteries, and replacing carbon dioxide filters. "A scuba diver uses a tank of air in sixty minutes; in Apollo an equivalent amount of oxygen lasted fifteen hours," George Low explained. "Oxygen was not simply inhaled once and then discarded: the exhaled gas was scrubbed to eliminate its CO_2, recycled, and reused. At the same time, its temperature was maintained at a comfortable level, moisture was removed, and odors were eliminated. That's not all: the same life-support system also maintained the cabin at the right pressure, provided hot and cold water, and a circulating coolant to keep all the electronic gear at the proper temperature. (In the weightless environment of space, there are no convective currents, and equipment must be cooled by means of circulating fluids.) Because astronauts' lives depended on this system, most of the functions were provided with redundancy—and yet the entire unit was not much bigger than a window air conditioner."

They were now, truly, on their way into history. One pilot said, "You don't pass anyplace on the way [to the Moon]. It makes it seem a little magical or mystical getting there," with the journey through space being "a blackness that is almost beyond conception," alternating with more starlight in the sky than sky. Apollo 14's Ed Mitchell said that outer space was "just as beautiful and strange as anything conjured by a child's imagination. . . . There is a sense of unreality here, with the absence of gravity and the tapestry of blackness broken only by an overwhelming glitter of stars that surrounded our craft."

If astronauts and cosmonauts are considered particularly brave, it is not just because they were the first to travel into outer space, but also because they willingly endured what could only be described as the world's worst camping trip—perhaps one reason why so many NASA moonwalkers were onetime Boy Scouts. To begin with, while space food technology was considered thrillingly modern in 1969, that was a view held only by those who didn't have to eat it. The situation was so dire that John Young and Gus Grissom snuck a corned beef sandwich aboard their March 23, 1965, Gemini III flight, resulting in a full-out congressional investigation and a public reprimand for this scandalous breach of flight rules.

On Apollo 11, ham, tuna, and chicken salad sandwich spreads were squeezed out of a tube (as was, for cosmonauts, *shchi*—fermented cabbage stew). Freeze-dried entrees (a process that combined flash-freezing with vacuuming to remove all moisture) of roast beef, ham, potatoes, yam, bacon, applesauce, vegetable medley, and hash were vacuum-sealed in plastic and needed to be rehydrated with three shots of hot water and kneaded into a mush, which was then, again, squeezed out like toothpaste. After finishing, the crew would insert germicide pills into the empty packs to keep bacteria (and its gas) from contaminating their hermetic environment. Aldrin became fond of the shrimp cocktail (but not the "gritty" chicken salad), gourmand Collins liked the cream of chicken soup and the salmon salad, and Armstrong reported his favorite outer-space meal as spaghetti with meat sauce, scalloped potatoes, pineapple fruitcake, and grape drink.

Apollo crewmen could also raid a snack pantry of dried pears, peaches, apricots, caramels, bacon bits, barbecued beef bits, and peanut cubes. Apollo 11 featured, for the first time on a NASA mission, not only meals you could eat with a spoon, but coffee—instant, and drunk through a straw (fifteen blacks for Armstrong, sweets for Collins, and regulars for Aldrin). During the run-up to launch, the men had taste-tested the childhood favorite Tang, but decided on a grapefruit-orange punch instead. If the cabin was depressurized and the men had to live in their suits, there was a hole in their helmets they could slip straws through and survive on a liquefied mash, what NASA called the "Contingency Feeding System."

Water came from two spigots—a hot for food and a cold for drinking—with six-foot tubes and pistol grips. If a squirt of water escaped into zero-g, "you produce a most amazing array of tiny jeweled spheres, each glistening like a crystal lens as they scatter in all directions," ISS science officer Don Pettit said. The water was a fuel-cell by-product, but Apollo 11's hydrogen gas filters didn't work well, making every drink so bubbly that Aldrin joked, "we could have shut down our altitude-control thrusters and done the job ourselves." That epidemic of gas inspired one repeated comment: that the bravest hero of an Apollo mission was the navy frogman who first opened the ship's hatch after splashdown.

If it took courage to eat in outer space, it took even more bravery to excrete. The first question that King George asked Charles Lindbergh at his royal audience was, "How did you pee?", and zero-g versions of the process had never been successfully addressed by any NASA engineering triumph. Defecating was so troublesome that agency doctors prescribed foods that produced as little waste as possible, and at least one astronaut spent his entire mission on Lomotil to avoid the procedure entirely.

For those who did submit to the urge, a variation on an airplane vomit bag was first unrolled, and its adhesive lips were secured to the skin. The bag had an indentation where a finger could be inserted to swipe the floating-in-air feces away from the body. "Nothing goes to the bottom of the bag in zero gravity," remembered one astronaut. "Everything floats. So you've got to get this off and make sure everything's still in the bag." After this was accomplished, another germicide pill had to be thoroughly mixed with the end results before storage.

Urinating meant sliding into a condom-shaped plastic sheath, which was attached with a tube to a vacuum. At just the right moment, a valve was turned to begin suction: doing this too soon, and discomfort would cause irritation; too late, and bubbles of urine would leak out to float across the cabin. Solid waste was kept for postmission analysis, but urine was released into space, where it immediately froze into a shower of glistening ice crystals. One astronaut reported that the most beautiful thing he saw during his epic voyage to the heavens was a "urine dump at sunset."

This waste elimination procedure was, however, a great advance over that employed on the previous Gemini system, the instructions for which illustrate a sensibility unique to engineers:

1. Uncoil collection/mixing bag from around selector valve.
2. Place penis against receiver inlet—check valve—and roll latex receiver onto penis.
3. Rotate selector valve knob (clockwise) to the "Urinate" position.
4. Urinate.
5. When urination is complete, turn selector valve knob to "Sample."
6. Roll off latex receiver and remove penis.
7. Obtain urine sample bag from stowage location.
8. Mark sample bag tag with required identification.
9. Place sample bag collar over selector valve sampler flange and turn collar 1/6 turn to stop position.

10. Knead collection/mixing bag to thoroughly mix urine and tracer chemical.
11. Rotate sample injector lever 90 degrees so that sample needle pierces sample bag rubber stopper.
12. Squeeze collection/mixing bag to transfer approximately 75 cc. of tracered urine into the sample bag.
13. Rotate the sample injector lever 90 degrees so as to retract the sample needle.
14. Remove filled urine sample bag from selector valve.
15. Stow filled urine sample bag.
16. Attach the CUVMS to the spacecraft overboard dump line by means of the quick disconnect.
17. Rotate selector valve knob to "Blow-Down" position.
18. Operate spacecraft overboard dump system.
19. Disconnect CUVMS from spacecraft overboard dump line at the quick disconnect.
20. Wrap collection/mixing bag around selector valve and stow CUVMS.

Despite all the inconvenience it presented for normal routines, though, most astronauts fell in love with zero-g. "Feeling weightless . . . it's . . . a feeling of pride, of healthy solitude, of dignified freedom from everything that's dirty, sticky," Wally Schirra rhapsodized. "You think well, you move well, without sweat, without difficulty, as if the biblical curse 'In the sweat of thy face and in sorrow' no longer exists. As if you've been born again. . . . You can love the earth with all the love in the world; returning is regret, is sorrow."

"Contrary to [weightlessness] being a problem, I think I have finally found the element in which I belong," John Glenn said. But weightlessness did come with physical side effects—including a feeling of "fullness," with too much blood staying in the head instead of being pulled by gravity back into the body—as well as cosmetic ones. Michael Collins: "With no gravity pulling down on the loose fatty tissue beneath their eyes, [Neil and Buzz] look squinty and decidedly Oriental. It makes Buzz look like a swollen-eyed allergic Oriental, and Neil like a very wily, sly one."

Zero-g slumber, however, was rest at its finest. After the window shades were drawn and the cabin lights dimmed, each man floated in a sleeping-bag-like mesh hammock, tethered to keep his body from knocking into a dashboard switch, or one of his shipmates. Sleepers were untouched by any sense of weight, lulled by the sweet lightness of being. Yet that night, perhaps due to their minds' being focused on the enormous task ahead, while Collins got seven hours of rest, Aldrin slept only five, and Armstrong, a bare three.

They awoke to a breakfast of sausage, cinnamon toast, fruit cocktail, co-coa, coffee, and punch.

CapCom: Is that music I hear in the background?
Collins: Buzz is singing.
Aldrin: Pass me the sausage, man.

GET 22:30: With one eye covered in an eyepatch to avoid the fatigue of endless squinting, Collins peered through the ship's sextant to sight the positions and attitudes of two stars, and entered the figures into the navigation computer, which automatically combined the results with its onboard data of the ship's speed and position to gimbal and fire the Service Module's rocket and send Apollo 11 onto its correct course. This same engine would slow to allow the ship to be grasped by the Moon's gravity and fall into lunar orbit, and then fire again to release the Moon's grip and head back home. When its fuels (hydrazine and unsymmetrical dimethyl hydrazine) and oxidizer (nitrogen tetroxide) met in the thrust chamber, they ignited on contact. Nearly every element of this engine was duplicated to counteract mechanical failure. If one rocket had to operate successfully on Apollo 11, no matter what, it was this one.

When NASA first started making plans for interplanetary travel, there was more research available on using celestial navigation to get to Mars than there was on going to the Moon, with so little knowledge on spaceship navigation in general that many at NASA fretted it might prove impossible. Administrator Jim Webb harassed the legendary Charles Stark Draper, the man in charge of designing Apollo's navigation software at MIT's Instrumentation Laboratory, to the point that an exasperated Draper finally wrote deputy director Seamans "to volunteer for service as a crew member on the Apollo mission to the moon. . . . let me know what application blanks I should fill out." If NASA didn't believe MIT was up to the task, its director would fly the rocket himself.

George Low: "If you had to single out one subsystem as being most important, most complex, and yet most demanding in performance and precision, it would be Guidance and Navigation. Its function: to guide Apollo across 250,000 miles of empty space; achieve a precise orbit around the Moon; land on its surface within a few yards of a predesignated spot; guide LM from the surface to a rendezvous in lunar orbit; guide the CM to hit the Earth's atmosphere within a 27-mile 'corridor' where the air was thick enough to capture the spacecraft, and yet thin enough so as not to burn it up; and finally land it close to a recovery ship in the middle of the Pacific Ocean. G&N consisted of a miniature computer with an incredible amount of information in its memory; an array of gyroscopes and accelerometers called the inertial-measurement unit; and a space sextant to enable the navigator to take star sightings. Together they determined precisely the spacecraft location between Earth and Moon, and how best to burn the engines to correct the ship's course or to land at the right spot on the Moon with a minimum expenditure of fuel. Precision was of utmost importance; there was no margin for error, and there were no reserves for a missed approach to the Moon."

The centerpiece of Guidance and Navigation was a Sputnik-sized inertial measuring unit, or IMU, which used three gyroscopes spinning at right angles to one another to create a stable reference in the three dimensions of outer space. "It's like a horizon in an airplane," RETRO Chuck Deiterich (who would be responsible for Apollo 11's rockets on the way home) said. "The vehicle flies around this thing, so it's a little eight ball that tells them what their attitude is, which way they're pointing. Well, the thing that defines how that thing is oriented in space with respect to the stars is called a REFSMMAT. That's a Reference Stable Member Matrix, a mathematical set of numbers that tell that platform how it's aligned with the North Pole, for example, and you have to have that in the computer to tell them what to fly with that little eight ball so they're in the right attitude for reentry or for anything else they do."

After the gyroscopes were correctly set, Collins dialed in the position of two reference stars. The navigation system compared the ship's position with the IMU, and fired its thrusters to set it correctly on course. While that might sound like a straightforward operation, depending on the amount of sunlight hitting the ship, it could be nearly impossible to recognize stars without their surrounding constellations, composed, as they are, of dim (and frequently invisible) companions. The Apollo sextant included a computer that tried to find the needed stars; Collins discovered he often had to use this feature.

Columbia was equipped with two 17.5-pound Raytheon computers, while its docked Lunar Module, *Eagle*, had one. With a memory of 36K, each was less powerful than a modern cell phone. The MIT programmers called their Command Module computer program Colossus IIA, and the Lunar Module's, Luminary. Mike Collins: "As it was to flash crazy lights in Buzz's face all the way to the lunar surface, I guess it was aptly named." Guidance software support leader Jack Garman: "The computer interface in those days was a device called the DSKY, display and keyboard, and the paradigm used was verbs and nouns. There was actually a *V* key for verb and an *N* key for nouns. So maybe a verb would be displayed, and then a noun would be what you were displaying. No letters, of course. It's all numbers. But with two-digit verbs and two-digit nouns and there was, I think it was three five-character displays, that's it, three, because in navigation you were living in a three-axis world. So like velocity, there'd be X, Y, Z. There's always three of everything."

GET 26:46: The ship had now reached the outer edge of Earth's gravity, which slowed it to 3,400 miles per hour. From here on, lunar gravity would pull it faster and faster toward its target.

As the men of Apollo 11 would now discover, at a certain point, you can't see the Moon when you fly to the Moon. On the flight path, it is too forward to be seen out the windows, or it lies so close to the sun that solar glare blots it out. On day four, Collins turned *Columbia* so that the men could have a view of their destination—now 200,000 miles away from where it was located at launch—and suddenly, there it was, looming and enormous, the surface blue-white from the reflected shine of the Earth, which, with its clouds and water, reflects four times as much sunlight as the full moon. To the arriving astronauts, their target also seemed illuminated distinctly three-dimensional by the solar eclipse directly behind it, with the Moon's perimeter a ring of sunlight fractured by lunar craters, sparkling like a diamond necklace, while the great star's corona, invisible to human eyes except during such a blackout, cast five-million-mile-high spiking bands of red-neon gases into lunar halos. At the same time, the Apollo 11 crew discovered they could also see the stars again, in fact, more stars in the sky than sky, untold billions shining against the blackest of nights.

GET 34:00: For their first global television broadcast, Michael Collins showed viewers watching at home how he could use a flashlight hanging in midair to help him look through the food packs, and then pointed the camera out the window while Neil Armstrong gave a geography lesson. The wives became more and more impatient; they wanted to see their husbands, not Greenland. Finally Mike appeared fully on-screen, shocking and dismaying Pat: "He's growing a mustache!"

GET 55:10: The guys were supposed to start their second TV relay at 56:20, but decided to back it up an hour. No one told their families, who were spending the afternoon at Joan Aldrin's Nassau Bay home for a pool party, and who missed the broadcast entirely.

While their kids splashed and screamed in the pool, the wives strategized about how to deal with the press, compared notes about how their children were reacting, and worried about putting on weight. Pat Collins had been deluged with generous offerings from friends, neighbors, and even unknown admirers—deviled eggs, shrimp casserole, multiple cakes, lasagna, assorted cheeses, bean salad, paté, a case of Champagne, and thirty pounds of ice— which led her to remark, "It reminds me of a wake."

The wives worried about how their young children were being affected by the mission, but none of the astronauts' kids seem to recall anything particularly negative about this time in their lives. Buzz's son, Andrew, said, "In the community I lived in when I was a kid, everybody's dad was an astronaut. Three of the five houses adjacent to us were all astronauts, so it was no big deal. I thought my dad was cool because he could pole-vault. . . . When it got to the mission, it was the crowd of press in our front yard that interested me, because these guys had doughnuts and wanted to play football."

"I didn't know much about my father's job, to be honest with you," remembered Frank Borman's son, Edwin. "I knew there were dangers and risks, but they were always training, and they were always upbeat, and they never showed their fears."

Is there a father alive who doesn't want to be a hero to his children? That would seem to come automatically with the job of NASA pilot but in this life, nothing is certain. Astronaut Dick Gordon tells a story: "When my eldest son, Rick, was eight or nine years old in 1965, [he asked a boy at school], 'What does your dad do?' The other kid said 'He's a sheriff,' and Rick went, 'Wow, he's a sheriff? Does he wear a badge? Does he wear a gun?' and got all excited. Then the other kid asked Rick what his did and he replied, 'Aw, he's just an astronaut.'"

"One Saturday morning we were having breakfast at a long table we had," Gemini IV/Apollo 9's Jim McDivitt said. "And so, we finally got to this dramatic moment and I said, 'Kids, I'm going to tell you something really important. You know that Dad's an astronaut and the astronauts fly in space. I just want to let you know that I'm going to fly in space soon.' And my older boy, Mike, who was probably seven or eight, says, 'Oh yeah, Dad, I heard that at school.' And then my daughter Ann said, 'Oh yeah, Dad, I heard that at school, too.' And my son Patrick said, 'Dad, there's a fly in the milk bottle.'"

Instead of watching his father on television, Mike Collins, Jr., only wanted to play with his pet bunny, Snowball. When someone asked what he thought of his father's making history, Mike said, "Fine. . . . What is history, anyway?" Mike Aldrin, meanwhile, was more excited about getting an autographed picture from *Gunsmoke*'s "Doc" than he was in following his own dad's mission. After paying little attention to almost everything connected with Apollo 11, however, Marky Armstrong looked up from what he was doing at one point, realized that it was his dad who was on the TV screen, and ran over to hug it.

Buzz Aldrin: "I remember one day picking up a copy of *Life* magazine with a story on us in it . . . I remember reading the story and thinking, 'if only it was like that.' Here were all the happy contented wives and children smiling out from happy backyards with husbands standing proudly by. Well, the fact is that the husband probably flew halfway across the country to pose for the picture, the kids were half-strangers to him, and the wife was scared to death about any number of things. . . . My kids had been forced to reconcile . . . the father they saw on television with the one they saw at home . . . often inattentive, tired, and asleep on the den sofa by nine o'clock."

On almost every night of the mission, meanwhile, the nightly news reported that yet another couple had named their baby Apollo. Memphis, Tennessee's Eddie Lee McGhee took a slightly different path and blessed his baby girl with the name Module.

GET 61:40: Apollo 11 was now 33,822 nautical miles away from the Moon. Since the Earth is eighty-one times the size of its satellite, the two bodies' neutral gravity point is about 90 percent of the distance between the two. From that point on, the ship could rely on the Moon's pull to help with its mileage.

GET 75:32: As the Moon travels at a speed of 2,287 miles per hour, it takes great precision to get from Florida to lunar orbit. NASA decided to hedge its trajectory procedure by using two rocket firings. In *Columbia*, Collins would eventually need to achieve a circular orbit in order to successfully dock with Armstrong and Aldrin in *Eagle* when they returned from from the lunar surface, but when Apollos 8 and 10 had tried this maneuver, instead of circular sixty-nautical-mile orbits, they fell into ellipses of between fifty-four and sixty-six nautical miles over the Moon. Agency engineers had not been able to overcome formula problems with the Moon's mascons, or concentrations of matter, which make its gravitational pull (and NASA's trajectory calculations) erratic.

Apollo 11's first burn in its lunar approach lasted six minutes and two seconds and slowed the craft to 2,917 feet per second, allowing it to fall into the pull of the Moon's gravity. The gravitational warp in space time then threw it like a slingshot to the dark side, where radio contact with the Earth was not possible. There, the crew would have to decide if they could use the second burn to enter a sixty-mile-altitude lunar orbit and continue their mission as planned, or abort out of lunar gravity and use the Earth's draw to return home. Until radio contact resumed, Mission Control would not be aware if all systems checked out and Apollo 11 had entered lunar orbit, or if the procedure had, once again, failed, and the mission had to be aborted.

Jan Armstrong synchronized her watch with NASA time so she would know immediately what the crew's decision had been. If AOS (acquisition of signal) came around 12:40 p.m. CST, it would mean the men had not fired the retrorockets for the six-minute period needed to slow the craft to 2,000 miles per hour to enter orbit; they'd decided they had to give up and come home. Waiting for her squawk box to burst with the news, Jan mental-telepathied luck to her husband: *Don't you dare come around. Don't you dare!*

Losing radio contact was so unnerving to everyone back home that a full complement of the brotherhood was standing by at Mission Control: Anders, Lovell, Haise, Glenn, Cernan, Scott, Worden, Swigert, and Stafford. The world followed the announcements by public affairs officer Jack Riley:

> And we've had loss of signal as Apollo 11 goes behind the moon. We were showing a distance to the moon of 309 nautical miles at LOS, velocity 7,664 feet per second. Weight was 96,012 pounds. We're 7 minutes 45 seconds away from the Lunar Orbit Insertion number 1 burn, which will take place behind the moon out of communications. . . . With a good Lunar Orbit Insertion burn the Madrid station should acquire Apollo 11 at 76 hours 15 minutes 29 seconds. . . .
>
> We're 24 and one-half minutes away from acquisition of signal with a good burn. . . . We are past the burn acquisition now and we have received no signal. . . . It's very quiet here in the Control Room. Most of the controllers seated at their consoles, a few standing up, but very quiet. . . .
>
> We are 4 minutes away now. . . . There are a few conversations taking place here in the Control Room, but not very many. Most of the people are

waiting quietly, watching and listening. Not talking. . . . That noise is just bringing up the system. We have not acquired a signal.

We're a minute and one-half away from acquisition time. . . . 30 seconds . . . Madrid AOS! Madrid AOS!

At almost exactly 12:50, Madrid's radio dish had caught the craft's signal, and everyone now knew: they were "Go" for lunar orbit.

Collins: The accuracy of the overall system is phenomenal: out of a total of nearly three thousand feet per second, we have velocity errors in our body axis coordinate system of only one tenth of one foot per second in each of the three directions.

Armstrong: That was a beautiful burn.

Collins: Goddamn, I guess . . . I don't know if we're at sixty miles or not, but at least we haven't hit that mother.

Aldrin [keying in the computer for its numbers]: Look at that, 169.6 by 60.9. [They were looking to get 170 by 62.]

Collins: Beautiful, beautiful, beautiful, beautiful!

Armstrong: Now the flight plan says we roll 180 degrees and pitch down 70 degrees.

Collins: Why are we pitching down for? I don't know what we're doing.

Armstrong: We're going to roll over and pitch down so that we can look out the front windows down at the moon!

Aldrin [looking out the window]: I have to vote with the Apollo 10 crew that the surface is brown.

Collins: It sure is.

Armstrong: It looks tan to me.

Houston now asked if they could identify their landing path.

Armstrong: The pictures and maps brought back by Apollos 8 and 10 have given us a very good preview of what to look at here—it looks very much like the pictures, but like the difference between watching a real football game and watching it on TV. There's no substitute for actually being here.

Collins [to Armstrong]: There go Sidewinder and Diamondback. God, if you ever saw checkpoints in your life, those are it.

Armstrong: But we don't get to see them.

Collins: You don't?

Armstrong: No, we're yawed face-up.

Their arrival coincided with a lunar sunrise (the speed of their orbit meant each lunar "day" was all of two hours), and the strong light threw a harsh contrast of glare and shadow across the terrain.

Aldrin: Boy that sure is eerie looking.
Armstrong: Isn't that something?

On arrival, many Apollo fliers found their first close encounter with the Moon disturbing—one saying, "It was a totally different moon than any moon I'd ever seen before. It was in this eerie shadow, no motion, utterly silent, it gave one a feeling of foreboding and didn't seem a very friendly or welcoming place."

"There was this big black void, black hole," Bill Anders said. "And that was the moon. That was the moon shielding the stars and yet not illuminated. It was as black as I've ever seen black. So that was the only time in the flight the hair kind of came up on the back of my neck a little bit." Neil Armstrong, however, observed, "I was sure that it would be a hospitable host. It had been awaiting its first visitors for a long time."

Mike Collins said of that moment, "I feel that all of us are aware that the honeymoon is over and we are about to lay our little pink bodies on the line. . . . We have not been able to see the Moon for nearly a day now, and the change is electrifying. . . . The Moon changes character as the angle of sunlight striking its surface changes. At very low Sun angles close to the terminator at dawn or dusk, it has the harsh, forbidding characteristics which you see in a lot of the photographs. On the other hand when the Sun is more closely overhead, the midday situation, the Moon takes on more of a brown color. It becomes almost a rosy-looking place—a fairly friendly place so that from dawn through midday through dusk you run the whole gamut. It starts off very forbidding, becomes friendly and then becomes forbidding again as the Sun disappears."

From 11's windows, the satellite's dark side looked especially impressive, and hazardous.

Armstrong: What a spectacular view!
Collins: Fantastic. Look back there behind us, sure looks like a gigantic crater; look at the mountains going around it. My gosh, they're monsters. . . . You could spend a lifetime just geologizing that one crater alone, you know that?
Armstrong: You could.
Aldrin: There's a big mother over here, too!
Collins: Come on now, Buzz, don't refer to them as big mothers; give them some scientific name.

Aldrin: It sure looks like a lot of them have slumped down.

Collins: A slumping big mother? Well, you see those every once in a while.

Astronomers over the centuries had reported various and mysterious lunar glows, perhaps from volcanoes, and 11's current orbit would take it near one site notable in that respect: Aristarchus crater. After CapCom Bruce McCandless asked them to have a look and report what they saw, Collins confirmed that "there's an area that is considerably more illuminated than the surrounding area," and Aldrin guessed, "It does seem to be reflecting some of the earthshine"—which in time would be proved to be the case; a part of the crater had high albedo, or reflectivity. When McCandless asked the crew to look for more details through the monocular, Armstrong shut down the experiment, telling Houston: "We'll give it a try if we have the opportunity next time around. We're in the middle of lunch."

That day, Jan Armstrong and her synchronized swimming team made seventy dollars from a lemonade stand serving the press on her lawn for their trip to a conference in Toledo. Later, when Pat Collins went to have her hair done, she found three journalists in the waiting area, each with an appointment. Besides their bravery in the face of their husbands' mission and their staunch fortitude in the face of a media assault, one little-known and remarkable aspect of NASA wives was their determination to understand what, exactly, their husbands did for a living. Jan Armstrong spent three evenings with astronaut Dave Scott meticulously going over a copy of the flight plan so he could explain every element that, after hours of study, she hadn't been able to grasp on her own.

On Saturday, July 19, the ship's cameras began picking up detailed photographs of the lunar surface.

Houston: We are getting a beautiful picture of Langrenus [crater] now with its rather conspicuous central peak.

Collins: The Sea of Fertility doesn't look very fertile to me. I don't know who named it.

Armstrong: Well, it may have been named by a gentleman whom this crater was named after. Langrenus was a cartographer to the king of Spain and made one of the early reasonably accurate maps of the moon.

CapCom: Roger, that is very interesting.

Armstrong: At least it sounds better for our purposes than the Sea of Crises.

Jan Armstrong laughed out loud hearing this: "So that's what he was doing with the *World Book* in his study. . . ."

Lunar cartography began over three hundred years ago, with Galileo's drawings of the terrain he saw through his telescope. Then in 1651, Jesuit astronomer Giovanni Battista Riccioli attempted to combat Galileo's and Copernicus's cosmic heresies with *Almagestum Novum* ("The New Almagest"), which included a detailed lunar map that established the distinction between mountains (highlands, of white anorthosite) and plains (maria, seas of black dust), and introduced many names still in use today, including Apollo 11's landing site: the Sea of Tranquility, Mare Tranquilitatis.

Riccioli also began the tradition of naming craters and peaks for astronomers and philosophers, generously including his ecclesiastical foes Copernicus, Kepler, and Galileo, alongside countless obscure Jesuits. (In honor of his family, Collins named a crater in the Foaming Sea KAMP, for Kate, Ann, Michael, and Pat.)

Before Apollo, scientists argued for millennia over whether the Moon's topography was the result of meteor strikes or volcanic eruptions. After Apollo, we know that the answer is both. Early in its life, the Moon was bombarded by so many meteors that its surface effectively melted, with the biggest strikes crashing out enormous basins. Half a billion years later, after the Moon had cooled, volcanic eruptions threw lava across its surface, smoothing those once-vast impact craters into "oceans" or maria.

Though a quarter of Earth's diameter and an eightieth of its mass, the Moon is so large—of 150 moons in the solar system, ours is the largest in relation to its host—that many believe it should be properly considered a planet, and that together we form a double-planet system. The lunar mass pulls a dimple on the surface of the Earth which, as it spins, creates twice-daily high tides. It also both slows our rotation infinitesimally and stabilizes that rotation—otherwise, the poles would turn erratically on the Earth's axis—providing for consistent agricultural seasons. A similar effect of the Earth on the Moon caused it to stop rotating entirely.

We think of the Moon as white, or silver, when it is actually various shades of black, charcoal, and asphalt; all of the lunar dust returned by Apollo 11 would be repeatedly described with one phrase: "black as coal." The "hostility" that many astronauts felt upon their first close-up view is borne out by surface conditions: with temperatures of just under four hundred kelvins in sunlight (260° F, 127° C) and minus one hundred kelvins in darkness (−279° F, −173° C), a lunar day is 327.5 hours of darkness followed by 327.5 hours of light (fourteen earth days), and two lunar days make a lunar year.

"It's hard to wrap your mind around a place where nothing ever happens," NASA astromaterials curator Carlton Allen said. "But the Moon is that place." For all of human history, the lunar surface has remained the same, and this condition reminds us how alive the Earth really is in the weather of its atmosphere, the tides of its seas, the seasons of its tilting axes, and the geology of its volcanoes, tectonic plates, and erosion.

We think of the Earth and Moon as round, but the Earth is somewhat pear-shaped, while the Moon is something like a vertical football. We think of their pairing as eternal, but every year, the Moon moves 1.5 inches away from the Earth. The crust on the Moon's near side, which always faces the Earth, is much thinner than the far side, which has far fewer plains of maria and far more craters. It is now believed that water ice may exist in deep craters at the lunar poles, and NASA will one day send a reconnaissance drone into lunar orbit to see if this is so; in 2008, scientists at Brown and Carnegie Institution used secondary ion mass spectrometry on the volcanic glass beads brought back by Apollo to discover evidence of water deep within the Moon itself. The seismometers left by Apollo missions have counted over one hundred moonquakes in the 1970s alone, several a day occurring halfway to the core, 750 miles down. The Moon may appear dead, but to scientists, even after forty years of research based on Apollo, "quite a lot of the darned thing is still quite mysterious," as Canadian astronomer Kimmo Innanen summed up.

Before Apollo, there were three theories accounting for the birth of the Moon. One, a notion called fission theory, developed in 1878 by George Darwin, son of Charles, and England's preeminent expert on tides, held that it was originally part of the Earth, and was torn away somehow. Another, Thomas Jefferson Jackson See's 1909 "capture" idea, argued that the Moon originated elsewhere, and was caught by Earth's gravity. A third, Édouard Roche's 1873 coaccretion model, proposed that the Earth and its Moon were formed at the same time and are like twin planets, forever conjoined. This last idea held particular sway, as it is not exactly true that the Moon orbits around the Earth, but more accurately, that they are two bodies orbiting each other around a point within the Earth, 2,900 miles from its center, the point that marks their joint-path ellipses around the Sun.

With Apollo, everything changed. The lunar samples brought back by missions 11 to 17 revealed that the Moon had the same isotope ratios as Earth. This discovery inspired William Hartmann in 1974 to offer a Darwin variant, the Giant Impact Theory: Four and a half billion years ago, the Earth was molten, and without form. A heavenly body, as big as Mars, smashed into our planet. The collision ejected pieces of molten rock which, captured by Earth's gravity, formed into a ring of debris, spinning around the globe, much like Saturn's. Time and the force of Earth's gravity then coalesced this ring of liquid rock and stone into one beautiful creation—the Moon. At a 1984 lunar conference in Hawaii, the majority of selenologists accepted Giant Impact as the default theory explaining the Moon's origins.

15

The *Eagle* Has Wings

July 20, 1969, the day of Apollo 11's descent to the lunar surface, was a Sunday in America. At the White House, Frank Borman recited the Bible passages everyone had heard the previous Christmas Eve aboard Apollo 8. At Buzz Aldrin's Webster Presbyterian Church in Houston, the service was so overwhelmed by prayerful attendants that rows of folding chairs had to be added to the back and sides of the congregation. Still, the atmosphere was reserved, as everyone knew the success of this mission was not preordained.

The morning began with a bad omen. Sim Sup Dick Koos—the man who had devised the test that would eventually save the flight—had just bought himself a beautiful Triumph TR3. On his way to Clear Lake City he flipped the car over, but was able to brush himself off and come in to work anyway. Kranz honored him with a seat at the Trench.

Mission Control woke the crew with a tip from the Chinese, "asking that you watch out for a lovely girl with a big rabbit. An ancient legend says a beautiful Chinese girl called Chang-O has been living there for four thousand years. It seems she was banished to the moon because she stole the pill of immortality from her husband. You might also look for her companion, a large Chinese rabbit, who is easy to spot since he is always standing on his hind feet in the shade of a cinnamon tree." The CapComs passed along the morning's news, which included a *Pravda* reference to Armstrong as the "Czar of the ship." Mike Collins thought it would be amusing to use this as the boss's new title, but gave it up after a couple of mentions when Armstrong's pronounced lack of amusement became clear.

CapComs Haise and McCandless: "And in Italy, Pope Paul VI has arranged for a special color television circuit in his summer residence in order to watch you—that's even though Italian television is still black and white."

After thirteen orbits around the Moon, Armstrong and Aldrin began day five of their flight by dressing in their lunar EVA suits, which included the diaper and waist-bag urinal system, the liquid cooled garments (white cotton long johns laced and water-chilled by a filigree of brown capillary tubes connected to their backpacks' circulation pumps, a necessity as sweat does not cool human bodies in a vacuum), and their pressure suits, made of rubber and glue, similar to those worn by scuba divers, but sealed and pressurized with oxygen from their backpack. Both the oxygen flow, and the amount of undergarment cooling, could be controlled by a chest-mounted remote control. They then zipped each other into their twentieth-century suits of armor—the white, thirteen-layered, Mylar-and-Teflon-coated beta-cloth Integrated Thermal Meteoroid Garments, or space suits. Their Snoopy communications skullcaps were also connected to radio transmitters and receivers in the backpacks, as were body sensors ferrying medical data back to Mission Control. Because of the terrible problems Gene Cernan had encountered while spacewalking, they sprayed their Lexan polycarbonate helmets with defogger to keep from being blinded. There was a snack bar included inside the helmet, as well as a bag of water draped around the neck and linked to a straw for drinking. Collins had to suit up as well, since, if there was an emergency and the docking procedure failed, he might have to rescue his teammates by maneuvering *Columbia* and *Eagle* side by side and opening both hatches—the sole reason, besides matching Soviet achievements, that Apollo astronauts had been trained in spacewalking.

Buzz Aldrin and Neil Armstrong then took their positions in *Eagle*. The LM (pronounced *lem*) was a peculiar craft. Pete Conrad described it as "an ugly and unearthly bug, with windows for eyes, a front hatch for a mouth and antennae jutting from its shiny skin. It has the innards of a flying machine—but an aerodynamic shell would be useless on the airless moon and therefore has been dispensed with. Each piece of gear has simply been put in the right place to the right job—aesthetics be hanged."

"The complexity of [the other lunar landing designs] kept making the Lunar Module look more and more attractive," Max Faget said. "So we finally gave it a good hard look and [it] solved all our problems. Not only did it solve the problem of being able to get there with one launch of the Saturn, but it solved problems that we didn't think about before. It made one spacecraft that would be nothing but a space flight, and we wouldn't have to operate [except] in space. The LM didn't have to have any heat protection, no aerodynamic considerations at all in the design of the LM. It only had to be designed to do one thing, and that is land on the moon, on a planet that had one-sixth the gravity of our planet. That's what it had to do."

From the outside, the LM looked exactly like a T4 bacteriophage virus; it

was, after all, a machine engineered to attach itself to a foreign body and release its alien DNA—Armstrong and Aldrin. Just under twenty-three feet
high and just over fourteen feet wide, with 18,000 pounds of fuel to land,
5,200 to return to *Columbia*, and 600 to maneuver, *Eagle* was outfitted with
thirty miles of wire, eighteen rockets, legs that extended via explosives, eight
kinds of radio, and two kinds of radar.

At each of its corners was a set of peroxide thrusters, their bells a match
for the collapsible plastic cups used on camping trips by Boy Scouts. There
were no seats or couches inside, since no one reclined in zero-g, and forgo-
ing two hundred-pound couches meant a fuel savings of 6,500 pounds. Instead, *Eagle* was flown standing up, like a trolley, with a dashboard stick
that controlled yaw, pitch, and roll, by crewmen with boots Velcroed to the
carpet, their waists tethered to the walls. The LM's nickel-steel skin was as
thick as three sheets of aluminum foil—.0000833 inches. In flight, however, with its façade covered in black, orange, silver, yellow, red, and gold
aluminized-plastic insulation to shield against solar rays, *Eagle* shone like a
star. In keeping with NASA's dinghy paradigm, the ship would be used just
once, to land on the Moon and then return to dock with *Columbia*, its poste-
rior (with the descent engine) left on the lunar surface, and its main cabin
(with its ascent engine) eventually released to drift in space.

After his colleagues were settled in their moon ship, Collins maneuvered
across the Foaming Sea, rebuilt the connecting tunnel's probe and drogue
assembly, disassembled the electrical connections tying the two ships together, and was supposed to set up a camera, but instead told Houston:
"There will be no television of the undocking. I have all available windows
either full of heads or cameras, and I'm busy with other things." Mission
Control could tell by his tone that this was not up for negotiation, and replied, "We concur."

The three then ran a series of tests to check, once again, *Eagle*'s systems.

Collins: So we're about 3 minutes ahead of the printed flight plan; it might be
   wise to try to SEP about 3 minutes early, and we can give them a GET of
   SEP that's precise whenever they want it. . . . I have 5 minutes and 15 sec-
   onds since we started. Attitude is holding very well.
Aldrin: Roger, Mike. Just hold it a little bit longer.
Collins: No sweat, I can hold it all day. Take your sweet time. How's the Czar
   over there? He's so quiet.
Armstrong: Just hanging on—and punching [computer buttons].
Collins: All I can say is, "Beware the revolution."
Armstrong: [No response.]

Collins: You cats take it easy on the lunar surface; if I hear you huffing and
   puffing, I'm going to start bitching at you.
Aldrin: Okay, Mike.

GET 100:39: At 1:47 p.m. CST on July 20, during their thirteenth revolution of the Moon, Collins fired explosives that separated his *Columbia*
from Armstrong and Aldrin's *Eagle*. Armstrong then swiveled the LM around
so that Collins could confirm, through the windows, that its rockets had
fired and its legs had telescoped correctly. At the same time, Aldrin and Armstrong could see, for the first time during the flight, their mother ship's
polished-mirror skin, and its black nozzle tail.

This procedure took place on the lunar far side, meaning everyone back
home on the ground had to wait once again for AOS to know if the maneuver
had been a success, or if the mission would now be aborted. CapCom Charlie Duke repeatedly murmured into his headset, *"Eagle,* Houston. We're
standing by. . . . *Eagle,* Houston. We're standing by. . . ."

Finally, there came a response.

Armstrong: "The *Eagle* has wings."

The whole of Mission Control seemed to take a giant breath, like one living creature. There was a pause, and then shouts of congratulations and
whoops of success.

Collins [to Armstrong]: You've got a fine-looking flying machine there, *Eagle,*
   despite the fact you're upside down.
Armstrong: Somebody's upside down.
Collins: OK, *Eagle* . . . you guys take care.
Armstrong: See you later.

Armstrong's mother, Viola: "The tense period of the separation of the LM
from *Columbia,* when Buzz and Neil left Mike, really pulled on my heart-
strings. We felt sorry for him because he was alone and could not go along
with the other two. We were so proud of him because he was so faithful."

Though a mere sixty miles from the Sea of Tranquility, Michael Collins
would be one of the few human beings unable to witness history's first Moon
landing.

Joan Aldrin said to a guest in her home: "It's like a dramatic television
show, but it seems unreal. There are just no words. Don't you agree?"

What no one realized at that moment—not any of the crew, nor anyone on
the ground—was that a tiny puff of air in the separation procedure pushed

Columbia and *Eagle* farther (and faster) apart than NASA had calculated. "We didn't know this until after the mission, but the crew had not fully depressed the tunnel between the two spacecrafts," Gene Kranz said. "They should have gone down to a vacuum in there, and they weren't. So when they blew the bolts, when they released the latches between the spacecraft, there was a little residual air in there, sort of like popping a cork on a bottle. It gave us velocity separating these two spacecrafts."

This tiny puff of unvented air added incrementally to *Eagle*'s speed, throwing NASA's carefully considered landing plan into disarray. Instead of a computer-controlled touchdown, Armstrong would have to land on the Moon himself. And there were other problems with the process, many that would not be fully understood until it was too late. "On Apollo 11, we wound up five miles off target because of the navigation errors coming around the Moon, and we didn't mathematically model the Moon all that well," mission planning and analysis division chief Floyd Bennett said. "And there was some— we've assumed it to be homogenous gravity, and it wasn't. It had what they call mass concentration, or mascons. So that caused us to be off when we started in the descent, and subsequently we were; we were off at the landing. Armstrong wound up over a boulder field."

16

One of Those Sad Days When You
Lose a Machine

Just as astronauts competed aggressively for crew slots, so, too, did mission ground teams. The winner for Flight of Apollo 11's lunar landing was Gene Kranz, the very essence of a no-nonsense man, an emotional bear of a guy whose feelings were never in question, a man who, before ulcers were known to be caused by bacteria, drank endless cartons of milk to calm his churning stomach, and a person who held a profound appreciation of the effort that had led to this moment in time: "Everything we needed to go to the Moon with, we had to create—and this joy of creation was a marvel to behold," he said. Even so, Kranz's nickname was "General Savage"; during missions, he was noted for wearing white and silver-brocade vests sewn by his wife, Marta, and beginning each day with the rousing marches of John Philip Sousa.

Gene Kranz: "I've got probably thirty or forty records, tapes. Every time you see an opportunity and the kids want to get anything for Christmas or Father's Day or any special event, it's always another—a new march record. At this time also we had eight-track recorders, so I had them in the car. Every place I'd go, I'd have John Philip Sousa. And this is the way I get up to speed, get the energy, get the adrenaline flowing." Kranz replaced Mission Control chief Christopher Columbus Kraft in this slot, which Kraft described as: "I am Flight and Flight is God."

Actually, as Apollo 11 would prove, Flight was one god among many. Gene Kranz:

> The day that Cliff Charlesworth came into the office and said, "You're going to do the lunar landing," I just ricocheted around the office virtually all day, and I don't think my secretaries ever saw me as happy. I had the white team. Cliff Charlesworth [who had the most Saturn experience and so oversaw launch] had green. [Glynn] Lunney had black [which would oversee lunar liftoff and rendezvous], Gerry Griffin had gold. . . .

I always find that when time came for a mission, when time came to do something, there was just an incredible degree of just solitude. You just felt so comfortable. I'm trying to find the right word for this thing here, but basically it's just peace. You're at total peace with yourself, and when you reach this total peace, you're ready to go. It's interesting. The adrenaline's pumping, but you have this incredible confidence in your team and in yourself as a result of this training Sim Sup's given you. He's given you your confidence. It's sort of like the first time you solo an airplane. It's the same doggoned feeling. This instructor has given you this absolute confidence in your ability to get the job done. You never think about, "If I get airborne, I've got to get back to ground. I may crash." It doesn't come across that way. It's just, "Certainly." It's just peace in the business. I used to fly early supersonic fighters, and you'd get out to the end of the runway, and, boy, when you're cruising down the runway at 250 miles an hour and something goes wrong, you've just got to have confidence that you're going to be able to pull it off. . . .

Down in the trench I've got Jay Greene, who's a Brooklyner, and he's got the Brooklyn accent that, I mean, it almost drenches you with this thing here. You feel you're walking the streets in the Northeast. And cocky. He is so cocky, it's incredible. He and I were the ones that were crashing spacecrafts on the Moon. Sitting right next to him, on his left, is Chuck Deiterich, who's a Texan, great big brush mustache, right on down the line with the drawl. Chuck is RETRO [retrofire officer]. He's also the guy that prepares a bunch of messages for them. To Greene's right is Steve Bales. He's one of the original computer nerds. I mean, he looks like one. He's got these big owlish-type plastic-rimmed glasses you got in there. I don't think any of them—they all look like they never needed to shave. I mean, they're baby-faced kind of people.

CapCom is Charlie Duke, and he's probably the best of the best of the best from a standpoint of the astronauts. He was personally requested by Neil Armstrong to be the CapCom for this mission, and you've got to respect Armstrong [Duke, famous for his Carolina drawl, also had an M.S. from MIT in aeronautics]. You've got to respect [Deke] Slayton, because Slayton also has to concur on this thing here. And Charlie Duke was just absolutely a master of timing. It seemed when we were in the pits, Duke always had the right words to say to just pull this team up and convince this team that we'd get it together.

Kranz, like the great majority of NASA employees, had been working such hard and long hours that he rarely had time to watch television or even read the paper, and was only vaguely aware, if at all, of the tumultuous events of the 1960s whirling outside the Texas and Florida compounds. Thanks to this insularity and its military heritage, despite its state-of-the-art technology,

NASA during the 1960s often seemed like a 1950s television drama that had never been canceled, starring crew-cut engineers, sundressed astronaut wives, and *Life* magazine encomiums. But such a characterization would be a caricature, for even General Savage himself, Gene Kranz, would by the launch of Apollo 11 exchange the Sousa eight-track tapes in his car for *Hair*'s "Aquarius." Ground-pounding engineers, whose average age at the time of Apollo 11 was twenty-six, were as distinct from one another as were the astronauts, ranging in style from USAF Minuteman officers, to slide-rule-wielding genius introverts, to those who, in their polo shirts and Joe Namath haircuts, were as stylish as any flier. All of them, however, were familiar with the essential tools of the trade during a mission: nail-biting, desk-drumming, and pencil-snapping.

NASA's Clear Lake City, Texas, facility, known as the Manned Spacecraft Center and then the Johnson Space Center, looks like nothing so much as a wildly successful community college, its well-tended lawned campus hosting midcentury-modern rectangular office buildings of black windows and calcium-white exteriors made from crushed Gulf of Mexico seashells instead of common sand . . . buildings made from the sea, to go to the Moon. Building 30, at the heart of the enterprise, has two wings. One, for mission planning, analysis, flight control, and support, has windows overlooking golf course–style ponds that are home to ducks and swans. The other has no windows whatsoever, for it held two auditoriums surrounded by tiny offices with cathode ray tubes. These were the MOCRs, Mission Operations Control Rooms, where Mission Control did its work, modern caves of industrious silence, blinding fluorescence, and the reverberant odors of pizza, smoldering ashtrays, burned-to-the-bowl coffee, and greasy Mexican takeout.

Four ascending rows of desk consoles—a Greek amphitheater for technowizards—featured tiny desktop TVs refreshing, once per second, data streams from the building's computers, as well as from the DSKYs aboard both modules of Apollo 11. "We didn't have computers on the console, because there weren't computers that would fit on the console," flight dynamics officer Jay Greene explained. "The computers [two green Univacs and five blue IBMs] were in the ground floor of Building 30, and in order for us to make inputs to the computer, we'd have to call down to a guy who would do the actual typing for us."

"[The men and women of Mission Control] were in their twenties," Jack Schmitt said. "They just came out of engineering school. They had just been hired from Rensselaer and Tulane and LSU and VPI and places like that [and they] really believed that they were doing the most important thing they were ever going to do with their lives. And it wasn't to beat—initially, it may have been to beat the Russians to the Moon—we sort of sensed that over here. And that everybody believed that this was something we ought to do. We ought to go to the Moon. And that's why they worked those sixteen-hour days and eight-day weeks. . . .

"Those people were the reason that you could get almost anything done. There was never a paucity of ideas. Imagination was rampant, and most of it very good imagination on how to solve problems. And a group of people could get around a table, work together, and in a noncompetitive—it seemed non-competitive, at least at the time—and the sum of the output of that table was far greater than just the individual parts that were there. It was really an exciting time to be involved. And that's why Apollo 13 was saved. That's why Apollo 11 landed at the time it did. It's really why any of the in-flight emergencies were dealt with successfully, is because people could get together and figure out how to solve the problem."

At the front of the MOCR's stage was a twenty-by-ten-foot screen flanked by twin ten-by-ten companions, each displaying relevant data at various points of the mission—four trajectory plots during launch, and the Earth, or its Moon, during orbits. For the lunar landing, the proscenium hosted a giant green Moon with two tiny *Columbia* and *Eagle* models blinking their way across its surface.

The stadium's front rows were known as the Trench, home to the first order of controllers, each wired to multiple communication loops through a headset and mic plugged into his dashboard, which was arrayed with big square plastic buttons, a cathode-ray computer display, and an embedded dial phone. The Trench got its name from the fact that the floors of Mission Control were linked by pneumatic tubes of compressed air through which memoranda-stuffed aluminum cylinders shot back and forth—a technology from the space age of 1806. During one particularly bad mission, flight dynamics officer (and ex-marine) John Llewellyn eyed the growing piles of cylinders around him and announced, "I think I am back in the trenches again, surrounded by empty 105 howitzer canisters."

For mankind's first lunar landing, FLIGHT Gene Kranz was assisted by the platinum-haired, Alabama-drawling CONTROL Bob Carlton (in charge of navigation and propulsion); pipe-smoking Brooklynite FIDO (flight dynamics officer, who plotted the actual flying of the ship at each crucial burn) Jay Greene; Lunar Module team leader and laconic Oklahoman TELMU Don Puddy (who oversaw the LM's electrical, communications, and life-support systems); the mustachioed Texan RETRO Chuck Deiterich (in charge of bringing the spacecraft home); and the twenty-seven-year-old GUIDO (guidance, or navigator), Steve Bales. Next to them sat the honey-voiced CapCom (Capsule Communicator) from the Carolinas, Charlie Duke; flight surgeon John Zieglschmid (who monitored both the crew's and the controllers' health, prescribing "whoa and go" speed and sleeping pills); EECOM (electrical, environmental, life support, and communications) John Aaron; GNC (guidance, navigation, control) Buck Willoughby; Booster (a Saturn V expert from Huntsville who only stayed from launch to translunar injection); CControl (LM's GNC); O&P (operations and procedures—an expert in mission rules who additionally ran the screens at the front of the stage); the flight activities officer (an expert in astronaut training ready to translate if there was an issue between the ground and the crew); and the network controller (in charge of the worldwide ground and sea stations acquiring data for Houston). The back row had consoles for the public affairs officer, the Pentagon, and NASA executives; behind them was a glassed-in room seating seventy-four guests: astronauts, politicians, NASA management, and assorted celebrities famous enough or connected enough to get a ringside seat at the greatest show in the universe. During key moments, however, communications to that room were silenced, and only those with a need to know heard, through their headsets, what was happening aboard Apollo 11.

The communication loops included the air-to-ground transmission; FLIGHT's private line; and links to back-office cubicles manned by NASA and subcontractor experts. For the sake of speed, everyone spoke, like navy flight crews, in phrases instead of sentences, and with as many acronyms as possible.

FIDOs like Jay Greene were so wrapped up in their rocket's performance that they often imagined themselves the ship's pilots, if not the ship itself. Apollo 8 FIDO David Reed: "You take on such a literal involvement that you picture yourself as, say, the S-IVB rocket—you say to yourself, 'If I bend that hard, I will break.'" At the FIDO desk was the "bat handle" lever, which could send a signal of two flashing lights to the spaceship ordering the astronauts to abort. Chuck Deiterich said that FIDO, GUIDO, and RETRO "were a team to do trajectory operations, or trajectory control, and we each had our own little areas that we worked in. The retro was more concerned about recovering from abort situations and in normal reentries and deorbits, returning from lunar trajectories, and things like that. So the retro was always worried about aborts and just bringing the crew back, either nominally or in an abort situation. The flight dynamics officer was worried about getting into orbit and doing rendezvous and doing lunar landings. And the guidance

officer was worried about the flight software and the onboard computer."
Gene Kranz:

> The Apollo 11 mission, like many missions, started off quite easy. Every-
> thing was normal. No major challenges. In fact, [Black Team Chief Glynn]
> Lunney, in his log, was tracing what he calls "nits." We finished my last shift
> before the lunar landing, and I had a thirty-two-hour period until my next
> shift came up. [And that's when] it finally sinks in that this next mission is
> what the whole program's been about: landing on the Moon.
>
> Once we got to the point where we were getting ready to land on the Moon,
> there were only three options that day: you were either going to land, you're
> going to abort, or you're going to crash. And, you know, those options are
> pretty awesome when you think about it, that, "Hey, we're not only in this par-
> ticular mode of operation now. We're going to be doing it in front of the entire
> world."
>
> This room is bathed in this blue-gray light that you get from the screen, so
> it's sort of almost like you see in the movies kind of thing. But you also get this
> feeling that this is a place something's going to happen at. I mean, this is a
> place sort of like the docks where Columbus left, you know, when he sailed
> off to America or on the beaches when he came on landing. So it's a place
> where you know something is going to happen.

As the morning of July 20 got under way, George Low whispered to Chris
Kraft, "I've never never seen things so tense around here."

At her home, the normally unflappable Pat Collins finally sighed out loud,
"Oh God, I can't stand it." In Nassau Bay, Mike Aldrin watched the television coverage by himself in a room upstairs. His mother, listening to the
squawk box, got so nervous that at one point, she leaned against the living-
room mantel and rested her head in her arms to calm herself. Astronaut
Rusty Schweickart was visiting the Aldrins with his wife, Clare, who made
him promise to truthfully interpret the squawk box's technical information,
even if the news was not good.

Bill Anders was at the Armstrongs', where Jan finally grew so annoyed
with the melodramatic TV commentary that she went to the bedroom to
listen to the squawk box. Before launch, she had warned Deke Slayton that
she would not be treated on this mission the way she had been on Gemini
VIII when, after her husband's ship spun out of control and the crew's life

was in danger, NASA turned off her home box. She immediately drove to Mission Control, but they refused to let her in or even tell her anything.

To begin the day when they would, for the first time in history, attempt to land two men on the Moon, Kranz ordered his team to dial into the assistant flight director intercom loop, which was a private line solely for their own use. Because of that privacy, his words to his men were not recorded, but he later struggled to remember them word-for-word: "Today is our day, and the hopes and dreams of the entire world are with us. This is our time and our place, and we will remember this day and what we will do here always. In the next hour we will do something that has never been done before—we will land an American on the Moon. The risks are high, but that is the nature of our work. We have worked long hours and had some tough times, but we've mastered our work. Now we are going to make that pay off. You are a hell of a good team; one that I feel privileged to lead. Whatever happens, I'll stand behind every call you make. Good luck, and God bless us today!"

Kranz: "The next thing I do is I have the doors of Mission Control locked; we do this for all critical mission phases. Steve Bales was probably one of the most vocal about it, saying, 'You know, you don't really know what you're do- ing when you've got a twenty-six-year-old kid in this room and basically you're going to write in the history books whatever happened today. And then you lock those doors, and I realize, I can't leave anymore! I can't say, "Hey, I don't want to do this job! Okay? It's too much for me."' Then we go to what we call 'battle short,' where we physically block all the main building circuit breakers in there. We would rather burn the building down than let a circuit breaker open inadvertently and cause us a loss of power."

There will forever be disagreement over who, exactly, flies NASA ships—the engineers on the ground, or the spacemen in the capsules. Later in life, as most of his colleagues continued to claim the pilot role, Aldrin sided with the engineers: "Gemini and Apollo were computerized and pre- planned, so the era of the pilot in command, having the creativity to decide what he wants to do—that's gone. Only in an emergency is it apparent. And in an emergency, like Apollo 13, they had no idea what went wrong. It was like, 'We got a problem, all the lights are comin' on!' And it was up to the ground to figure out what the problem was." Given that Aldrin was a fist-clenched second-in-command during Armstrong's hair-raising lunar touchdown, when the computer that was supposed to land the ship had to be overridden, it's puzzling that he didn't take that experience into consideration when arriving at this conclusion.

—————

A patch in the southwest corner of the Sea of Tranquility had been selected for the first lunar landing, since fly-by photos had indicated that it was relatively free of boulders and craters; it was beneath a possible orbit path for *Columbia*, and it could be approached with the sun at the crew's backs and at a ten-degree angle, which was optimal for viewing surface details without too much shadow or glare. This meant that there was only one day in each one month of lunar cycle that would be ideal for landing based on the sun's position, which was how the date of July 20 had been selected.

Eagle fell to an altitude of ten miles, and then fired its rocket to slow itself out of orbit and into descent. The craft's computer used its landing radar (the Primary Navigation, Guidance, and Control System, PNGS or "pings") to drop to around five hundred feet, at which point the Raytheon computer commanded *Eagle* to gimbal its rocket and fire its thrusters to approach the Sea of Tranquility portholes down, so that the ship's commander could assess the terrain. If Armstrong decided that the computer's designated target site was unacceptable, he could use the ship's landing radar—the Abort Guidance System, AGS or "aggs"—to land the LM himself. This was considered the most difficult piloting job ever: flying a craft different from anything ever built, using controls devised from scratch, in one-sixth gravity, with no atmosphere, on a landscape whose features weren't known, and only enough fuel for one attempt. And of course, there was the time that Apollo 11's commander, in the lunar module training vehicle, had almost gotten himself killed.

Neil Armstrong: "It was my twenty-first flight in the Lunar Landing Research Vehicle. [Powered by hydrogen peroxide rockets, this simulator contraption looked, as Armstrong described it, like "a kind of cross between an old-fashioned Stanley Steamer and a calliope," or like a "flying bedstead."] I lifted the vehicle off the ground and reached an altitude of about five hundred feet in preparation for making the landing profile. I had been airborne for about five minutes and was down to about two hundred feet when the trouble began. The vehicle began to tilt sharply. Afterward this incident was reported as an explosion, but that was erroneous. It's just that there are all the exhaust products of those rocket engines going off, and since there were a lot of engines firing at once people on the ground thought they were seeing an explosion. . . ."

"The frightening films show that he escaped death by just two-fifths of a second," Chris Kraft said. "Winds were gusting that day, something that can't happen on the airless moon, but Armstrong was fully in control for the first five minutes. He took it up several hundred feet and was ready to practice a nearly vertical descent and landing. Then the machine suddenly dropped. He steadied it and climbed back up another two hundred feet. Then the LLTV began to bounce around in the sky. It pitched down, then up, then sideways. Its stabilization had failed and it was clearly out of control. A ground controller radioed Neil to bail out. He activated the ejection seat with only a fractional second of margin. Neil's parachute opened just before he hit the ground. He wasn't hurt, but the LLTV was demolished in a fireball."

"Offhand, I can't think of another person, let alone another astronaut, who would have just gone back to his office after ejecting a fraction of a second before getting killed," Alan Bean recalled. "He never got up at an all-pilots meeting and told us anything about it. That was an incident that colored my opinion about Neil ever since. He was so different than other people."

Neil Armstrong: "It's one of those sad days when you lose a machine."

GET 102:27: The first sign of trouble for Apollo 11 appeared when Mission Control lost radio contact with *Eagle*. The engineers jerry-rigged a solution by patching communications from *Eagle* to *Columbia*, and then to Houston. This added a new element of pressure, for in the event of an emergency, there would be additional delays in communications. "Immediately, as soon as we acquire telemetry, we're in trouble because the spacecraft communications are absolutely lousy," Gene Kranz said. "We can't communicate to them; they can't communicate to us. The telemetry is very broken. We have to call Mike Collins in the command module to relay data down into the lunar module, and immediately this mission role has come into mind because it's decision time, go/no go time."

Whenever a serious issue arose during a mission, Kranz polled the whole of his immediate team, a process known as "going around the horn." For additional advice, he could turn to the horde of NASA and subcontractor engineers sitting in cubicles around the outskirts of MOCR. Now, confronted with the radio problems, he asked, should they continue with the mission?

Kranz: OK all flight controllers, Go/No Go for powered descent [landing on the rocket's tail]. Retro?

Deiterich: Go.
Kranz: FIDO?
Greene: Go.
Kranz: Guidance?
Bales: Go.
Kranz: Control?
Carlton: Go.
Kranz: TELCOM?
Puddy: Go.
Kranz: GNC?
Willoughby: Go.
Kranz: EECOM?
Aaron: Go.
Kranz: Surgeon?
Zieglschmid: Go.
Kranz: CapCom, we're go for powered descent.
Duke: *Eagle*, Houston. If you read, you are go for powered descent.
[Silence.]
Collins: *Eagle*, *Columbia*. They just gave you a go for powered descent. . . .
   Eagle, do you read *Columbia*?
Aldrin: We read you.
Duke: *Eagle*, Houston. We read you now. You're go for PDI.
Aldrin: Roger.

GET 102:33: While Armstrong, at the controls, looked out the window to assess the approaching landscape, Aldrin, monitoring the displays and gauges—all of which were additionally watched by Mission Control—called out his data readings. Aldrin, in fact, was so busy with the dashboard that he never took a single glimpse out the window to watch the landing. Although they were plummeting at 7,614 feet per second, neither man could feel any sensation of descent. "Scooting in at 3,000 mph at 47,000 feet [altitude], you are really hauling mail at that point in time," Apollo 10's Gene Cernan said.

One astronaut remembered that "When [the LM] pitches over and you get your first look, there's nothing but nine million craters. You get this terrible sinking feeling, because you don't recognize a thing, even after studying all those photographs."

"Here around the Earth you get used to an orbit, you look at rivers, you see interstate highways, when the weather's right, and cities. There's some judging of distance," Tom Stafford said. "Up there, there were no roads, no section lines. There was nothing, no cities to look at in trying to judge distance. So we had the map that we were going down, we'd put that out in front of us, put the thumb down and look at these awesome craters and boulders. Those boulders were what really amazed me. . . . You see those big boulders that I thought were as big as a three- or four-story building. Well, they were bigger than the Astrodome down in Houston."

As Armstrong studied the landscape of the landing path known as "US 1"—Maskelyne A crater, followed by the Italy-shaped Boot Hill, Rima Maskelyne 1 canyon, Last Ridge hill, and their designated landing spot, in Moltke crater—it became clear that they were going to overshoot their original target by almost four miles. At MOCR, Steve Bales's readouts showed why: they were flying twenty feet per second faster than planned. If that number reached forty, flight rules would call for an abort. As Armstrong considered this, the Houston–*Eagle* com link failed yet again. Through *Columbia*, it was recommended that Aldrin switch their antenna to slew, which seemed to fix their transmission somewhat.

"We now get to the point where it's time to start engines," Gene Kranz said. "We've got telemetry back again. As soon as the engine starts, we lose it again. This is an incredibly important time to have our telemetry because as soon as we get acceleration, we settle our propellants in the tanks, and now we can measure them, but the problem is, we've missed this point. So now we have to go with what we think are the quantities loaded prelaunch. So we're now back to nominals. Instead of having actuals, we've got our nominals in there. We got down to the go/no-go for start-up powered descent (this is done about four minutes prior to the landing point), and again we—there's no reason I had to wave off. The team was working well. So, we made the go to continue. And as soon as we gave them the go to continue, we lost communication. So, we couldn't even call the crew! So again we relayed—Charlie Duke relays through Mike Collins down to the Lunar Module that they're go to continue. Here we're getting ready to go to the Moon, and we can't even talk to the crew directly!"

GET 102:38: At 33,500 feet, Aldrin engaged AGS, the ship's landing radar. It conflicted with PNGS by 2,900 feet. Then, as *Eagle* fell 132 feet per second, a yellow warning lamp flashed, and the DSKY computer said PROG, meaning a program error. The master alarm clanged in the headsets. Armstrong reported this to Houston and then queried DSKY, which replied: "12 02." Even with all their training and onboard documentation, Aldrin and Armstrong could not decipher this code. Aldrin: "I was completely surprised when that [alarm] happened and I felt an incursion or you know, somebody was getting in our way, they were disturbing what we were trying to do, it was a nuisance. . . . We just didn't wanna stop doing what we could, what we were doing so they could pick up the book and look up 1202 and see what it meant and what we're gonna do about it, the best thing to do is to get advice of the people on the ground."

CapCom Charlie Duke: "When I heard Neil say '12 02' for the first time, I tell you my heart hit the floor. . . . Communication dropouts were a nuisance more than a danger, but a computer problem was a showstopper." MIT space guidance and analysis instrumentation lab director Richard Battin: "[On] the Apollo guidance computer, the memory was a unique system called a core-rope memory . . . little magnetic cores, and one core could represent an entire word, and you'd have sixteen sense wires. The thing that NASA didn't like about it is the fact that although the memory was reliable, you had to very early on decide what the computer program was going to be, because somebody had to put the wires through all these little cores, and it was literally done using the LOL method—Little Old Ladies—who sat there with something that looked like a weaving machine. They would push a probe through, and they had to, in essence, thread those cores by hand. . . .

"The final descent to landing was far and away the most complex part of the flight. The systems were very heavily loaded at that time. The unknowns were rampant."

Then the ship's alarm flashed again, this time reporting a "12 01" error. Buzz Aldrin: "During the descent, when we started having problems with the computer, my attention was focused entirely inside the cockpit looking at the displays and trying to relay the information on the computer and also on the altitude and altitude rate meters to Neil so he could use this with his out-the-window determination as to where we should go to find a suitable landing place. Things were happening fairly fast and it was just a question of making sure the most correct thing was done from my standpoint at that instance. Not much time was allowed for reflection on the situation."

Kranz had guidance officer (GUIDO) Steve Bales research the problem. Should the landing be aborted? It was Bales's call. Bales turned to MIT software engineer Jack Garman, who remembered a history with these alarms: "[The simulation trainers] would think up the problems, the failures to cause. You have to picture in those days that as you got close to one of these 'going where no man has ever gone before' kind of flights, they didn't want to put in failures that you couldn't recover from. That would be both demoralizing and it would make the papers. I mean, even simulations made the papers.

"Because I was a backroom guy, they didn't think it was cheating to come to ask us for what kind of failures they could put in to make the computers not work, because, again, there weren't a lot of people that knew about computers, much less the onboard computers. So we would help make up failures and then pretend we didn't know what they were."

Gene Kranz:

[During Lunar Module simulator runs for Apollo 11] the training people looked, and they saw one entire area that wasn't treated in the rules. It was associated with the various alarms that are transmitted from the spacecraft computer down to the ground. And on the final day of training, which I would—I had expected would sort of like be the graduation ceremony, they'd give us some problems, they'd give us tough problems, but they wouldn't give us anything that would kill us. Well, that wasn't their approach to doing the job. We started our final day of training, and about midway through the day, we had done more aborts, and I was really starting—it was starting to get irritating to me, because what I wanted to do was practice the landing, continue to refine the timing of the landing, but we were aborting when I really felt—and I was really seething, I mean just really frustrated at Sim Sup, but there's no—I mean, he's the boss during training. He's going to call the shots.

On the way down, we started seeing a series of alarms coming from the spacecraft. And there are two types of alarms: one of the alarms said, "Hey, I'm too busy to get all of the jobs done. So, I'm going to revert to an internal priority scheme; and I'll work off as many things as I can in this priority scheme until a clock runs out, and then I'll go back and recycle to the top of this priority listing." And it's going to get the guidance job done. It's going to get the control. But it may not be updating displays. It may not. And then if these type alarms continue for a sustained period of time, it goes now to a much more critical alarm, which we call poo-due. A due program zero zero, where the computer will go to halt and await further instructions. Well, if this happens up and away, you're not going to land on the Moon that day. We'd never seen these before in training. We'd never studied these before in training. My guidance officer, Steve Bales, looked at the alarms and decided we had to abort.

We aborted, and I was really ready to kill Koos at this time, I was so damned mad. We went into the debriefing, and all I wanted to do is get hold of him at the beer party afterwards and tell him, "This isn't the way we're supposed to train," and in the debriefing we thought we'd done everything right.

Koos comes in to us, and he says, "No, you didn't do everything right. You should not have aborted for those computer program alarms. What you should have done is taken a look at all of the function. Was the guidance still working? Was the navigation still working? Were you still firing your jets? And ignored those alarms. And only if you see something else wrong with that alarm should you start thinking about aborting." We told him he was full of baloney. In the meantime, I gave an action to Steve Bales to come up with a set of rules—now we're literally about two weeks from launch—a set of rules. We've had our final training run. A set of rules related to program alarms.

He says, "Hey, Flight, I'm going to look at this overnight. I'm going to call together a bunch of people from MIT [Massachusetts Institute of Technology] Draper Labs, and we'll find out what we should've done here." Well, I got a call about ten o'clock that evening that said, "The training people were right. We had made the wrong decision." And they wanted to do some more training the next day.

Jack Garman:

Gene Kranz, who was the real hero of that whole episode, said, "No, no, no. I want you all to write down every single possible computer alarm that can possibly go wrong." He made us go off and study every single computer alarm that existed in the code, ones that could not possibly happen, they were put in there just for testing purposes, right down to ones that were normal, and to figure out, even if we couldn't come up with a reason for why the alarm would happen, what were the manifestations if it did. . . .

The onboard computers ran in two-second cycles. The notion of navigation and flight control is such that kind of like if you were walking down the street, you open your eyes to see where you are once every two seconds, and you see the hallway around you and the ceiling and the road ahead, and then you shut your eyes and then decide where to put your feet. Okay? . . . The software was designed so as it progressed through any cycle, it would memorize by writing data in this erasable memory where it was in such a way that if it got lost, it could go back to the prior step and redo it. . . .

When one of these alarms came up, it would ring what was called the master caution and warning system, like having a fire alarm go off in a closet. One of these in the earphones, lights, everything. I gather their heart rates went way up and everything. You know, you're not looking out the window anymore. . . .

Bales said, "What is it?" So we looked down at the list at that alarm, and, yes, right, and if it doesn't reoccur too often, we're fine, because it's doing the restarts and flushing. So I said, on this backroom voice loop that no one can hear, I'm saying to Steve, "As long as it doesn't reoccur, it's fine."

Bales is looking at the rest of the data. The vehicle's not turning over. You couldn't see anything else going wrong. The computer's recovering just fine. Instead of calculating once a second, every once in a while it's calculating every second and a half, because it flushes and has to do it again. So it's a little slower, but no problem.

By this point Armstrong was growing disturbed enough to again ask Houston: "Give us a reading on the 12 02 program alarm." Bales: "In the Control Center, any more than three seconds to reach a decision during powered descent is too long; and this took about ten to fifteen minutes."

Finally, CapCom Duke replied: "Roger. We gotcha. We're Go on that alarm."

Neil Armstrong: "The concern here was not with the landing area we were going into but, rather, whether we could continue at all. Consequently, our attention was directed toward clearing the program alarms, keeping the machine flying, and assuring ourselves that control was adequate to continue without requiring an abort. Most of the attention was directed inside the cockpit during this time period and, in my view, this would account for our inability to study the landing site and final landing location during the final descent. It wasn't until we got below two thousand feet that we were actually able to look out and view the landing area. . . .

"My own feeling was, as long as everything was going well and looked right, the engine was operating right, I had control, and we weren't getting into any unusual attitudes or things that looked like they were out of place, I would be in favor of continuing, no matter what the computer was complaining about. . . . [Besides,] aborts were not [a] very well understood phenomenon—no one had ever done an abort. You were shutting off engines, firing pyrotechnic separation devices, igniting other engines in midflight. Doing all of that in close proximity to the lunar surface was not something in which I had a great deal of confidence."

On the ground, however, there was still a lingering worry. Charlie Duke: "We were concerned, very concerned, at the time. Suppose we had to lift off a couple of hours after touchdown? The computer is busier during ascent than it is during descent! So here we are with a computer that seems to be saturated during descent and my gosh, we might be asking it to perform a more complicated task during ascent."

GET 102:39: At 25,000 feet, PNGS tilted the ship upright to its feet-first landing position. The Earth rose to shine through a porthole. The computer then began to throttle the rocket up from 10 percent to 100 percent, for a slow, smooth descent.

This, in fact, was just one of the Lunar Module rocket's unique features. Instead of pumps that might break down, its plumbing was charged with high-pressure gas; instead of a coolant system, its combustion chamber burned up and melted away, just like the heat shield on the Command Module. But it was the throttle that was the signal breakthrough. Rocketdyne had tried, and failed repeatedly, to make a throttle rocket that worked. Then a scientist at NASA's Jet Propulsion Labs, Glenn Elverum, teamed up with TRW for the solution. They used a protruding pipe and bell made of columbium to spray propellant in the middle of the combustion chamber, where it could directly strike the flow of oxidizer. The throttle was a little sleeve that changed the size of the injector's holes, like tiny shutters on a camera. It took months of hard engineering labor, however, to make this notion succeed in the material world.

As *Eagle* descended to four thousand feet of altitude at one hundred feet per second velocity, Kranz called for the ultimate decision from his team at Mission Control:

> Kranz: All flight controllers, Go/No-Go for landing. Retro?
> Deiterich: Go.
> Kranz: FIDO?
> Greene: Go.
> Kranz: Guidance?
> Bales: Go.
> Kranz: Control?
> Carlton: Go.
> Kranz: TELCOM?
> Puddy: Go.
> Kranz: GNC?
> Willoughby: Go.
> Kranz: EECOM?
> Aaron: Go.
> Kranz: Surgeon?
> Zieglschmid: Go.
> Kranz: CapCom, we're go for landing.
> Duke: *Eagle,* Houston. You're go for landing.
> Aldrin: Roger. . . . Understand . . . Go for landing . . . 3,000 feet . . . Program
>   alarm . . . 12-01 alarm.
> Duke: Roger. 12-01 alarm.
> Bales: Same type. We're Go, Flight.
> Duke: We're Go. Same type. We're Go.

From a height of two thousand feet while traveling twenty feet per second, Armstrong tried to orient himself to the landscape below from his studies of the Apollo 10 photographs, but *Eagle* was now too low to the ground, and too far off-course. He switched to semiautomatic, with the computer controlling throttle of the descent velocity, while he ran the attitude and flying speed. Gene Kranz: "We see Neil take over manual control, and he uses an input with his hand controller that redesignates the landing point. He's got a grid in the lunar module window that's sort of like a gun sight. And throughout the mission, it's basically oriented that if 'I don't do anything different right now this is where I'm going to land.' So basically, he's redesignated. And all of a sudden you start becoming intensely aware of the clock that says, in most of the training runs, we would've landed by now, and we haven't landed."

Buzz Aldrin: "He slowed our descent from twenty feet per second to only nine. Then, at three hundred feet, we were descending at only three and a half feet per second. As *Eagle* slowly dropped, we continued skimming forward. Neil still wasn't satisfied with the terrain. All I could do was give him the altimeter call-outs and our horizontal speed. He stroked the hand controller and descent-rate switch like a motorist fine-tuning his cruise control. We scooted across the boulders."

Neil Armstrong: "As we dropped below a thousand feet it was quite obvious that the system was attempting to land in an undesirable area in a boulder field surrounding [West] Crater. I was surprised by the size of these boulders; some of them were as big as small motorcars. And it seemed at the time that we were coming up on them pretty fast; of course the clock runs at about triple speed in such a situation.

"I was tempted to land, but my better judgment took over. We pitched over to a level attitude which would allow us to maintain our horizontal velocity and just skim along over the top of the boulder field. That is, we pitched over to standing straight up. . . . That was when I took command of the throttle to fly the LM manually the rest of the way. . . . I was quite concerned about the fuel level . . . we had to get on the surface very soon or fire the ascent engine and abort."

Mission Control, meanwhile, could not understand why Armstrong was taking so long to touch down. Mission planning and analysis division chief Floyd Bennett: "We kept watching his forward velocity and his altitude rate and, like I said, he was going around like this [gestures], but his forward velocity was still going forty feet per second forward or something, and you can't land at those speeds. And I said, 'What is he doing?' We didn't know what he was doing. He didn't have time to tell us it was a rock field out there."

"I see the vehicle going across the surface of the moon like I have never seen it do in simulations," Steve Bales remembered. "I say, 'What has gone wrong? What is going on?' It's going five times as fast horizontally, it's never supposed to do that, it's just supposed to gently hover down." Don Lind: "At the end, all we knew was that the LM was descending at one foot per second and scooting across the surface at forty-seven feet per second, with only about sixty seconds' worth of descent fuel left. My heart was pounding so hard I was afraid they'd kick me out of the Astronaut Corps."

"I was in the room off the Mission Control Center, the Sun Room," Goddard Theoretical Division director Wilmot Hess recalled. "I was hearing the conversation between mission control and the astronaut. And remember, the spacecraft was silent. And the mission control guys said, 'It's going sideways. It's going sideways. What's it doing that for? Where's it going? It's only got forty seconds of fuel left.' And they were very nervous. They were really out of their minds, because instead of just coming down and landing like this, like the plan had it, he comes down and he starts going like this. . . . What he was doing was overflying a crater, but he didn't say that. He didn't talk at all during this period of time—Neil Armstrong. And so people were really, really upset. It was fun."

"Deke Slayton is sitting next to me," CapCom Charlie Duke said. "We're both glued to the screen on my console, and I'm just talking and talking and telling them all this stuff, and Deke punches me in the side and says, 'Charlie, shut up and let them land.' So I got real quiet, and the tension began to rise in Mission Control."

Gene Kranz: "Carlton calls out in hushed tones, 'Attitude hold.' I acknowledge, 'ATT hold,' then silence. The crew is searching for a landing site. Duke, in even more hushed tones, states, 'I think we better be quiet from here on, Flight!' I respond, 'Rog, the only call-outs from now on will be fuel.' My voice loops become silent, the atmosphere electric as we hang on to each of the crew's words and wait for Carlton's call. We are within five hundred feet of the surface and continuing the descent. We watch displays that the crew cannot see and listen for sounds yet to be uttered. If anyone so much as clears his throat, twenty other voices shush him."

Armstrong [to Aldrin]: OK, how's the fuel?
Aldrin: 8 percent.

Neil Armstrong: "For some reason I'm not sure of, we started to pick up left translational velocity and a backward velocity. That's the thing I certainly didn't want to do, because you don't like to be going backwards, unable to see where you're going. So I arrested this backward rate with some possibly spasmodic control motions, but I was unable to stop the left translational rate. As we approached the ground, I still had a left translational rate which made me reluctant to shut the engine off while I still had that rate. I was also reluctant to slow down my descent rate any more than it was, or stop [the descent], because we were close to running out of fuel. We were hitting our abort limit."

LM flight controller Bob Carlton:

Not only did you have these thrusters that were fine-tuning your attitude, but the big main engine, the landing engine, could gimbal. So if this thruster fails, it's going to cause the spacecraft to rotate. Now, the other thrusters will turn on, granted, but the gimbal of the engine, when it sees this rotation, that big engine is also controlling the attitude, so—whup—it'll gimbal the engines.

The sim guys, they threw us a failed thruster problem. For some reason the ground systems was messed up and my event lights were not working that day. Well, this was the day to really work this over, and they threw me failure after failure. Oh, that was a nightmare sim day.

They just kept failing thrusters and failing thrusters and failing thrusters.

Kranz said, "Carlton, what's wrong with you?"

There was nothing I could do. I said, "As long as they fail the thruster, we're going to crash the mission. I just don't have any way to detect it."

But about that time, one of the guys in the back room—I don't remember which one of them it was—said, "Hey, Bob, look at what the gimbal's been doing."

And he got to looking, and what would happen is, this thruster turns on—whomp—you know, the spacecraft starts to rotate, the big gimbal, and that gimbal would go cattywampus off way to one side, torquing against the thruster, and then it would land. The instant you saw it, you knew, "Well, why didn't I think of that before?"

Come Apollo 11 and we're landing on the Moon, we're two thirds of the way down the trajectory landing on the Moon, and my event light popped up, a failed thruster. About the time I saw that, I saw him take over the hand controller, and then all the thrusters started blinking. I'm sitting there using my logic, but I went to the gimbal and looked down there, and the gimbal was just as steady as a rock. Now, what did that tell me? It says there's no thruster failed. If there's a thruster failed, that gimbal would be off in the left corner.

Now I go back and every time I see the sim guys, I tell them, "Boys, you saved my life. You saved my life."

As fuel gauges dwindled, nerves began to tighten even further. At seventy-five feet of altitude, Houston finally warned *Eagle* that Armstrong had only sixty seconds of fuel remaining. Buzz Aldrin: "When sixty seconds and the [low fuel] light came on, and we were still not real close to the ground, we were maybe a hundred, hundred and twenty feet, then I guess I was getting a little concerned. But what could I do? Could I say, 'Neil, hurry up, get it on the ground'? That would just excite him a little bit more, so I couldn't say that. . . . I was trying to do whatever body English I could to nudge Neil to get that machine on the ground. . . .

"We were in the so-called dead man's zone, and we couldn't remain there long. If we ran out of fuel at this altitude, we would crash into the surface before the ascent engine could lift us back toward orbit."

"Yes, it was touch-and-go there at the end," Floyd Bennett said. "At the experimental test pilots' meeting in Los Angeles after they came back, I asked [Armstrong] about that, and he said, 'Well, I was just absolutely adamant about my God-given right to be wishy-washy about where I was going to land.' Now, I was allowed to put that in this AIAA [American Institute of Aeronautics and Astronautics] paper, but when I did a NASA paper on it, they wouldn't let me. They said that wasn't technical enough. I said, 'Well, that's what he felt.'"

Buzz Aldrin: "In simulations, someone's training you to give a certain response. So you want to do the right thing in the simulation. When it's not a simulation, you want to do the right thing to get the mission done."

Bob Carlton then announced, "Low level." He was referring to the telemetry that meant "running on fumes," a measure of 5.6 percent fuel remaining, which gave Armstrong perhaps a minute to land, or crash. Gene Kranz: "I never dreamed we would still be flying this close to empty. Inside the tanks we don't have a gas gauge like you have in a car, even an aircraft. Once you get at the point, you have a cylindrical tank that's got a round dome at the top and the bottom. The fuel is sloshing around back and forth in this tank, and you have what they call a 'point sensor,' and this point sensor says that we have 120 seconds of fuel remaining if we're at a hover throttle setting. This is roughly around 30, 35 percent throttle.

"Bob Nance is looking at a recorder which is tracing out actually the throttle position that Armstrong's using, above hover throttle, below hover throttle, above hover throttle, below hover throttle, and he is mentally integrating now how many seconds he is above hover throttle and subtracting that from the minutes below hover throttle, trying to give us a new number for how many seconds of fuel we've got. . . . Armstrong has to pick out a landing site, and he's very close to the surface. Instead of moving slowly horizontal, he's moving very rapidly, and ten and fifteen feet per second, I mean, we've never seen anybody flying it this way in training."

Engineering Analysis Office Chief Milton A. Silveira:

> Just before [Apollo 11], we had run a test where we were concerned about what happens to the descent engine on the LM if you're firing it against a solid surface. The shock gets swallowed by the engine, and then the engine would blow up. So we ran this test. We got close to the plate that we've simulated the LM [engine approach], and sure enough, the engine blew up. So I went to George Low and I said, "We've got some bad news."
>
> George said, "Well, you ought to go talk to Neil about this."
>
> I said, "Okay," so I went and had a one-on-one with Neil. Neil was always a funny individual. You weren't sure whether he was listening to what you were saying or not. I said, "You've got to be careful. This thing might blow up."
>
> "Oh, okay."
>
> So when the actual landing was occurring on the Moon . . . everybody was worried about if he was going to run out of fuel; I was worried that he would not run out of fuel.

Goddard liaison Bill Easter remembered that at that moment, MOCR "was packed. You could have heard a pin drop. The room was designed for probably 60 people, and it probably had 150 in there. . . . That, to me, probably was the most exciting time I've ever had at NASA, when they started landing. There was no other feeling like that. It was like watching a man, some snake trainer, put his hand on a cobra, anything can happen any minute, and probably will."

Neil Armstrong: "I changed my mind several times, looking for a parking place. Something would look good, and then as we got closer it really wasn't so good. Finally, we found an area ringed on one side by fairly good sized craters and on the other side by a boulder field; it wasn't particularly big, a couple of hundred square feet—about the size of a big house lot."

> Aldrin: Five percent. Quantity Light. [A second "fuel tank empty" sign.]
> Carlton: Coming up on sixty. . . . Sixty!
> Kranz: Sixty seconds.
> Duke [to *Eagle*]: Sixty seconds.

At home, Jan Armstrong watched and listened, hugging her son Ricky close to her. Joan Aldrin supported herself by leaning against a door, her hands shaking, her eyes wet with tears.

Aldrin: "Forty feet, down two and a half [feet per second]. . . . Picking up some dust. . . . Thirty feet, two and a half down." Jack Garman: "We'd watched hundreds of landings in simulation, and they're very real, and on this particular one, the real one, the first one, Buzz Aldrin called out, 'We've got dust now,' and we'd never heard that before. You know, it's one of those, 'Oh, this is the real thing, isn't it? My God, this is the real thing.' And you can't do anything, of course. You're just sitting down there. You're a spectator now."

Aldrin: Four forward. Four forward. Drifting to the right a little.
Nance: Thirty seconds.
Carlton: Thirty.
Kranz: Thirty seconds.
Aldrin: Twenty feet, down a half.
Duke: Thirty seconds.
Aldrin: Drifting forward just a little bit; that's good.

With scant hypergolics left, his nose and throat dry and cold from *Eagle*'s tanked air, Armstrong finally saw a clear spot, and righted his ship into vertical landing position. The flight surgeon noted that Aldrin's heart rate was 125 and Armstrong's 156.

The craft shuddered from its thrusters firing to maintain trajectory.

The descent kicked up so much dust that, at thirty feet, the ground below was a roiling cloud.

Neither man could feel it happen, but then, with about seventeen seconds of fuel remaining, the dashboard's blue light ignited, meaning that one of the LM's foot sensors had made contact.

Aldrin: Contact light!
Armstrong: Shutdown.

GET 102:45:41: Kranz: "Carlton was just ready to say, 'Fifteen seconds,' and then we hear the crew saying, 'Contact.' We have a three-foot-long probe stick underneath each of the landing pads. When one of those touches the lunar surface, it turns on a blue light in the cockpit, and when it turns on that blue light, that's lunar contact, their job is to shut the engine down, and they literally fall the last three feet to the surface of the Moon. So you hear the 'lunar contact,' and then you hear, 'ACA [Attitude Control Assembly] out of Detent [out of center position].'"

Neil Armstrong: "I was absolutely dumbfounded when I shut the rocket engine off and the particles that were going out radially from the bottom of the engine fell all the way out over the horizon, they just raced out over the horizon and instantaneously disappeared, you know, just like it had been shut off for a week. That was remarkable. I'd never seen that. I'd never seen anything like that. And logic says, yes, that's the way it ought to be there, but I hadn't thought about it and I was surprised."

Aldrin immediately turned off the engine's power and keyed "413" into the computer, to store the ship's location of 0.71 degrees North, 23.63 degrees East.

GET 102:45:58 (July 20, 1969, 3:17 p.m. CST): Armstrong: "Engine arm is off. (Pause) Houston, Tranquility Base here. The *Eagle* has landed."

In NASA history, filled with so many great moments, this was the greatest of them all.

"The only thing that was out of normal throughout this entire process, that we had never seen in training, was the people behind me in the viewing room start cheering and clapping and they're stomping their feet," Gene Kranz remembered. "And our instructors are over in the room to the right of the room, again behind a glass wall, and they're all cheering. And you get this weird feeling. It's chilling that it soaks in through the room; and I get it, and say, 'My God! We're actually on the Moon!' And I can't even relish that thought because I got to get back to work. Because we have to make sure, almost instantaneously, whether the spacecraft is safe to leave on the surface of the Moon or should we immediately lift off? We go through what we call our T-1 stay/no-stay decisions. So that within sixty seconds of getting on the Moon, I have to tell the crew, 'It's safe to stay on the Moon for about the next eight minutes.' And I don't have any voice. I'm clanked up.

"I was so hung up by this cheering coming in from the lunar room that I could not speak, and pure frustration, because I had to get going on the stay/no stay. I just rapped my arm down on the console there, just absolutely frustrated. I broke my pencil, the pencil flies up in the air. Charlie Duke's next to me, and he's looking and wondering what the hell has happened here. And finally I rap my arm on the console and break my pen, and I finally get going. Get back on track again. And in a very cracked voice say, 'Okay, all flight controllers, stand by for T-1 stay/no-stay.'"

Finally CapCom Duke was able to reply: "Roger Twan . . . Tranquility, we copy you on the ground. You've got a bunch of guys about to turn blue, we're breathing again, thanks a lot. . . . Be advised there are lots of smiling faces in this room, and all over the world." Later, Duke would remember: "We were so excited I couldn't even pronounce 'Tranquility.' It came out 'Twanquility'!"

Aldrin: "There are two of them up here."

Collins: "And don't forget one in the command module. . . . You guys did a fantastic job."

Walter Cronkite (while pulling off his glasses to mop the sweat from his forehead): "Phew! Wow, boy! Man on the Moon!"

Aldrin and Armstrong, their faces grizzled with four days of stubble, looked at each other, grinned, shook hands and clapped each other's shoulders. "So far, so good," said Armstrong. "Let's get on with it." To Houston he announced, "Okay. We're going to be busy for a minute."

Buzz Aldrin: "If there was any emotional reaction to the lunar landing it was so quickly suppressed that I have no recollection of it. We had so much to do, and so little time in which to do it, that we no sooner landed than we were preparing to leave, in the event of an emergency. I'm surprised, in retrospect, that we even took time to slap each other on the shoulders."

Armstrong later would reveal his profound feelings at this moment, that successfully landing on the Moon was "a real high in terms of elation. It marked the achievement that a third of a million people had been working for a decade to accomplish."

"The first thing that happens when you land is you experience the most quiet moment in your entire lifetime," Gene Cernan said. "I mean when you're coming down everything's dynamic: it's shaking, the engines are running, you're flying, you're landing and you get close and you hear [the Lunar Module pilot] talking, you're listening. You've got dust and all of a sudden you shut down. And wow, you are now where no man has ever been before. It's quiet, it's still. There's nothing moving, there's no wind, there're no trees. I mean, you look around, and it's almost like science fiction."

"As I went out of the Control Center, the Moon was up, and it was quite a feeling to stand there outside of the Control Center and look at the Moon and say, 'There you are,'" director of flight operations technical assistant Wayne Koons remembered, while Lunar Module systems engineer Robert Heselmeyer said, "I can remember during that time, though, walking from the Control Center back to my office and looking up at the Moon and saying, you know, 'That's where those guys are.' It was across where the ponds are on the JSC campus, and back then the deer would show up by the ponds, and so there you are at night with the deer around the ponds and the Moon up there and thinking, 'This is really kind of surreal.' For a guy who was trained as an engineer for technical stuff to start thinking surreal thoughts was a breakthrough."

Apollo communications chief Ed Fendell:

The first thing you have to remember, I don't know what you've heard from other people, but most of us didn't believe we would land on Apollo 11. Have people told you that, in their opinion, that they thought we would make it the first time? I didn't believe we would ever land the first time, but we did. Okay. And when it started getting down close, I don't think I was touching my chair. I actually believe I was levitating somewhere over that chair. That's the way I felt. I know I wasn't levitating, because I can't do that, but that's the way I felt. It was so intense that I don't think most people really fully realized what we did. I know I didn't.

I went home and I slept for a couple of hours, I got cleaned up, and I was going back to work and I stopped to eat some breakfast. And between Monroe and Edgebrook in those days there was a Dutch Kettle, you know, one of these little coffee shops with the round stools. I walked in there, and I knew we had landed on the Moon, and I was proud and all that and everything, but because I wasn't out there with the public when it all happened, I really wasn't that jived as to what the real effect was going on in the world. You know what I'm saying? You know, there were people going crazy all over the world. You'll see these pictures on the movies and the newsreels, you'll see thousands of people standing in Times Square watching this stuff and so on, you know. So you weren't into all that, you were so intense in what was going on and what you were looking at and so on.

I'm sitting there reading the paper and so on, and two guys walk in and sit down on the two stools next to me. They are from the Exxon or Enco or whatever the gas station was down at the corner down there, and they're in their gas station clothes and they've got the grease under the fingernails and so on. They were a little bit older. They sit down and they get their coffee and they're waiting for their breakfast. They start talking.

One of them says to the other one, he said, "You know, I went all through World War II. I landed at Normandy on D-Day." And he said, "It was an incredible day, an incredible life, and I went all the way through Paris and on into Berlin," wherever the heck it was he was talking about. He said, "But yesterday was the day that I felt the proudest to be an American." Well, when he said that, I lost it. It all of a sudden hit me as to what we had done, you know. And I just threw my money down, grabbed my paper, and walked out and got in the car and started to cry.

"I was with Joan Aldrin that night," Nurse Dee O'Hara said. "I had gone over that evening, and several of her friends and other wives were there, and it was probably the only time I've experienced a surreal moment. I don't know quite how to describe it, but it was truly unreal. I saw the TV flickering and the LM was there, and we were told it was the Moon, and the LM was on the surface of the Moon. Of course, we were so relieved that they had landed. But it just simply wasn't real. Intellectually you know it is, and, of course, Joan, as was everyone, was terribly relieved that they had landed. I remember sitting there. I kept shaking my head. I thought, this can't be real, it just can't be. Here we are, on another planet. It was goose bumps all around. It really was just unreal."

Joan Aldrin: "My mind couldn't take it all in. I blacked out. I couldn't see anything. All I could see was a match cover on the floor. I wanted to bend down and pick it up, and I couldn't do it. I just kept looking at that match cover."

Janet Armstrong tried to avoid, as much as possible, watching the mission on television; instead, she listened to the squawk box: "The speculation by the TV commentators—the drama of things that could happen if there was a problem along the way—I didn't need to hear all that. That just drove me nuts." But even she was watching TV for her husband's moment in history. A huge mob of friends and relatives had shown up at her house: "I was paying attention to the flight, and that was most important. It was not a social occasion as far as I was concerned. Well, it was and it wasn't. It was a great tense time."

Houston that evening was being drenched by a summer thunderstorm. The media huddled outside the Aldrin, Armstrong, and Collins houses, their plastic slickers and broad umbrellas useless against the soaking rain. Of the three wives, Joan Aldrin was the last to open her front door and offer a statement to the horde destroying her lawn. She had been emotionally overwhelmed by the landing, and needed time to prepare.

Though Buzz had told her repeatedly that the most dangerous part of the mission would be the launch from the Moon that would return *Eagle* to *Columbia*, Joan couldn't get the idea out of her head that the landing was terribly vulnerable, and had spent the entire procedure in a state of profound anxiety. Rusty Schweickart had been with her, explaining, as he had promised, the technical details as revealed by the squawk box. But as the craft ran out of fuel, she became more and more upset.

When she finally came out to speak to the press and a reporter asked, "What were you doing when they landed?" she could only whisper, "I was holding on to the wall. I was praying." When the journalists then continued with the standard-issue and somewhat inane media queries of that era—Will the children be allowed to stay up past their bedtimes to watch their fathers walk on the moon?—Joan could not contain herself. "Listen!" she shouted. "Aren't you all excited? They did it! They did it!"

17

"Mr. President, the *Eagle* Has Landed"

While the world and especially the United States stood by, enthralled at the wonder of what had just happened, the crew of *Eagle* had no time to savor this moment for themselves. Immediately after landing, Armstrong and Aldrin began a 03:12 countdown, testing the ship's systems to determine whether they could remain on the Moon, or would have to immediately abort and launch back to *Columbia*. That urgency was necessary because a longer stay would require waiting two hours for the mother ship's next orbit. The tests were successful, and they were "stay" for a minimum of those two additional hours.

Checking their own readouts, Mission Control realized that the Moon's chilly surface had frozen a dollop of propellant, clogging one of *Eagle's* pipes. As the plug refused to budge, it caused a building up of heat and pressure, and threatened an explosion. Grumman engineers theorized that the engine could be "burped," dislodging the plug, but that solution might have the side effect of toppling the LM over.

"Right after they landed on the Moon, they had to dump the pressure out of the helium pressurization system, called supercritical helium," White Sands aerospace engineer Ray Melton explained. "Well, that venting of that pressure forces ultra, ultra-cold helium gas through a heat exchanger, which is full of rocket fuel. But the rocket fuel is no longer flowing, and so it would freeze. . . . So they were quite worried on the very first flight, Apollo 11, when they saw this temperature and pressure in this short section of line rising, rising, rising, and they were perhaps even worried that that pressure and temperature might get so high that that section of line would explode. And if it were to explode, it would destroy [not only] the LM descent stage, but also the ascent stage on top. So the astronauts even had to wait inside the ascent stage for this pressure/temperature problem to be resolved in case it started getting out of hand."

"If it got up about four hundred degrees," Grumman LM engineering manager Thomas Kelly said, "the fuel could explode, go unstable. There wasn't much fuel in there, just vapor, but still, we just didn't like the idea and we were nervous about it. So we had some very hasty consultations with the NASA people and our own propulsion people, and we finally decided we were going to burp the engine. We were going to ask the astronauts to flick the engine on and then off right away, just to relieve the pressure in that line. We didn't think it would start up enough to [cause] any problem. The Cap-Com, the capsule communicator, was just about to tell the astronauts about it when the problem solved itself. The ice plug in the heat exchanger melted by itself, and all of a sudden the pressure dropped down to zero, so no problem, it went away. But we sweated that out for about ten minutes right after the landing."

"The night that Apollo 11 landed," Ray Melton continued, "I was a mixture of excited, and I was awestruck. In fact, the hair is still up on my arms right now, even remembering it, because I knew of the problems that we still had, and I kept thinking, 'We haven't totally resolved this 400-cycle pressure oscillation in the LM descent engine, and that might be a real problem.' [But the astronauts] were, 'Let's go.' They're going to fly it anyway. There's all sorts of little things that the public didn't know weren't totally, totally resolved. We [knew we didn't have] everything worked out. But when those guys landed, I cried. I was so happy, I just cried."

After their landing run-through was completed, Armstrong could finally take a breath and explain to Mission Control what had happened during the approach: "Hey, Houston, that may have seemed like a very long final phase. The AUTO targeting was taking us right into a football field–sized crater, with a large number of big boulders and rocks for about . . . one or two crater diameters around it, and it required us going in P66 and flying manually over the rock field to find a reasonably good area."

Later, he said: "That [West Crater] was a big dude! See, Buzz could not see it, because he was on the right side. On the LM, between the right and left crew positions, there's a panel that goes out so that either crew member can only see forward and out his side, but not to the other side. So we came over the [north side of] big West Crater, which looked enormous, humongous. . . . To me it looked like a football stadium—you know, a circular football stadium—about that size. And he couldn't see it! . . . He couldn't appreciate that at all! We were coming right into the northeast slopes of it. And there were a lot of big boulders on that slope. A generally undesirable landing area for the first one. Geologically interesting, but not a good place to land."

———

Buzz Aldrin: "We'll get to the details of what's around here, but it looks like a collection of just about every variety of shape, angularity, granularity, about every variety of rock you could find. The colors—well, it varies pretty much depending on how you're looking relative to the zero-phase point. There doesn't appear to be too much of a general color at all. However, it looks as though some of the rocks and boulders, of which there are quite a few in the near area, it looks as though they're going to have some interesting colors to them."

Collins: "Sounds like it looks a lot better than it did yesterday at that very low Sun angle. It looked rough as a cob then."

Armstrong: "It really was rough, Mike. Over the targeted landing area, it was extremely rough, cratered, and large numbers of rocks that were probably some, many larger than five or ten feet in size."

Collins: "When in doubt, land long."

Then, after asking "every person listening in, whoever and wherever they may be, to pause for a moment and contemplate the events of the past few hours, and to give thanks in his or her own way," Buzz Aldrin took from his personal preferences kit a small chalice, a few drops of red wine, a host, and a notecard. He had decided that, should he and Armstrong successfully make their landing, he would take Holy Communion.

Originally, he'd wanted to conduct his ceremony on open mic, but atheist activist Madalyn Murray O'Hair was suing NASA over Apollo 8's Bible reading broadcast, and Deke Slayton had asked to keep it private. Aldrin was very close to his minister at Webster Presbyterian, Dean Woodruff—the two had just recently marched together on Palm Sunday in honor of Dr. King, and in 1996 Aldrin would even play Woodruff in a TV movie about Apollo 11. Following the reverend's suggestion, Aldrin read from John 15:5–12: "I am the vine, you are the branches. Those who abide in me and I in them bear much fruit, because apart from me you can do nothing. Whoever does not abide in me is thrown away like a branch, and withers; such branches are gathered, thrown into the fire, and burned. If you abide in me, and my words abide in you, ask for whatever you wish, and it will be done for you. My Father is glorified by this, that you bear much fruit and become my disciples. . . . This is my commandment, that you love one another as I have loved you."

While Buzz conducted his ceremony, Neil began his own lunar report: "I'd say the color of the local surface is very comparable to that we observed from orbit at this Sun angle, about ten degrees Sun angle, or that nature. It's pretty much without color. It's a very white, chalky gray, as you look into the zero phase line; and it's considerably darker gray, more like an ashen gray as you look out ninety degrees to the Sun. Some of the rocks in close here that have been fractured or disturbed by the rocket engine plume are coated with this light gray on the outside; but where they've been broken, they display a very dark gray interior—could be country basalt. . . . Out my overhead hatch, I'm looking at the Earth. It's big and bright and beautiful.

"[The view from] the window is a relatively level plain cratered with a fairly large number of craters of the five- to fifty-foot variety and some ridges, small twenty, thirty feet high, I would guess, and literally thousands of little one- and two-foot craters around the area, we see some angular blocks out several hundred feet in front of us that are probably two feet in size and have angular edges. There is a hill in view, just about on the ground track ahead of us, difficult to estimate but might be half a mile, or a mile. . . . The sky is black, you know. It's a very dark sky. But it still seemed more like daylight than darkness as we looked out the window. It's a peculiar thing, but the surface looked very warm and inviting. It looked as if it would be a nice place to take a sunbath. . . . It's some kind of lighting effect, but out the window the surface looks much more like light desert sand than black sand."

GET 105:00: As a safeguard against bad press and exhausted fliers, NASA's original plan was for the astronauts to land, test their craft, and then go to sleep for four hours before stepping out onto the surface on July 21 at 01:00 Houston time. Armstrong and Aldrin asked to postpone this rest period and immediately move forward with the first moonwalk, and Mission Control agreed . . . a decision that had the additional benefit of a primetime television audience at 8:00 p.m. CST. Neil Armstrong: "We had thought, even before launch, that if everything went perfectly and we were able to touch down precisely on time; if we didn't have any systems problems to concern us; and if we found that we could adapt to the one-sixth gravity lunar environment readily, then it would make more sense to go ahead and complete the EVA while we were still wide awake—but, in all candor, we didn't think this was a very high probability."

The men disconnected themselves from the LM's ports and clipped on the harnesses of their personal life support system (PLSS) backpacks (batteries, radio, oxygen, exhaust, cooling water for the long johns, and a remote control panel that was attached to the front of the suits for radio, camera, and ventilation control), slung on their oxygen purge systems (which would be used if the PLSSes failed), and screwed in the PLSS's ducts to their suits' various outlets. They stuffed their pockets with data cards, scissors, and penlights, and finally attached their boots, overshoes, and gloves to the suits' various metal cuffs.

The EVA prep was supposed to take two hours; instead, it took three, due to problems with the cooling units and with depressurization, and due to the fact that two large men in bulky suits and enormous backpacks that stuck out a full foot were attempting to operate within the confines of a tiny spacecraft. Buzz Aldrin: "We felt like two fullbacks trying to change positions inside a Cub Scout pup tent."

NASA scientists knew that when men landed on the Moon they would be bombarded by cosmic rays, solar radiation, and the one-in-ten-thousand chance of a micrometeorite—the size of a fingertip, they would strike at 64,000 miles per hour—but because they weren't wholly certain what other perils might endanger them, they created outfits that seemed ready for any contingency. The Apollo 11 Extra-Vehicular Activity (EVA) suit had five hundred parts and was three layers thick (of Nomex and Beta cloth), with a fourth layer at every joint to safeguard from life-threatening rips. The indoor versions weighed thirty-five pounds, while the moon excursion suits ended up a massive fifty pounds, not including the PLSS backpack that pumped oxygen and air-conditioning. The helmet had two visors, one like a pair of mild sunglasses to combat glare, the outer a one-way mirror of gold electroplate which would reflect the worst of the sun's blinding rays. Also, in the unlikely event of a hostile encounter, its reflection would keep aliens from being able to peer into a human face.

It was very difficult to work in pressure suits, as Jack Schmitt explained: "The biggest problem is that the gloves are balloons. . . . Whatever it is, to pick something up, you have to squeeze against that pressure, in our case eight psi . . . that squeezing against that pressure causes these forearm muscles to fatigue very rapidly. Just imagine squeezing a tennis ball continuously for eight hours or ten hours, and that's what you're talking about.

"The other part of the glove that was a problem is that no matter how closely you cut your fingernails, every time you reach for something or moved in that glove, you would tend to scrape your nail against the bladder of the suit, the rubber bladder. I even wore liners, nylon liners, to reduce that, but still you would do that, and you'd gradually lift the nail off the quick. That is painful to some degree while you're working, but particularly gets painful later after you've gotten out of the suit and you then prepare for the next day."

What would happen if these suits got punctured, or torn? In 1965 a NASA accident resulted in a man's being subjected to a near-vacuum, similar to the conditions of space. He passed out in fourteen seconds, but before that happened, he could feel the spit on his tongue beginning to boil.

GET 109:04: Before Aldrin and Armstrong could commence with the extra-vehicular activity (NASA for "going outside"), the LM had to be depressurized. This did not work well, as the venting was drastically slowed by the use of a biological filter to keep earthly germs from infecting the Moon. Buzz Aldrin: "We tried to pull the door open and it wouldn't come open. We thought, 'I wonder if we're really gonna get out or not?' It took an abnormal time for it to finally get to a point where we could pull on a fairly flimsy door. You don't want to rupture that door and leave yourself in a vacuum for the rest of the mission."

"I think it's dangerous," Armstrong's grandmother, Caroline Korspeter, remarked. "I told Neil to look around and not to step out if it didn't look good. He said he wouldn't."

Jan Armstrong had her own explanation for the delay: "It's taken them so long because Neil's trying to decide about the first words he's going to say when he steps out on the moon. Decisions, decisions, decisions!"

For months, Armstrong had endured thousands of friends, coworkers, and complete strangers demanding to know one thing: What would be the first words of the first man on the Moon? Astronauts were not trained for such responsibilities—they were, after all, "just the facts, ma'am" kind of guys—and this particular one preoccupied Armstrong, who hadn't really considered it an issue until everyone started asking about it.

Though inspired by a heartfelt idea, the children's game "baby steps, giant steps," one of the most memorable quotes in history was, from the start, mired in controversy. Many insiders, from NASA employees to the press, assumed that United States Information Agency officer Simon Bourgin was involved, since he'd come up with the idea of Apollo 8's Genesis reading. Others up in the agency hierarchy tried to take control of the wording, until press officer Julian Scheer finally put a stop to the matter by publicly asking: Did Isabella tell Columbus what he would say in advance?

Then there's the missing "a." The most famous astronaut in history's most famous statement doesn't really make sense grammatically in the way it has been historically recorded. Armstrong has always insisted he said it correctly, but has also admitted that, given the way he speaks, it might not have come out, exactly. Neil Armstrong: "The a was intended. I thought I said it. I can't hear it when I listen on the radio reception here on Earth, so I'll be happy if you just put it in parenthesis." NASA transcribers first insisted that it was nowhere to be found on the tape, but thirty-odd years later, Australian computer programmer Peter Shann Ford was adamant that he'd discovered it, though many did not accept his conclusions.

After the hatch was finally opened, Aldrin guided his commander as he slowly crawled backward in his inflated suit and massive backpack through the narrow doorway onto *Eagle*'s "porch." The debarkation was far more difficult than originally anticipated, which made everyone involved grateful they'd come up with a different method of flying than direct ascent, with its ninety-foot ladder.

> Aldrin: Neil, you're lined up nicely. Toward me a little bit, OK down, OK make it clear . . . Here roll to the left. OK, now you're clear. You're lined up on the platform. Put your left foot to the right a little bit. OK, that's good. Roll left.
> Armstrong: That's OK? How am I doing?
> Aldrin: You're doing fine.

Getting out the door was so difficult that Armstrong forgot to turn the handle releasing the modular equipment stowage assembly, or MESA, hatch (NASA for "cabinet"), where everything they needed on the ground—including the TV camera for recording the first step—was stowed. When Houston reminded him of the oversight he pulled himself back up to release it, and then continued his climb downward. When Walter Cronkite asked on CBS why this was taking so long, Rusty Schweickart replied that Armstrong "doesn't have eyes in his rear end."

> Armstrong: Houston, the MESA came down all right.
> CapCom McCandless: This is Houston, roger. We copy. And we're standing by for your TV.
> Armstrong: Houston, this is Neil. Radio check.
> McCandless: Neil, this is Houston. Loud and clear. Break. Break. Buzz, this is Houston. Radio check, and verify TV circuit breaker in.
> Aldrin: Roger, TV circuit breaker's in, and read you five square.

There had been a problem with television transmission from California's Goldstone, the main facility for reception at this time of day and lunar position, and instead the video signal was being received by Honeysuckle Creek,

Canberra, Australia, while the audio feed continued from California. Would this jerry-rigging work to successfully broadcast the first human being to touch another planet?

Then, it happened: the ten-by-ten screen at Mission Control suddenly bloomed to life. The whole of MOCR—its engineers and controllers, its visiting astronauts, military officers, and VIPs—began to clap and cheer.

McCandless: Roger. We're getting a picture on the TV.

Aldrin: You got a good picture, huh?

McCandless: There's a great deal of contrast in it, and currently it's upside down on our monitor, but we can make out a fair amount of detail.

Aldrin: Okay. Will you verify the position—the opening I ought to have on the camera?

McCandless: Stand by. . . . Okay. Neil, we can see you coming down the ladder now. . . .

McCandless: Buzz, this is Houston. F/2—1/160th second for shadow photography on the sequence camera. [The ladder with Armstrong stepping down was facing away from the sun and in almost total darkness.]

Aldrin: Okay.

Armstrong [to Houston]: I'm at the foot of the ladder. The LM footpads are only depressed in the surface about one or two inches, although the surface appears to be very, very fine grained, as you get close to it, it's almost like a powder. Now and then it's very fine . . . I'm going to step off the LM now. . . .

Jan Armstrong [to her husband on television]: "Be descriptive!"

GET 109:24:26 (July 20, 1969, 9:56 PM CST): With six hundred million people—one-fifth the earth's population—watching or listening, Armstrong's foot touched the ground of another world for the first time in human history, and he said,

"That's one small step for (a) man. One giant leap for mankind."

Actually, the step was not so small. Armstrong had set down *Eagle* so gently that the legs' shock absorbers did not compress and deploy, leaving the ladder's last rung 3.5 feet above the surface. Before actually making the historic step, Armstrong dropped from the rung to *Eagle*'s footpad, and then jerked himself back onto the ladder, to make sure that after their EVA, he and Aldrin would be able to climb back aboard their ship. He then warned Aldrin about what "a long one . . . a three-footer" the small step actually was.

Aldrin, meanwhile, had to remember not to lock the cabin door after exiting *Eagle*, since the designers had neglected including a handle on the outside.

One NASA executive commented that, with "one small step," "Armstrong surprised everybody. Yes or no from him is a big conversation." Back in Houston, Neil's brother, Dean, was so thrilled he could barely contain himself. Finally he joked, "When we ask him about it later, he'll say, 'a piece of cake.'" Neil's twelve-year-old son, Rick, however, corrected his uncle, saying of his dad, "Usually when you ask him something, he just doesn't answer."

Neil Armstrong: "The surface is fine and powdery. I can—I can pick it up loosely with my toe. It does adhere in fine layers like powdered charcoal to the sole and sides of my boots. I only go in a small fraction of an inch, maybe an eighth of an inch, but I can see the footprints of my boots and the treads in the fine, sandy particles.

"There seems to be no difficulty in moving around as we suspected. It's even perhaps easier than the simulations at one-sixth g that we performed in the various simulations on the ground. It's actually no trouble to walk around. Okay. The descent engine did not leave a crater of any size. It has about one foot clearance on the ground. We're essentially on a very level place here. I can see some evidence of rays emanating from the descent engine, but a very insignificant amount."

When first seeing the Moon's surface from the height of lunar orbit, astronauts typically described a harsh, mottled, and barren landscape, a destroyed and empty battleground. "When I was in orbit around the Moon, I had a feeling I was in an animation," Apollo 17's Alan Bean said. "It was like orbiting a ball, because it's so much smaller than the Earth, and you can see the curvature and it just seemed magical that we would keep going around this little ball in this little spaceship, without drifting off into space. But I remember looking out and there'd just be craters all over the place and I'd be scared and I remember saying to myself, 'Well, I can't do my job scared,' so I'd look in and I'd pay attention to my computer screen, which looked just the same as in the simulator, and I calmed down. But I didn't want to miss the trip, so I'd look out and get excited again, then look in. . . . I remember looking out the window and saying to Pete, 'Wow, this is scary.' . . .

"Astronauts aren't fearless . . . it's a case of trying to find a way that you can manage your fear and still do the job at hand. And the thing to remember is that we couldn't do this when we began; we gradually learned to. I tell young people all the time, as part of their education, 'You're not born brave; you gotta learn to be brave. You got to find a way.'"

With their lunar landings, however, the Americans discovered a landscape of overpowering beauty, lit by a sun unfiltered by any atmosphere, with one-sixth gravity that inspired hardened pilots to act like eight-year-old boys, and a top dust as soft as the powdered beaches of Cancún. Aldrin heard Armstrong remark, as they were taking their first steps, "Isn't it fun?" But what Armstrong remembers saying is, "Isn't it fine?"—meaning the remarkable lightness of the soil. Buzz Aldrin: "The surface of the moon was like fine talcum powder. When you put your foot down in the powder, the boot-print preserved itself exquisitely. When I would take a step, a little semicircle of dust would spray out before me. It was odd, because the dust didn't behave at all the way it behaves here on Earth. On Earth, you're sometimes dealing with puffy dust, sometimes with sand. On the moon, what you're dealing with is this powdery dust traveling through no air at all, so the dust is kicked up, and then it all falls at the same time in a perfect semicircle."

Neil Armstrong: "[The Moon has] a stark beauty all its own, like much of the high desert of the United States. It's different but it's very pretty down here."

At the time, an awestruck Buzz Aldrin could only describe what he saw as "magnificent desolation." Across the monochrome lunar vista of white anorthosite highlands and black cratered seas were colored shots of sparkle, which would later be identified as either volcanic glass, or particles produced by the poundings of meteorites. The other unique quality about being on the Moon, both men later reported, was that it was obvious you were standing on an orb since the horizon had a pronounced curve to it. Armstrong: "The horizon seems quite close to you because the curvature is so much more pronounced than here on earth. It's an interesting place to be. I recommend it."

Just as dramatic was the vast difference between lunar shade and light. "The light was sometimes annoying, because when it struck our helmets from a side angle it would enter the faceplate and make a glare that reflected all over it," Neil Armstrong said. "As we penetrated a shadow we would get a reflection of our own face, which would obscure everything else. Once when my face went into shadow it took maybe twenty seconds before my pupils dilated out again and I could see details." Aldrin: "Stepping out of the LM's shadow was a shock. One moment I was in total darkness, the next in the sun's hot floodlight. From the ladder I had seen all the sunlit moonscape beyond our shadow, but with no atmosphere, there was absolutely no refracted light around me. I stuck my hand out past the shadow's edge into the sun, and it was like punching through a barrier into another dimension."

With no atmosphere, lunar vistas were brilliantly clear, like the Earth's after a drenching rain, but even more striking, to the point where distances could not be estimated, and the landscape seemed to roll on and on forever. Returning lunar visitors were always filled with a near-religious feeling, incapable of fully describing their experience—"serenity . . . peacefulness . . . unreal clarity." One astronaut was flabbergasted by the "unbelievably beautiful naked charcoal ball," while another tried to capture the experience with "The serenity of it. Pristine purity," and a third said that, while the Moon is gray, "until you've been there, you have no idea how many shades of gray there are" (Aldrin called it "ash-cocoa"). There were two kinds of "daylight": blinding klieg-lamp sunlight, and blue-white earthlight—the Earth being eight times brighter on the Moon than any full moon is for Earth. Two very different kinds of light, playing in contrary ways across the terrain.

"I think the feelings I had the whole time was the feeling of awe," Charlie Duke recalled. "The Moon was the most spectacularly beautiful desert you could ever imagine. Unspoiled. Untouched. It had a vibrancy about it. And the contrast between the moon and the black sky was so vivid, it just made this impression of excitement, and wonder." Even the austere Gene Cernan would say that, on the Moon, "What I was feeling at that time, science and technology has no answers for."

Overhead was no blue sky, but one infinite expanse of velvet black. "The sun itself was brighter than any sun that I had ever seen, in New Mexico or anywhere else, in a desert-like landscape," Jack Schmitt said. "But most hard, I think, to get used to was a black sky, an absolutely black sky. The biggest problem I think photographers have in printing pictures from space is actually finding a way to print black, absolute black. Certainly slides that you show will have a little bit of blue in that background, and you're just never going to get the contrast that we had visually on the Moon, because the sky was black."

At one moment, Armstrong realized that he could extend his fist and, using only his thumb, blot out the Earth. Asked later if this made him feel like a giant, he said, "No, it made me feel really, really small." Other astronauts echoed this sentiment. "[The Earth is] very delicate," Bill Anders said. "It reminded me of a Christmas tree ornament." Stu Roosa remembered: "It's the abject smallness of the Earth that gets you." Another said, "Everything that I know, my family, my possessions, my friends and my country, it's all down there on that little thing. And it's so insignificant in this great big vastness of space."

"From space there is no hint of ruggedness to [the Earth]; smooth as a billiard ball, it seems delicately poised on its circular journey around the Sun, and above all it seems fragile," Michael Collins wrote. "Is the sea water clean enough to pour over your head, or is there a glaze of oil on its surface? . . . Is the riverbank a delight or an obscenity? The difference between a blue-and-white planet and a black-and-brown one is delicate indeed."

Working in their three-part space suits was something like wearing a house, so protected as to be divorced from any sense of the environment. The experience of being extraterrestrial is muted—no touch, no smell, no sound, and even vision is necessarily tinted. "When you're on the moon, there's very little audio around you, only the sounds of your suit—the hum of pumps circulating fluid," Buzz Aldrin said. "But you don't hear any amplified breathing inside your mask; that's a Hollywood contrivance. The name of the game on the moon is stay cool and don't exert too much so you're never out of breath."

So many science fiction writers have imagined the first human setting foot on another planet, but none imagined that, when it actually happened, the rest of the world would be watching it all, live on television . . . or that the first men on another planet would be like little children, stumbling about in their inflated snowsuits, hopping with glee in the reduced gravity, to-do lists sewn into their sleeves. Because the space suits' tinted visors hid the faces of the men inside, making them anonymous, anyone watching, no matter his gender, race, or nationality, could imagine himself, a fellow astronaut, exploring the Moon.

To the estimated six hundred million (and others have guessed over one billion) earthlings watching it all in their living rooms, the sight was at the same time dreamlike and spectral. The 240,000-mile transmission was drained of color and detail, the visual equivalent of a radio station heard roughly through static. Today, the films of Apollo 11 look like moments from the dawn of cinema, silent and jerky, grainy and blurred. If the Hasselblad's exposures could never capture the otherworldly contrasts of the black sky, the overpowering white-hot light of the sun and the blue-white glow of the Earth, or the variegated landscape of our black Moon, there is even so, a joy and wonder in everything about those films and photographs. Everything about them looks so unbelievable that it's not difficult to understand why so many Americans now believe that the entire episode was all a sham, that we never made it to the Moon at all. At the same time, it was the most extraordinary thing anyone had ever seen in human memory, a revolutionary moment of history that seemed to bathe the world in a manifest sense of achievement, in a future of unlimited potential.

Because the Manned Spaceflight Network tracking station in Carnarvon, Australia, had no television reception, the Australian Broadcasting Corporation arranged for a set to be hooked up at the local movie theater. Every seat was taken. Those in the balcony watched the lunar landing with binoculars.

The global audience was so engaged that Italy's crime rate sank to a new low.

Anyone in an American city who wasn't watching at that moment would have known something had happened: there were screams of joy and triumph, the pop of celebratory firecrackers, and a note with a bouquet left at the Arlington gravesite of John F. Kennedy that read, "Mr. President, the *Eagle* has landed."

18

To Rediscover Childhood

There were so many last-minute changes to the Apollo 11 mission that a number of crucial chores couldn't even be included in the task lists sewn to the arms of Aldrin's and Armstrong's space suits. One of these was to show the TV camera that, bolted onto *Eagle's* bottom stage, was a plaque engraved with the signatures of Collins, Armstrong, Aldrin, and President Nixon, with the inscription:

Here Men from the Planet Earth First Set Foot Upon the Moon
JULY 1969 A.D.
We Came in Peace for All Mankind

"There was a committee formed to determine how to celebrate the first moon landing," technical services department director Jack Kinzler remembered, "and I said, 'Well, the first thing I'm caught up in is the idea we should have a plaque. We ought to have something with words on it indicating the crew's names and when they landed and where they came from.' I returned to the shops, and we went ahead and we designed a stainless steel plaque that would fit nicely on the center leg of the descent ladder. Whenever they would come down from the moon and step off, that plaque would be mounted right there in the immediate vicinity, and it would be there forevermore because it was mounted on the descent stage, which stays on the moon." Public affairs assistant administrator Julian Scheer:

Peter Flanigan [from the White House] called. There was one thing the President wanted changed—the plaque to be left on the lunar surface, which read "We Came in Peace for All Mankind." The President wanted "Under God" inserted after the word "Peace."

"Peter," I said, "there is no universal god. We do not want to offend any religion . . ."

"Julian," he said, "the President was insistent."

I protested again.

"Dammit, Julian, the President wants that change. The president is big on God."

"What?"

"Julian, Billy Graham is here nearly every Sunday. The President wants 'God' on the plaque!"

There was nothing left to do but say "yes."

It occurred to me that in the rush of events, no one would remember. That worked out.

Using the rope and pulley known as the "Brooklyn clothesline" that extended from the LM's hatch to the ground, Aldrin passed a bag with the EVA Hasselblad down to Armstrong, whose first assignment was to immediately grab a rock, just in case there was an emergency abort. Instead, he became so engrossed in taking pictures with the camera pinioned to the front of his suit (the two would in their limited time on the Moon capture 857 black-and-white and 550 color photographs) that Mission Control had to nag him three times about the sample. "I'm going to get to that just as soon as I finish these picture series," the commander finally said, irritably.

"As I recall the story, Al Shepard bought the first Hasselblad in a shop down in Florida just before his suborbital flight," Jack Schmitt said. "NASA was not going to supply him with a camera, which shows how every once in a while you come into little glitches of thinking. After that, Victor Hasselblad, the owner of Hasselblad in Sweden, volunteered to adapt it in various ways for more easy use by the astronauts. He introduced for the first time a motorized winding system, a trigger system so that we could operate it on the surface. Large magazines that they hadn't had before. The rissole plate that gives the little plus marks on the pictures; that was put into it for geometric control of the developed film. All of these things came as a result of Hasselblad's personal effort." In time, the black skin of Apollo 11's Hasselblad would become the preferred color for all professional cameras, a design element still in use forty years later.

When it was his turn to take pictures, for some reason, Aldrin took only four of Armstrong, each portraying a tiny figure in a huge panorama featuring the LM. Today, there is only one decent photograph of Armstrong on the Moon—one he took himself, reflected in Aldrin's gold visor. Buzz Aldrin: "It wasn't until we were back on earth . . . that we realized there were few pictures of Neil. My fault perhaps, but we had never simulated this in our training." Photographic technology lab manager Richard Underwood:

On Apollo 11, nobody wondered why we never released any pictures of Neil Armstrong on the Moon. Because there weren't any. But we were told, "Don't mention it." And nobody in the news media picked this up.

I can't figure that out to save my life, why every picture released was Buzz Aldrin, because Buzz was mad at Neil, didn't take his picture? Got hundreds of the other eleven guys walked on the Moon, none of number one.

Even PAO [Public Affairs Office] for a while thought of, "Why don't we say this picture by the flag is Armstrong? How do you know? You can't see his face or anything."

I said, "Well, there's some nine-year-old kid out there who's a space groupie and he knows every aperture and wire and seam in a space suit. The day after you publish it, *The New York Times* is going to have a letter from a nine-year-old kid saying, 'No, you're wrong. That's Buzz Aldrin.'"

"Well, don't mention it." So that's the way that sort of worked for years. Nobody brought that idea up. A lot of things weren't mentioned and got away with from that standpoint.

After handing the camera over to Aldrin, Armstrong started collecting his geologic samples, using a collection tube to burrow into the ground. "This is very interesting," he said to Houston. "It's a very soft surface but here and there where I plug with the contingency sample collector, I run into a very hard surface but it appears to be very cohesive material of the same sort. I'll try to get a rock in here. Here's a couple." When he then tried using the scoop to put the contingency sample in his leg pocket, however, the space suit's pressurized rigidity made the job impossible, and Aldrin had to help.

While Aldrin set up the equipment needed for in-depth sample collecting, Armstrong placed the television camera on its tripod, and turned and paused it every few degrees to send a panorama back home. Aldrin then unfurled the solar wind collector, a sheet of aluminum on a stand (like a home movie screen), which would become impregnated with noble gases—ions of argon, neon, and helium—blown to the Moon by solar winds. After a timed exposure, the sheet would be rolled up and sent for diagnosis to a Swiss lab, back on Earth. Aldrin had trouble getting the feet into the ground, which, at around four inches deep, turned out to be solid rock.

Armstrong later confessed to Alan Bean that their next task was the most difficult and daunting of them all: raising the American flag. It turned out that, contrary to many geologists' conjecture, the Moon's surface (at least in the Sea of Tranquility, where *Eagle* had landed) was a very thin sweep of dust covering hard, dense, impenetrable rock (just like the deserts of Arizona). After pounding and sweating away at the task for much too long, Aldrin and

Armstrong could only get their flagpole sunk a few inches deep, at best. Both were convinced that, live on television with billions watching, they would step back from the flag, only to see it topple over into the dust. Armstrong tried patting a mound of lunar soil at the base to stabilize it, but it remained so precariously upright that he and Aldrin spent the rest of their moonwalk carefully avoiding going anywhere near the flagpole.

"We wanted the flag to be able to suspend itself nicely, and we knew there's no atmosphere on the moon to speak of, so what we did was sewed a hem in the top of the flag, and we made an aluminum flexible tube that slipped through the hem, and you could take hold of it and you could pull it by hand until you extended the top out, and there's a latching effect," as technical services department director Jack Kinzler described it. "Well, as it turned out, when they landed on the moon and they went out and started to deploy it, they saw the rippling effect that it had if you left that extender there slightly short. So they decided to take the picture that way, and I'm so glad they did, because it makes it look more realistic, like it's fluttering in the wind."

The flag was another Apollo 11 element that originated in controversy. Since the UN had passed legislation forbidding territorial conquest in outer space, Congress had argued over whether the men should plant a United States or a United Nations flag, with some insisting that a Christian flag be included. This issue was being debated just as NASA's appropriations bill for the following year was under consideration, and finally a bill was passed decreeing that "the flag of the United States, and no other flag, shall be implanted or otherwise placed on the surface of the moon, or on the surface of any planet, by members of the crew of any spacecraft . . . as part of any mission . . . the funds for which are provided entirely by the Government of the United States. . . . this act is intended as a symbolic gesture of national pride in achievement and is not to be construed as a declaration of national appropriation by claim of sovereignty." NASA agreed that Apollo 11 would carry the American flag on June 10, so late in the mission process that it had to be transported to the Moon strapped to the LM's ladder, protected by a steel-and-Thermoflex casing from the 2,000-degree temperatures at touchdown. NASA had kept secret the manufacturer of the flag, insisting that it had been bought anonymously. But the president of flag-maker Annin learned that it had been bought at Sears, an exclusively Annin retailer, and begged the agency's Public Affairs Office to publicly acknowledge this. Eager to avoid any appearance of NASA or its astronauts making commercial endorsements, however, NASA refused, saying, "We don't want another Tang."

When the broadcast showed their digging finally accomplished and Aldrin smartly saluting the American flag, everyone in MOCR gave the men a loud and long standing and cheering ovation. Joan Aldrin was sobbing with joy. "Being able to salute that flag was one of the more humble yet proud experiences I've ever had," Aldrin has said. "To be able to look at the American flag and know how much so many people had put of themselves and their work into getting it where it was. We sensed—we really did this almost mystical identification of all the people in the world at that instant."

Collins [from *Columbia*]: How's it going?
Joan Aldrin: He doesn't know what's going on. Poor Mike!
McCandless: The EVA is progressing beautifully. . . . I guess you're about the only person around that doesn't have TV coverage of the scene.
Collins: That's all right, I don't mind a bit. How is the quality of the television?
McCandless: Oh, it's beautiful, Mike. It really is.
Collins: Oh, gee, that's great!

"[At that moment] I didn't have any great feeling of 'We've done it!'" Collins later said. "We've done part of it. I was a lot more worried I guess about getting them up off the Moon than I was about getting them down onto the Moon. The motor on the Lunar Module was one motor, and if something went wrong with it, they were dead men. There was no other way for them to leave."

At that moment, Houston patched through a call to the Moon from Richard Nixon. It turned out that Armstrong knew that the president might be calling, but hadn't bothered mentioning it to Aldrin.

"Hello, Neil and Buzz, I am talking to you by telephone from the Oval Room at the White House, and this certainly has to be the most historic telephone call ever made. I just can't tell you how proud we all are of what you . . . For every American, this has to be the proudest day of our lives, and for people all over the world, I am sure they, too, join with America in recognizing what an immense feat this is. Because of what you have done, the heavens have become a part of man's world, and as you talk to us from the Sea of Tranquility it inspires us to redouble our efforts to bring peace and tranquility to earth. For one priceless moment, in the whole history of man, all the people on this earth are truly one. One in their pride in what you have done, and one in our prayers that you will return safely to earth."

"Thank you, Mr. President," Armstrong said. "It's been a great honor and privilege for us to be here representing not only the United States but men of peace of all nations."

Meanwhile, Earth's geologists, watching Armstrong and Aldrin spend precious moments raising the flag, jumping around to test their suits, talking on the phone, putting on a television show, and taking each other's pictures, grew more and more impatient. Selenologist Harold Urey finally erupted at the TV, "Oh, hurry up and get the samples!" When NASA had asked which geologic laboratories would have the wherewithal to study lunar rock, 150 had signed up, and they were all now anxiously waiting for them.

As they began their geological digging, Aldrin and Armstrong found that it was, again, much harder work than expected. Aldrin was trying to get a core sample of eighteen inches, but the Moon was so hard that, even hammering until there were dents in the collection tube, he could only strike five inches deep. The men did, however, find it a great pleasure to work in the one-sixth gravity of the Moon, even though the first-generation space suits were not well designed at the joints. Knees did not bend easily, and walking was mostly a matter of using the ankles and toes; stopping or changing direction took some practice. But the rapturous lightness of being . . . The lunar space suit and PLSS backpack weighed 180 pounds on Earth. On the Moon, the combined 360 pounds of man, suit, and survival box evaporated into a mere wisp of 60 pounds. Walking became a series of floating hops, and every step released an eruption of powder that lifted with geometrical precision and settled like tiny blasts. The most efficient method of getting around turned out to be the bouncy gait that became known by future moonmen as the "kangaroo hop."

"To fall on the moon is to rediscover childhood," Dave Scott said. Neil Armstrong: "You can actually just fall over on your face like a dead man, right down to the surface, and push yourself back up. . . ." Said Apollo 12's Alan Bean, "Every time I threw something, whatever it was, it looked like it went as high and far as the best punt you ever saw in the NFL."

Aldrin said it was "a unique, almost mystical environment . . . The moon was a very natural and very pleasant environment in which to work. It had many of the advantages of zero-gravity, but it was in a sense less lonesome than zero-g, where you always have to pay attention to securing attachment points to give you some means of leverage. In one-sixth gravity, on the Moon, you had a distinct feeling of being somewhere, and you had a constant, though at many times ill defined, sense of direction and force. . . ."

Armstrong went to investigate a nearby crater, where he found a number of unusual rocks, but he was uncertain whether he could climb down and back up the fifteen-to-twenty-foot slope of its walls in the time NASA had allotted. "We got excellent sampling, we got excellent photography, the procedural stuff went very well; but until Apollo 17 we really did not get very much good, solid descriptive work, with one exception—Neil Armstrong, for a very short period of time," said geologist astronaut Jack Schmitt. "He was probably the best observer we sent to the Moon, in spite of very limited training; he just had a knack for it. . . .

"There was one incident involving Buzz Aldrin, where during his description of the Moon he referred to what he was seeing sparkling on the surface as 'looking like mica.' That's all he meant: it looked like mica. Unfortunately a bunch of the scientists—bless their black little hearts—criticized Buzz, and it came out publicly that they criticized him, for saying something looked like mica. There couldn't be mica on the Moon because there is no water; but it looked like mica, and that's all we wanted Buzz or anybody else to say! These little pieces of glass were sparkling just like he'd seen mica sparkling in streams and rivers. It was exactly what we wanted. And unfortunately, the public comments—not a lot, but the public comments that the astronauts heard, criticizing Buzz for saying that—turned them off from giving us good descriptive information."

Aldrin's EVA work included setting up the rest of the Early Apollo Surface Experiment Package, which included a lunar seismometer to measure moonquakes, and a Laser Ranging Retro-Reflector—a mirror of black quartz crystals that, by reflecting a laser beam shot from Earth, gave an absolute measure of Earth-to-Moon distance. Even though these small and primitive instruments would prove to be very useful, especially the LRRR, which would be so accurate as to provide support for Einstein's Strong Equivalence Principle (that all forms of matter accelerate at equivalent rates under the force of gravity), the relative power of NASA's engineers over its scientists was underscored by the limited amount of scientific work the *Eagle* crew was allowed to do. The Manned Spacecraft Center's science director, Wilmot Hess, was so frustrated that he would resign ten days later. Don Lind:

> One component of the scientific community got General Sam [Samuel C.] Phillips to agree that no science would be done on the first mission that couldn't produce results in ten minutes. Because if you have to abort after you've been there for twenty minutes, and you've started to set up a science station that's going to take ninety minutes to set up, you've aborted and got nothing to show for it. So they got this ten-minute rule, which obviously means that the only thing you can do is pick up rocks. Guess who got that rule through?
>
> Well, the rest of the scientific community rose up in righteous indignation, and said, "Hey, that cuts out everything except geology." General Sam did the logical thing. He said, "Well, if you can come up with any science that you can

do in ten minutes, it can go on Apollo 11." The scientific community decided collectively that the most important single piece of equipment they could send up there was a seismometer, because that's how we understand the interior of the Earth [by detecting] earthquake [generated] seismic waves [as they] pass through the Earth. We did not expect tectonic earthquakes up there, but we did expect meteor-impact moonquakes, and so a seismic station was extremely important. A [normal] seismic station is the size of a small room in the basement at some university. [Bendix Aerospace Group up in Ann Arbor, Michigan] had miniaturized that into a twelve-inch-diameter can sixteen inches high. It was an absolutely magnificent piece of scientific equipment.

But this had to be set up in ten minutes. This was so late in the procedures that Buzz, who was [going] to deploy this, had never seen a deployment. So I flew down to the Cape. I said, "Level it with a bubble level; you know what they look like. Line it up with a gnomon. You know what they look like. And then there's a handle that looks like a cane. Reach under the handle and squeeze the trigger. It's going to start doing things. Now, don't move, because if you move, you'll bust it. It's going to start deploying solar cells all the way around you, and when it quits moving, then just back out slowly and go away."

When he got back, he said, "Oh, Lind, that was fantastic. I wish we'd had the television set up to see that."

This deployed about thirty feet from the Lunar Module, and during the first night on the Moon, the crew [was] sleeping before the takeoff the next morning. And the seismic station picked up something. They didn't think it was a meteor impact at a distance. It was something really close.

So the science team called the Mission Control Room and said, "Did some relief valve just go off, or did some mechanical [operation] go on in the Lunar Module?"

They went down the line and quizzed every single [controller.] The last one on these surveys was always the flight surgeon. When they got to him, he said, "Oh yes. Exactly at that moment, Neil turned over in the hammock."

Aldrin continued taking pictures of the panoramas, and one of his bootprint in the lunar soil, both with the Hasselblad and with the "Gold" Surface Close-Up 3-D stereoscopic camera, named for radio astronomer Thomas Gold, whose theory of an immensely dusty lunar surface NASA was disproving with every step taken by the Apollo 11 crew.

"The one thing that gave us more trouble than we expected was the TV cable," Armstrong said. "I kept getting my feet tangled up in it. It's a white cable and was easily observable for a while. But it soon picked up the black dust which blended it in with the terrain, and it seemed that I was forever getting my foot caught in it. Fortunately, Buzz was able to notice this and keep me untangled. Here was good justification for the two men helping each other. There was no question about that, either; he was able to tell me which way to move my foot to keep out of trouble. We knew this might be a problem from our simulations, but there just was no way that we could avoid crossing back and forth across that cable. There was no camera location that could prevent a certain amount of traverse of this kind."

One viewer at home shared Armstrong's fears about that cable. "When I watched my dad bouncing about on the moon, I was scared," Buzz's son Andrew remembered. "Kids that age are worried about their parents embarrassing them, and that was me too. All I could think was: 'There's a bunch of cables on the ground, he's going to trip and fall and he's going to lie there like a bug on his back in front of three billion people and every one of my classmates.'"

Houston finally warned Aldrin that his time was up and he had to get back to the ship, and nagged Armstrong to make sure that everything on his agenda had been completed.

> CapCom McCandless: You have approximately three minutes until you must commence your EVA termination activities. . . . Neil, this is Houston. Did you get the Hasselblad magazine?
> Armstrong: Yes, I did. And we got about, I'd say, twenty pounds of carefully selected, if not documented, samples.
> McCandless: Well done.

"You know how it is—you have this gung-ho, can-do attitude and you don't want to admit that they are asking too much of you," said Skylab astronaut and solar physicist Ed Gibson. "But I really think they should allow the crewmen up there to set the pace. You can really get yourself into a box by trying to do too much."

Climbing back aboard *Eagle*, Aldrin was reminded by his commander that he hadn't yet left behind their cache of ceremonial memorabilia: Patches honoring Apollo 1's Grissom, White, and Chaffee; Soviet medals for Komarov and Gagarin; a silicon wafer etched with goodwill speeches from seventy-three heads of state; and a gold olive branch that matched the three that the men would take back home for their wives.

On February 25, 1969, administrator Tom Paine had initiated a Committee on Symbolic Articles to poll NASA managers for their views on what ceremonial objects should be left behind by the first men on the Moon. Gilruth suggested "something symbolic of mankind"; von Braun wanted a proclamation from Nixon on the historic meaning of the event; Langley director Edgar Cortright offered materials from the Wright Brothers and Robert Goddard, copies of the Bible and the Koran, a Star of David, and the genetic code; Max Faget thought a statue of a man and a woman made sense; and others wanted a memorial for those who'd died in service to the exploration of space. Paine asked Washington's 136 ambassadors if their nations would like to prepare a message; seventy-three responded, and their comments were included, along with remarks from Eisenhower, Kennedy, Johnson, Nixon, Glennan, Dryden, Webb, Seamans, Paine, Phillips, and seventeen other NASA executives, all engraved onto a silicon wafer that could be read with a sixty-power microscope. Armstrong and Aldrin had hoped to conduct an actual ceremony, but they'd run out of time. Buzz pulled the pouch from a shoulder pocket, and lofted it gently down to the surface. The three had agreed to authenticate all the items collectively on both *Columbia* and *Eagle* as "carried to the Moon," so that at least Michael Collins wouldn't be left out of that part of the mission.

GET 111:55: During their walk, Aldrin's highest heart rate was measured at 125, while Armstrong's hit 160 during his last chore, when he rushed to load the rock samples and heavy equipment from the ground to the LM hatch with his "Brooklyn clothesline." At that time, he discovered a unique quality to its design: "The LEC was a great attractor of lunar dust. It was impossible to operate the LEC without getting it on the ground some of the time. Whenever it touched the surface, it picked up a lot of the surface powder. As the LEC was operated, that powder was carried up into the cabin. When the LEC went through the pulley, the lunar dust would shake off, and the part of the LEC that was coming down would rain powder on top of me, the MESA, and the SRCs so that we all looked like chimney sweeps."

Their time was now up. When American astronauts come back from outer space, they are always asked, "So, what was it like?" and besides the general lack of verbal firepower endemic to servicemen in general and test pilots in particular, another reason that most have little to say is that they work like dogs. "The awe and wonder is pushed into the background and your immediate occupation is a series of mundane chores that have to be done," Michael Collins said. "Too bad in a way. There you are suspended in the most incredible position and everyone asks you, 'Really, what was it like?' and it's a little frustrating not to be able to tell them. But my whole attention was riveted on the next job in line. I wasn't blocking it out, either, you know, ignoring it, not looking down—nothing would have been more pleasant than just to float around and look." There was in fact a reason for subjecting them to this strenuous activity, discovered by a NASA psychiatrist who spent two days in an isolation tank with scuba gear to see for himself the effects of forty-eight hours in space: "I thought a little, and then I stopped thinking altogether . . . incredible how idleness of body leads to idleness of mind. After two days, I'd turned into an idiot. That's the reason why, during a flight, astronauts are always kept busy."

"When you're part of the pioneering effort, there's a focusing of an individual's concentration and level of attention that is at the exclusion of a lot of other things," Buzz Aldrin said. "It's a kind of gun-barrel vision. . . . Our surface activity was limited to two hours and forty minutes, and every minute was busy. Both of us were concerned with the next item on the agenda to be accomplished. This took precedence over any reflective thinking we might have been able to do. We understood the significance of what we were doing. I felt like we were not alone—we had people listening and looking at everything we were doing and I had the impression of being on center stage during the entire operation."

Neil Armstrong: "It would have been nice from our point of view to have had more time to ourselves so that we could have gone out and looked around a little bit. . . . I do remember thinking, 'Gee, I'd like to stay out a little longer, because there are other things I would like to look at and do.'"

Back inside *Eagle*, Aldrin and Armstrong removed their backpacks, outer suits, gloves, and boots, and along with used filters and food packs, the cameras, the visors, and their urine bags, threw it all out of the hatch to leave behind on the Moon's surface in order to lighten their ship's weight for the return flight to *Columbia*. They stowed forty-one pounds of lunar samples into their aluminum cargo boxes, and reattached themselves to the LM's support system.

> CapCom: We observe your equipment jettison on TV and the passive seismic experiment reported shocks when each PLSS hit the surface. Over.
> Armstrong: You can't get away with anything anymore, can you?

The two men ate Vienna sausages, drank fruit punch, and answered questions posed by scientists and engineers standing by at Mission Control. Neil Armstrong: "I don't know just what the temperatures were outside. I've heard it guessed that they were only zero to one hundred degrees Centigrade [32–212 degrees Fahrenheit]. I really wasn't aware of any temperatures inside the suit. And at no time could I detect any temperature penetrating the insulated gloves as I touched things—the LM itself, things in the shadow, things in the sunlight, the tools, the flagpole, the TV camera, the rocks that I held.

"Inside the *Eagle*, once we had repressurized the cabin and removed our helmets, there was one more little surprise. There was a decided odor in the cockpit. It smelled, to me, like wet ashes in a fireplace. It seemed presumptive to jump to the conclusion that the lunar material we had taken into the cabin was causing this odor, but we had to guess that to be the case."

The two astronauts had worked for over twenty-two hours without a pause, and needed to sleep. "We cleaned up the cockpit and got things pretty well in shape," Armstrong said. "This took us a while, and we planned to sleep with our helmets and gloves on for a couple of reasons. One is that it's a lot quieter with your helmets and gloves on." Aldrin: "We wouldn't be breathing all that dust."

Buzz took a hammock; Neil sprawled out on the floor. But, as Aldrin noted, "It was very chilly in there. After about three hours, it became unbearable. We could have raised the window shades and let the light in to warm us, but that would have destroyed any remaining possibility of sleeping." Armstrong: "One is that it's noisy; and two is that it's illuminated. We had the window shades up [that is, covering the windows] and light came through those window shades like crazy. The next thing is that there are several warning lights that are very bright and can't be dimmed. The next thing is that there are all those radioactive illuminated display switches in there."

The commander found that his position put him directly in the line of sight of the craft's telescope, which at that moment was focused on the Earth. For the exhausted but restless flier, it seemed as if a huge, unblinking blue eye was staring down at him. "I was on the engine cover with a loop that I'd rigged up of some kind to hold my legs, hanging from something up there," he said. "And my head was back to the rear of the cabin and there was a glycol pump or a water pump or something very close to where my head was. But the temperature control was probably the most troublesome."

There was a constant racket. "When Neil Armstrong and what's-his-name landed on the moon and we haven't heard anything about it since then, they were scared to death every time a switch went on and a little pump went on in the lunar module," joked spacecraft manager Ernie Reyes. "They were up there on the moon on a strange planet, and all of a sudden, the air-conditioning goes on or the heater comes off or something. If I was on the moon, if you're on the moon, you're going to worry all night! Every time something goes click, you're going to say, 'What was that? Oh my God, what is that?'"

Aldrin ended up sleeping a mere four hours, and Armstrong scarcely three. That same night, Russian probe Luna 15 approached its landing site about five hundred miles east of Tranquility Base, in the Sea of Crises, with Tass reporting that it "left orbit and achieved the Moon's surface in the preselected area." It was to have scooped up lunar samples and returned to Earth before Apollo 11, giving the Russians another win in the Space Race. Instead, it crashed into the Moon and was lost.

Far above *Eagle* Michael Collins continued in his glorious solo orbit, and got a very good night's sleep. "Not since Adam has any human known such solitude as Mike Collins is experiencing during this forty-seven minutes of each lunar revolution when he's behind the moon with no one to talk to except his tape recorder aboard *Columbia*," NASA's press office said of the Command Module pilot. "While he waits for his comrades to soar with *Eagle* from Tranquility Base and rejoin him for the trip back to earth, Collins, with the help of flight controllers here in Mission Control, has kept the command module's system going 'pocketa-pocketa-pocketa.'"

Many had predicted that flying alone over the far side of the Moon, out of touch with all human contact, would be excruciating, but Collins did not find it any more unnerving than when he flew F-86s across the eternally desolate winters of Greenland. And there was a rhythm to lunar orbits of blinding sun, soft earthshine, and impenetrable blackness that echoed, in its alien way, a kind of life, of dawn and dusk and the pitch of night.

On his return home, Collins received a letter from the ultimate solo pilot, Charles Lindbergh, that said: "I watched every minute of the walk-out, and certainly it was of indescribable interest. But it seems to me that you had an experience of in some ways greater profundity . . . you have experienced an aloneness unknown to man before." Collins himself had a very different reaction, writing that, "I am alone now, truly alone, and absolutely isolated from any known life. I'm it . . . I feel this powerfully—not as fear or loneliness—but as awareness, anticipation, satisfaction, confidence, almost exultation. I like the feeling."

Actually, during this period of the mission Collins was indeed a little on edge. First he'd received a disturbing message from Mission Control: "We show your EVAP OUT temperature running low. Request you go to manual temperature control and bring it up. You can check the procedures in ECS MAL 17." This meant that the coolant temperature control was malfunctioning—something like a radiator about to blow, but on a mission-threatening scale—and that Collins was to perform "Environmental Control System Malfunction Procedure 17," which turned out to be an incredibly elaborate and complicated business. He instead decided to try a more direct approach: switching the control from automatic, to manual, and then back to automatic. In time, this solution worked, and the coolant temperature returned to normal.

For the rest of that day, Collins spent countless hours peering through the sextant, trying to pinpoint his crewmates' landing spot. He never could determine where *Eagle* was, exactly, nor could Houston. *Eagle*'s pings reported that it was north of the originally intended site; aggs indicated that it had landed in the middle; and the tracking stations back on planet Earth insisted it was to the south.

Using this data, NASA mapping scientist Lou Wade and his team plotted fourteen possible landing points, while a band of geologists, using Armstrong's descriptions of the terrain, insisted there was only one possible answer: West Crater. They sent NASA their reasoning, but it wasn't taken seriously at Mission Control. On July 21 at 8:30 a.m. CST, Collins radioed from *Columbia,* "You've given up looking for the LM, right?" and Houston replied, "Affirmative."

In fact, the geologist team would be off by a mere two hundred meters.

19

A Tenuous Grasp

G **ET 124:03 (T minus 19 minutes):** After a too-early wakeup call from Houston and a freeze-dried breakfast of bacon, peaches, and cookies, Aldrin and Armstrong began preparing for what most at NASA considered the most perilous moment in their voyage: lunar liftoff. Grumman lunar module program director Joseph Gavin: "The critical thing was the takeoff, because you had a limited time, you had to punch the button, and everything had to work. The ascent engine had to ignite. The explosive bolts had to explode. The guillotine had to cut the connections, and then it had to fly up. And this is something we never saw happen until the last mission. So it was all, well, hearsay. It's something we never could test for, because the conditions couldn't be duplicated on Earth."

Almost every feature of Apollo 11 had been designed with redundancy in case of failure, except, because of weight restrictions, one significant element: *Eagle's* ascent engine. Here was rocket travel as the utmost in minimal design—hypergolic fuels, the only moving part a ball-valve allowing fuel to enter the injector and thrust chamber, and one control—an on/off switch. LOR evangelist John Houbolt called it "one of the best-tested engines in the universe," having been put through its paces numerous times on the ground and on Apollos 9 and 10. Even so, many of those tests had proven the machine to be maddeningly error-ridden, with bomb test instabilities (where bombs are exploded within the rocket engine, to see how it recovers) of 1 in 667—a very poor record.

The procedure was simplicity itself. Armstrong only needed to flip the switches marked "Abort Stage" and "Engine Arm," and Aldrin would then push the button "Proceed," and they would be launched. The rocket was so powerful, and *Eagle* so light, that on ignition, it would propel the ship from zero to three thousand miles per hour in two minutes. But rockets have a way to them; there's something moody about explosives.

"What if that burner would not ignite?" Amstrong's mother, Viola, wondered.

"What if it simply would not ignite? I was silently in deep concentration with our Lord. It seemed to take a while for the boys to get ready to leave. They seemed to be making sure they were doing it all just right. There were many last things to do."

In an emergency the best Mike Collins could do was lower *Columbia* to fifty thousand feet, but since some of the Moon's mountains were known to be thirty thousand feet high (and since there was a great deal of lunar terrain whose height was unknown), that was as low as he could safely approach. Collins had been trained in eighteen different methods of emergency rescue rendezvous, yet: "I have skimmed the Greenland ice cap in December and the Mexican border in August; I have circled the earth forty-four times aboard Gemini 10. But I have never sweated out any flight like I am sweating out the LM now. My secret terror for the last six months has been leaving them on the moon and returning to earth alone. If they fail to rise from the surface, or crash back into it, I am not going to commit suicide; I am coming home, forthwith, but I will be a marked man for life and I know it. . . . I would do everything I could to help them. But they know and I know, and Mission Control knows, that there are certain conditions and malfunctions where I just simply light the motor and come home without them."

Alongside the first launch of Alan Shepard, the difficult homecomings of Carpenter and Glenn, Apollo 8's return of radio contact from the far side of the Moon, and the Moon landing itself, this would be the tensest moment in the history of Mission Control.

GET 124:05 (T minus 17 minutes): After CapCom Ron Evans told the men they were go for takeoff, Aldrin replied, "Roger, understand, we're number one on the runway." To prevent the same computer overload problem that had plagued descent from happening on ascent, Houston had Aldrin turn off his rendezvous radar. Armstrong tested the attitude thrusters, and Aldrin fired two small charges that opened the engine's helium tanks. On the LM's control panel, however, the tanks' pressure gauge did not move. Aldrin tried again, and the same thing happened. They continued with the countdown, hoping that the problem was a stuck gauge, and not with their actual fuel.

GET 124:20:58 (T minus 1 minute 2 seconds): Aldrin armed the explosive bolts that would separate the two halves of *Eagle* and began to reach for the switch to arm the rocket that would bring her back to *Columbia*. "As I look down on the corner of the floor I saw a small black object and I immediately recognized what it was," he said. "It was the end of a circuit breaker that had broken off. So I looked up to see these rows of circuit breakers, see which one had broken off, and it was the one that armed the engine to ignite." The Command Module's key switches and dials had covers to keep them from being accidentally switched and turned by bulky spacesuits and backpacks operating in tight spaces. The same safeguards had not been included on the Lunar Module, but after this incident—the one switch that absolutely needed to work to get Armstrong and Aldrin back to Collins had broken in half—NASA would have them installed on all future missions.

Aldrin found a felt-tip pen, crammed it into the switch's remaining space, and pushed it, hard. It worked.

GET 124:22:00 (liftoff): The bolts connecting *Eagle* with its descent engine and tanks exploded. At the same moment, the 3,500-pound-thrust ascent rocket began to burn. The rickety *Eagle* began to rise.

When Joan Aldrin found out that the ascent engine had worked as planned, she was so giddy she had to lie down on the bed and kick her legs in the air. Listening in on her squawk box, Jan Armstrong collapsed in relief at this news, knowing that "As long as that thing is lit, Mike will come and get them, wherever they are. Wherever they are, he would come. Nobody told me that. Nobody ever had to. He would." When someone on the television broadcast complained that the astronauts weren't talking enough, Pat Collins argued back at the screen: "They may not be much on show biz, but the deed speaks for itself." Clare Schweickart pointed out, "Astronauts get along so well because they don't talk." Another viewer complained of CBS's melodramatic coverage, referring to their lead anchor as "Walter Crankcase," who "always likes to get you psyched up for tragedy, and then when it doesn't happen everything is all right."

One of the guests at the Collins house was Dave Scott's wife, Lurton, who brought the news that the Apollo 11 men and their wives would be leading parades in New York, Chicago, and Los Angeles, and then have dinner with President Nixon and First Lady Pat at the White House. Pat Collins said that she had expected a great public reception, something like what had happened to John Glenn, but she was sure that, by the end of the year, their lives would be back to normal. Lurton told her that, no, in fact, her life would never be the same.

Eagle's ascent was far from smooth. Computer navigation made constant adjustments to the thrusters over the 465-second rocket fire, producing a shake and a wobble. "In space, say you wanted to pitch up, you would use a down-pointing rocket in the front and an up-pointing rocket in the back," Neil Armstrong explained. "But we didn't want any rockets firing up when we're accelerating away from the Moon, because that would be wasting fuel. So we would only use the down-pointing rockets because they would be adding to our velocity, would be fuel-efficient. But the result of that is that there's a substantial rocking motion. As you pitch forward, the pitch-up thruster fires, lifts your nose up, then it stops, then the nose falls down again and the rocket fires as though you're in a rocking chair." Behind them, the ascent engine's exhaust threw debris across Tranquility Base, toppling the American flag into the dust.

Eagle soon reached a lunar orbit of 5,537 feet per second at 46.7 by 9.4 nautical miles, while Columbia was in orbit at 63.2 by 56.8. Eagle's ascent engine burned again, for forty-five seconds, raising it into Columbia's path and to a speed of 130 feet per second.

On Columbia, the guidance system tuned the sextant to where it thought Eagle might be; the mission plan was for Eagle to rise and catch Columbia, but there was an emergency procedure that would enable Collins to intercept the LM if the original procedure failed—more redundancy. It took two rotations about the Moon to line up the ships for docking, and it was only then that Collins, who had needed to keyboard 850 punches to bring his crewmen back home, was able to look through his sextant and locate Eagle for the first time since the three men had separated. As he watched the LM rise slowly to meet him, Collins would remember this as the mission's happiest moment.

To use the sextant, Collins had to fly sideways; now, he pitched Columbia over to mate. A series of very small burns lined up the two craft in the darkness on the far side of the Moon, where Armstrong could finally see his mother ship's homing light. He flipped Eagle over to reveal its drogue in Collins's reticule, so that the two ships could once again be aligned for head-to-head docking.

Armstrong: "You're not confused which end to dock with, are you? Looks like you are making a high side pass on us, Michael."

GET 127:55: After the two ships were connected by the light first-stage latches, Collins threw the hard dock controls, but then: "all hell [broke] loose. . . . The docking process begins when the two vehicles touch and the probe slides into the drogue. Neil made the first maneuvers to point his drogue at my probe, and then I took over—probably at a distance of twenty or thirty feet. . . . To make the combination rigid you fire a little gas bottle that activates a plunger which literally pulls the two vehicles together. At this point the twelve capture latches fire mechanically and you are held together very strongly.

"Just as I fired the charge on the gas bottle we got a quite abnormal oscillation in the yaw axis. We had eight or ten rather dubious seconds then, when I really thought we were outside the boundaries for a successful retract and that I was going to have to release the LM and go back and dock all over again."

The two ships were not correctly mated. As hard docking began, both tried to realign themselves, and each attempt threw off the other one, leaving them to thrash at each other. It was just this kind of moment that had been described by astronaut Dick Gordon, who said that rocketing into space was "just like the old fighter pilot's life. Long periods of boredom punctuated by moments of stark terror."

Neil Armstrong: "I allowed the platform to go into gimbal lock. [i.e., moving the ship into a position which prevented the navigation platform's gyroscopes from moving correctly, freezing the system and upending autopilot. Gemini never had gimbal lock since those capsules were equipped with four gimbals to avoid it; Apollo engineers, following a 'not invented here' philosophy, ignored that history and only provided Apollo ships with three—a significant annoyance to astronauts who knew better.] In aligning the *Eagle* with *Columbia* I was looking out the window, flying the LM to the proper roll attitude for docking, [not paying attention to the attitude indicator] and didn't notice that we were getting close to gimbal lock.

"Since we were flying digital autopilot, as soon as we went into gimbal lock the autopilot was no longer useful. It doesn't work without a stable platform reference. So that meant we had no attitude control system . . . so I switched to AGS, the abort guidance system, and flew on the alternative flight control system. I don't believe it has been done since then, and I don't recommend it."

This moment must have reminded Armstrong of the most frightening of all of the American Gemini missions, the astronaut's sole prior (and catastrophic) experience in space. Ten hours after liftoff, while 130 miles over South America and flying at 18,000 miles an hour, Armstrong and Dave Scott successfully pushed Gemini VIII's nose guide into the V-funnel of an Agena rocket (the same upper stage that launched Corona satellites) and completed the first orbital docking in NASA history. Twenty-seven minutes later, however, Scott noticed that he and Armstrong were rolling.

Armstrong used his thrusters to control the rotation, but the minute he stopped applying counterthrust, the docked pair of ships began spinning again. Armstrong: "A test pilot's job is identifying problems and getting the answers. We never once doubted we would find an answer—but we had to find it fast. We first suspected that the Agena was the culprit. We had shut our own control system off, and we were on the dark side of the Earth, so we really didn't have any outside reference, or very good reference. Neither of us thought that Gemini might be the culprit, because you could easily hear the Gemini thrusters whenever they fired. They were out right in the nose, in the back. Every time one fired, it was just like a popgun, 'crack, crack, crack, crack.' And we weren't hearing anything, so we didn't think it was our spacecraft. When the rates became quite violent, I concluded that we couldn't continue, that we had to [separate from the Agena]. I was afraid we might lose consciousness, because our spin rate had gotten pretty high, and I wanted to make sure that we got away before that happened. Although we had no way of knowing for sure, we were concerned that the stresses might be getting dangerously high that the two spacecraft might break apart. We discussed undocking, but we had to be sure that the tumbling rate at the instant of separation would be low enough to keep us from colliding moments later."

They separated, but becoming a much lighter craft only made the spinning worse. Scott then noticed their attitude fuel had fallen to 30 percent, meaning something was wrong with the engines. The fault was static electricity, leading to a short, which caused a capsule thruster to stick, firing over and over again. Armstrong: "The sun flashed through the window about once a second. The sensations were much like those you would feel during an aircraft spin. Neither Dave nor I felt the approach of loss of consciousness, but if the rates continued to increase we knew that an intolerable level would be reached." The rotation speed now rose to one rev per second, threatening to tear the ship into pieces. The disaster began at a moment between tracking stations and out of radio contact (each dish, moving into position, as the world turned), so the first words Mission Control heard were a disturbed Armstrong's: "We're tumbling end-over-end!"

The commander's solution was to switch from the craft's main thrusters to his reentry jets to calm the spinning bird, but because this procedure exhausted needed fuel, he then had to do the last thing any astronaut can bear—abort. Though the two astronauts were both still dizzy when they were safely brought home, the rescue otherwise went perfectly. Armstrong: "It was a great disappointment to us, to have to cut that flight short. We had so many things we wanted to do, and I know Dave wanted to do an EVA and try out the backpack and do all that kind of stuff. It was very disappointing to have to call it quits and come home." The abort meant no spacewalk for Scott, but it was surely an element leading to a significant future command position for Armstrong, who had gotten himself and his crew back home safely under extreme duress, the ultimate achievement for any test pilot/astronaut.

If a similar docking catastrophe had happened to Apollo 11, there was an emergency Plan B. Rails had been installed next to the hatches on both *Eagle* and *Columbia* so that Aldrin and Armstrong, carrying their rock boxes, could use them to spacewalk from one ship to the other—the only time, in fact, that spacewalking might ever be needed on an Apollo mission. For all the bravura shown during Gemini's out-of-capsule demonstrations to compete with similar Russian space feats, however, the procedure had a very poor history, since in outer space, if you want to use a screwdriver but don't have yourself fully tethered, you'll end up being the screw that turns. Buzz Aldrin: "Your body simply had to be anchored, because if it wasn't, flexing your pinkie would send you ass-over-teakettle."

As Michael Collins recalled of his spacewalk on Gemini X: " I grabbed the docking collar. It wasn't meant to be grabbed, and my momentum is still carrying me along, so I just slipped, and as I went by, then I went cartwheeling ass over teakettle, up and around and about, until I came to the end of my tether, and then it swung me in a great big arc." There wasn't enough handhold or toehold engineered to the side of the rocket to anchor a spacewalking astronaut, and Collins found himself doing double-flips, carried by his body's momentum in space. He was carrying a propellant gun, which could be fired to restore bearings in case of just such an emergency, but when he reached to pull it from its holster, he discovered that it had come loose from the pack. Mike fumbled to find the gun's power line, and finally pulled the gun back to him.

Collins and Armstrong both now tried to fix their misfires in the *Columbia-Eagle* docking. Mike Collins: "I instantly took action to correct the angle, and so did Neil in *Eagle*. . . . All the time the automatic retract cycle was in fact taking place, and we heard a loud bang, which is characteristic of those twelve big latches slamming home. And lo and behold—we were docked, and it was all over." When NASA released the Apollo 11 transcript, it removed Collins's comment at that moment: "all hell broke loose." A reporter had his own copy of the audio signal, however, and after it became public knowledge, the remark was reinstated into the record.

Mike Collins now had to dismantle the docking mechanism out of the tunnel connecting the two ships. ("Me, who couldn't repair the latch on my screen door. I hated that probe, and was half convinced it hated me and was going to prove it in lunar orbit by wedging itself intractably in the tunnel.") Finally the hatch was opened, and the three were reunited. After a few minutes of laughing and celebrating, Collins "grabbed Buzz by both ears and I was gonna kiss him on the forehead, I can remember that, and I got him to right about here and I said 'Eh, this is not a good thing to do somehow,' and I forgot, I clapped him on the back or shook his hand or something."

After the three transferred the two-foot-long, vacuum-packed safety-deposit boxes of lunar rock to *Columbia*, Aldrin and Armstrong then had to vacuum as much of the precious moon dust out of their clothing as possible, while Collins filled the LM with everything they wouldn't need for the trip home. He then released *Eagle*, which fell into lunar orbit, having to work solo as Armstrong and Aldrin did not want to be involved in orphaning their girl. In time, the world's first manned lunar spaceship would crash into the Moon, at a location still unknown.

GET 135:25: On the lunar far side, Apollo 11's computer now initiated the first step in coming home: Trans Earth Injection. Using five tons of gas in 2.5 minutes, the service module's rocket burned, boosting the ship's speed from 3,300 to 5,300 miles per hour—lunar escape velocity—rocketing itself out of orbit and onto its 3,625 miles per hour path toward Earth.

For the two and a half days it took to get back, besides executing the "barbecue" maneuver, Collins had to jiggle the craft to improve its antennae signal, not unlike viewers at home, fiddling with their TV's rabbit ears. The rotisserie swiveling that the Apollo Command Module performed created a dazzling effect, as Apollo 14's Ed Mitchell remembered: "The biggest joy was on the way home. In my cockpit window every two minutes: the earth, the moon, the sun, and a whole 360-degree panorama of the heavens. And that was a powerful, overwhelming experience. And suddenly I realized that the molecules of my body, and the molecules of the spacecraft, and the molecules in the bodies of my partners, were prototyped and manufactured in some ancient generation of stars. And that was an overwhelming sense of oneness, of connectedness. It wasn't them and us, it was 'that's me, that's all of us, it's one thing.' And it was accompanied by an ecstasy: 'Oh my God, wow! Yes!' An insight. An epiphany."

Otherwise, Apollo 11's trip home was completely uneventful . . . save for a mystery. At one point, Mission Control radio transmission picked up a disturbing series of whistles, and what sounded like bells. Houston asked Apollo 11, "You sure you don't have anybody else in there with you?" but the astronauts did not reply. At their first press conference back home, nearly a month later, the three would again be asked about this incident, and would finally admit that, before leaving, they had made a tape of sound effects and played it back as an alien prank on the ground controllers.

NASA told the crew that the lunar seismometer Aldrin had set up was working, but not the laser reflector, since they still hadn't determined exactly where *Eagle* had landed. For their next-to-last TV appearance from outer space, Buzz spread some ham salad on a piece of bread and then set the can spinning in zero-g to show how anything can be made gyroscopic, while Collins shot himself in the mouth with a water gun.

NASA meteorologists, meanwhile, were concerned about tropical storm Claudia, which was then 2,300 miles from the splashdown site, but could quickly move close enough to be a worry. The next day she was joined by typhoon Viola, which could converge with Claudia and threaten the mission. Viola would in time become a super typhoon, with 150-miles-per-hour winds killing eleven on Luzon in the Philippines.

GET 150:04: One last rocket blast aligned the CM with the U.S. Navy's waiting recovery operation of nine ships and fifty-four planes and helicopters. NASA had asked for aircraft carrier USS *John F. Kennedy* to take part as a tribute to the president's original vision; the Nixon White House gave them USS *Hornet* instead. The astronauts had one more TV broadcast to make, which included a pitch-perfect message of gratitude by Collins: "This trip of ours to the moon may have looked to you simple or easy. I'd like to assure you that that has not been the case. The Saturn V rocket which put us into orbit is an incredibly complicated piece of machinery, every piece of which worked flawlessly. . . . The SPS engine, our large rocket engine on the aft end of our service module, must have performed flawlessly or we would have been stranded in lunar orbit. The parachutes up above my head must work perfectly tomorrow, or we will plummet into the ocean. . . . All this is possible only through the blood, sweat and tears of a number of people. . . . All you see is the three of us, but beneath the surface are thousands and thousands of others, and to all those I would like to say: thank you very much." It had in fact taken 400,000 men and women, among NASA and its corporate subcontractors, to send the first men to the Moon.

Aldrin had other matters in mind: "We feel that [our mission] stands as a symbol of the insatiable curiosity of all mankind to explore the unknown. Neil's statement the other day upon first setting foot on the surface of the moon, 'This is a small step for a man, but a great leap for mankind,' I believe, sums up these feelings very nicely. . . . We've been particularly pleased with the emblem of our flight, depicting the U.S. eagle bringing the universal symbol of peace from the earth, from the planet earth to the moon, that symbol being the olive branch. It was our overall crew choice to deposit a replica of this symbol on the moon. Personally, in reflecting [on] the events of the past several days, a verse from Psalms comes to mind to me: 'When I consider the heavens, the work of Thy fingers, the moon and the stars which Thou hast ordained, what is man that Thou art mindful of him?'"

Armstrong then offered his own thoughts: "The responsibility for this flight lies first with history and with the giants of science who have preceded this effort; next with the American people, who have, through their will, indicated their desire; next with four administrations and their Congresses, for implementing that will; and then, with the agency and industry teams that built our spacecraft, the Saturn, the *Columbia*, the *Eagle*, and the little EMU, the space suit and backpack that was our small spacecraft out on the lunar surface. We would like to give special thanks to all those Americans who built the spacecraft; who did the construction, design, the tests, and put their hearts and all their abilities into those craft. To those people, tonight, we give a special thank you, and to all the other people that are listening and watching tonight, God bless you. Good night from Apollo 11."

GET 194:50: Collins shut down the Service Module's power and initiated the same jettison procedure that had separated the two halves of *Eagle*. He ignited a series of squibs, throwing a guillotine across the wires connecting the Command and Service Modules, and exploded the bolts holding the two craft together. With its Service Module disposed of, the Command Module capsule was now wholly on its own; at 12,250 pounds, it was one fifth of one percent of the full craft's original launch weight.

The last word from Houston was "Have a good trip . . . and remember to come in BEF"—"Blunt End Forward." As the ship struck the atmosphere at 39,000 feet per second (or 26,000 miles per hour—a rifle bullet travels 2,000 miles per hour) over the Solomon Islands, the crew was hit by a force of 6.3 g's and the craft by temperatures of 5,000 degrees Fahrenheit. Collins had to bring it in at an angle of between −5.5 and −7.5 degrees: too much tilt forward, and the ship would disintegrate in flames; too much tilt back, and it would bounce off the atmosphere and back into terrestrial orbit. There was not enough propellant or oxygen for a second reentry attempt.

GET 195:03: The Apollo 11 heat shield was a two-inch-thick honeycomb of carbon fiber and ablation compound, a phenolic glue resin that burned into char and evaporated away, taking the heat with it. The computer repeatedly swiveled the craft, to slow it further, and to alleviate the heavy g-forces being suffered by the crew. During the four-minute heart of reentry, radio communications were foiled by the fireball's ionizing air, which threw shots of aqua, violet, warning-lamp orange, and purple waves across the capsule—a comet tail, in reverse. As background to this dazzle of physics lay the black sky of space which, bit by bit, dissolved into the blue, and then white light of Earth. "You are literally on fire, your heat shield's on fire, and its fragments are streaming out behind you," Mike Collins said. "It's like being inside a gigantic lightbulb."

GET 195:12: At 23,000 feet, drogue parachutes were released, slowing the descent and providing visual contact to the first of the navy's tracking planes. At 10,000 feet and 175 miles per hour, the three orange-and-white-striped main chutes deployed, dropping Apollo 11 into an eighteen-knot-wind. "You have to understand that parachutes in a packed condition are a random assembly," Max Faget said. "No two will ever be packed exactly the same. You go through a packing sequence, and you try to get them exactly alike. But when you consider all the acres of cloth in a big chute, the risers, and the shroud lines, it's never going to come out exactly the same—no matter how hard you try. Another problem was that the chutes were highly loaded when they first opened. The parachutes are reefed when they first open—a line around the shrouds keeps the chutes from opening immediately. Little pyrotechnically operated cutters cut the reefing line, but all things have to work—the packing assembly, the reefing line, and the cutters—for the chute to deploy properly. The whole mission would be over, the capsule would be coming down, and people would be cheering. But for guys like me, who worried about the chute, we'd be saying, 'Don't cheer! Don't cheer yet! Wait until it's in the water!'"

Jan and Mark Armstrong played a game of "Go Fish" while waiting for splashdown, but it was too early in the morning and too dark for the television cameras to capture a visual. "On the Apollo missions that I was out on, I don't recall, ever, one of them landing far enough away from the ship that we weren't able to see it, which was pretty exciting," recovery systems chief John Stonesifer said. "Except Apollo 11. It was still dark when that landed. But the first clue we always had was the sonic boom, because when Apollo came back in, it was coming pretty much over the landing area, almost straight down, and you knew when you got that boom, boom."

In darkness, *Columbia* floated into a seven-foot-swelled Pacific Ocean at 13 degrees 19 minutes North latitude, 169 degrees 9 minutes West longitude. The force of the waves engulfed the parachutes, flipping the capsule upside down. As the 11 crew had lain on their backs for liftoff, now they hung from the ceiling on splashdown. Learning Frank Borman's lesson, they'd each taken a seasick pill at the start of reentry, and now they each took another, just in case. Collins deployed a set of airbags and a flotation collar to right the ship, while Armstrong called Air Boss to report on their condition. The three then waited, just as at liftoff, in complete silence.

Four of *Hornet*'s Sea King helicopters arrived, sending an underwater demolition team 11 member to rope the capsule with an anchor. Chippewa Falls, Wisconsin's navy frogman Clancy Hatleberg, in a biological isolation suit, leapt from Helicopter Swim Two into the water, reached the ship, and struggled to close the valves for filtering the capsule's air, which freed the hatch to open. He then passed three biological isolation garments inside to the waiting men. After they were dressed, the three inflated their Mae Wests and joined Hatleberg in a raft, where they scrubbed each other with iodine solution, and then again with bleach.

"The nation had established an Interagency Committee for Back Contamination, made up of the head of the CDC, a representative from the Department of Agriculture, the Department of Commerce," Stonesifer explained. "They were worried about contaminating the seas, the fisheries, and things. Department of Agriculture was concerned about the contamination or what you might bring back from the lunar surface that would destroy crops. The thought processes there were that the heat of reentry would possibly destroy anything that was coming back on the exterior of the spacecraft, which seemed logical. Or if when it landed into the water, the dilution factor of the ocean was another backup to the heat of reentry. But the spacecraft vented in the air. So there wasn't anything we could do about that.

"[The flotation collar] was put on primarily to give the pararescue men a platform to work from to assist astronauts out of the spacecraft. They had tanks of betadyne, and they sprayed the entire top deck of the Apollo spacecraft to decontaminate it, because that's where the ventilators were. Then a weakness in the system here, one of the pararescue men—these were Navy people, UDT, frogmen—opened the hatch and threw in what we called 'the bag of BIGs,' the bag of biological isolation garments. So here again, you open the hatch. What else are you going to do? We went through these processes with this Interagency Committee and got their approval, and we challenged anybody, 'If you can think of something else, have at it.'"

"I can remember this young frogman, Navy Seal, pulling the hatch back and poking his head in and then, with a shocked look kind of falling backwards," Apollo 8's Bill Anders said. "I didn't have time to contemplate that very much because we had to hop in the life raft which is now tied to the spacecraft and then hoisted by the helicopter onto the USS *Yorktown*. But later after we had been debriefed by the doctors and everybody, we had a chance to go out and look at the spacecraft and to meet the rescue crew. So here was these Marines, all lined up, you know, in their uniforms and I recognized the young corporal there that had first stuck his head in. And I asked him; I said 'Corporal, thanks a lot' and you know, we hadn't shaved, you know, we were dirty, I said, 'we really must have looked bad' and he said, without batting an eye, 'Sir, it wasn't how you looked, it was how you smelled.'"

The night before splashdown, Hatleberg had had a nightmare in which the astronauts had exited the capsule, but no matter how hard he tried, he couldn't close the hatch. The meaning? His anxiety that, whatever lunar contaminants had been brought home would make their way into the Earth's atmosphere, and it would be all Clancy's fault. This nightmare, in fact, came true: "I closed [the hatch], but it wouldn't lock. I just couldn't get it locked. I remembered that dream and I thought, *My God, everything is falling apart.* I motioned to the astronauts. Armstrong came back up, and we still couldn't get it closed. Then Collins came back and recycled the door, and we finally got it locked."

That open hatch was in fact a notable anomaly in the entire quarantine procedure—that, and the lack of filters on the Command Module's air vents. Since it couldn't be determined exactly how long it would take to get a landed capsule to the aircraft carrier, and the crew could not remain inside for a lengthy period, the open hatch was the less-than-ideal compromise. The BIGs, however, were not watertight, as they were supposed to be, and because they had no ventilation, they were both unbearably hot, and their faceplates kept fogging over, blinding the astronauts.

"The command module lands in the Pacific Ocean, and what do they do?" Mike Collins asked. "They open the hatch—you've got to open the hatch— and all the damned germs come out, right? They contaminate the whole Pacific Ocean! Then you've got these three guys in there. There's stuff on these suits, these BIGs. I mean, it doesn't make any sense. It was a huge flaw in the planning."

Another explanation was offered by future programs chief Jerome Hammack: "At one time the capsule was going to be lifted up with the crew intact and brought butt up against the Mobile Quarantine Facility. I began to worry about that because of the humidity, the heat. I thought it was more of a safety concern to do that. You've got to worry about the safety of the crew. You've got to worry about the lunar pathogens, but you've got to worry about the safety of the crew. What's the tradeoff?

"I hurt myself in several ways on that question, but I could not—what they wanted to do, the last straw was, they wanted to seal up the capsule. Didn't even want the filter to be working. They wanted to seal it up for a limited period of time. I know it wasn't long. But that was going to be entombing them out there, and so that's when we started objecting."

Collins, Aldrin, and Armstrong were then hoisted in baskets and carried by helicopter to *Hornet*, and before they left they gave the Apollo 11 emblems sewn onto their BIGs to the helicopter's crew. At the carrier, the men and their capsule rode an elevator to a below-hangar deck, where a modified Airstream trailer was waiting. Immediately, they felt the exhausting power of gravity. Buzz Aldrin and Mike Collins began a furious set of calisthenics. Armstrong did not join in.

The capsule was emptied of its cargo by engineer John Hirasaki through a plastic tunnel connected to the trailer (its wheels had been removed, and its air conditioning system was amended with biological filters). He, Dr. Bill Carpentier, Armstrong, Aldrin, and Collins would now spend three weeks in quarantine, an isolation that began for the fliers with a hot shower, a shave, and a cold martini.

Aboard *Hornet*, the men of 11 were greeted through their trailer window by an ebullient President Nixon, who seemed to dance a little jig while saying, "This is the greatest week in the history of the world since the Creation . . . the world is bigger infinitely and . . . as a result of what you've done, the world's never been closer together before. . . . Frank Borman says you're a little younger by reason of going into space. Is that right? You feel a little younger?"

Mike Collins: "We're a lot younger than Frank Borman." Oddly enough, their one portal to the outside world, the trailer's main window, was crotch high, forcing them to squat for any conversation, even for the national anthem.

It had been, to NASA and the crew's everlasting credit, a mission as perfect as anyone had ever known. "To me, the marvel is that it all worked like clockwork," Collins said. "I almost said, 'magic.' There might be a little magic mixed up in the back of that big clock somewhere. Because everything worked as it was supposed to, nobody messed up, and even I didn't make mistakes."

During the Apollo 11 splashdown, one navy helicopter carried a television camera, whose signal was sent back to Houston, where it was displayed on a giant MOCR screen. Next to the screen appeared these words:

I believe that this nation should commit itself to achieving the goal,
before the decade is out, of landing a man on the moon
and returning him safely to earth.
—John F. Kennedy, May 25, 1961.

And next to that (eight years, one month, and twenty-two days later), it said:

Task accomplished, July 24, 1969.

20

We Missed the Whole Thing

That night in every NASA directorate, one celebration after the next kicked off for every agency employee whose division could say: Mission Accomplished. "I was really impressed at some of the folks' chugging ability," Lunar Module systems engineer Robert Heselmeyer chuckled. "You know, that was the first time—even in college—but that was the first time I had seen a mug of beer consumed in probably a second. Amazing." The citizens of Huntsville carried Wernher von Braun on their shoulders for a parade across nearly the whole of the town. Flight dynamics officer Jay Greene:

> [The splashdown party for] Apollo 11 was the big blowout with the fabled piano in the swimming pool. I wouldn't know. I think I passed out before we got that far. . . . we landed early in the morning, and it didn't matter; we didn't know what time of day it was. The flight was over, so we went to drink, and there were a thousand people in the Flintlock, and at seven o'clock in the morning, Bloody Marys seemed like the right thing to do.
>
> We partied hard and partied into all hours, blowing off steam, and I don't remember losing anybody. We used to think that the curbs on the feeder of the Gulf Freeway were to keep you going in the right direction, and they did. I had one guy who got on the Broadway traffic circle. He was too drunk to figure out how to get off, so he just left his car and walked away. . . .
>
> One of the flight controllers was Bill "Big Shoes" Bucholz, an Air Force captain [with] eight children. The only affordable transportation for the entire bunch was an old large black Cadillac hearse that seated 12. . . . When they wanted to come home through the downtown Houston traffic, the controllers would form up their cars in a line behind Bill's hearse and turn on their lights mimicking a funeral procession. The ploy worked every time.

MSC chief Bob Gilruth woke up the day after splashdown to find his face sore and hurting. He'd been grinning too much. Another celebration bash ended with a speech by George Mueller:

We now stand at what is undoubtedly the greatest decision point in the history of this planet. Four billion years ago the earth was formed. Four hundred million years ago life moved to the land. Four million years ago man appeared on earth. One hundred years ago the technological revolution that led to this day began. All of these events were important, yet in none of them did man make a conscious decision that would change the future of all mankind. We have that opportunity and that challenge today.

There remains for mankind the task of deciding the next step. Will we press forward to explore the other planets or will we deny the opportunity of the future? To me the choice is clear. We must take the next step. Should we hesitate to exploit the first step? Should we withdraw in fear from the next step? Or should we substitute temporary material welfare for spiritual adventure and long-term accomplishment? Then will man fall back from his destiny. The mighty surge of his achievement will be lost and the confines of this planet will destroy it.

This is the time for a decision. This is the time for rededication to the spirit of our forefathers. A time for all men to move forward and together. The organization that brought men to the moon stands ready for the next step. The knowledge possessed by men is sufficient, the resources are adequate for the task of carrying out this next step. The will of the people of this nation and of the world will determine whether mankind will make the great leap to the planets.

In this moment of man's greatest achievement it is timely to dedicate ourselves to the unfinished work so nobly begotten by three of us. To resolve that this nation, under God, will join with all men in the pursuit of the destiny of mankind will lead to the way of the planets.

If Mike Collins, Buzz Aldrin, and Neil Armstrong imagined at this moment that they, too, had accomplished their mission, they would learn soon enough that their assignments were far from finished; that, in fact, they would have these jobs and be assigned these roles for the rest of their lives. A C-141 Starlifter cargo jet ferried the astronauts in their trailer from Hawaii to Texas, landing at Ellington AFB on July 27 at 1:00 a.m. CST. There the quarantined men were able to crouch down in their window and talk to their wives in a prisonlike reunion. Aldrin's first words to his wife Joan were to beg for fresh underwear. The trailer was then craned onto another flatbed truck, an operation that took hours because the logistics hadn't been properly worked out ahead of time. Once they were loaded, the men were paraded through Houston, appearing in their airstream at a welcome-home ceremony with Mayor Louie Welch.

Finally they were carted to the Lunar Receiving Laboratory at the Manned Spacecraft Center. Before leaving their ship forever, though, Collins couldn't help himself from sneaking back aboard and writing, next to the sextant, "Spacecraft 107—alias Apollo 11—alias *Columbia*. The best ship to come down the line. God Bless Her. Michael Collins, CMP."

What were astronauts like when they first returned from outer space? Nurse Dee O'Hara: "They have something, a sort of wild look, I would say, as if they had fallen in love with a mystery up there, sort of as if they haven't got their feet back on the ground, as if they regret having come back to us . . . a rage at having come back to earth. As if up there they're not only freed from weight, from the force of gravity, but from desires, affections, passions, ambitions, from the body. Did you know that for months John [Glenn] and Wally [Schirra] and Scott [Carpenter] went around looking at the sky? You could speak to them and they didn't answer, you could touch them on the shoulder and they didn't notice; their only contact with the world was a dazed, absent, happy smile. They smiled at everything and everybody, and they were always tripping over things. They kept tripping over things because they never had their eyes on the ground."

Earlier that year, Michael Crichton's bestseller *The Andromeda Strain* had terrified the world with its story of a crystal-based alien lifeform that turned human blood into ash and threatened global extinction—a life form brought to Earth by an American satellite. Though it seemed unlikely that such a creature would hitch a ride from the Moon somewhere in the bloodstreams of Armstrong and Aldrin, even so remote a risk could not be taken, and all three crewmen were quarantined for twenty-one days in the three-story suite of labs, offices, and living quarters that NASA called the LRL (Lunar Receiving Laboratory), directed by Persa Bell. If an alien life form did appear, the first job of the attending biologists was to figure out how to destroy it. "Even if it isn't harmful to the mice or the quail or anything else in the laboratories, we will have to be able to kill it, because if we let it out we may find that it will kill, say, all the oak trees in the world," exobiologist Gerald R. Taylor said.

"The LRL had a colony of I don't know how many white mice, and they exposed them to the moon rocks and to us," Mike Collins remembered. "Had there been some strange malady, maybe these mice's grandparents had some genetic defect or something, I don't know, but had the mice all sickened, jeez, I hate to think about it. We'd be in that building today."

With its pale yellow walls and furniture from the Sears, Roebuck catalog, the LRL became a temporary home for two cooks, a photographer, a NASA

PR officer, two doctors, a janitor, a researcher whose glove ripped while he was inspecting a lunar chunk, and a photo technician who'd accidentally touched dust on a film magazine. "We were worried about there being something toxic, some mysterious virus or what have you, so they were very uptight about handling any of the return samples, yet they overlooked the fact the film cassettes we brought back had lunar dust all over them, and we cleaned those and we had all the waste," science and applications technical assistant Fred Pearce said.

Another fellow inmate was NASA Public Information Office chief John McLeaish, which irritated the crewmen, who wanted to be both off-the-clock and off-the-record. They had complained about this plan to Slayton, who'd agreed with their position, but the four were overruled.

"Most of the crews who landed on the Moon and had to go into quarantine for three weeks really thought that was the best thing that ever happened to them, because it gave them a chance to unwind and to sit down and chronicle their thoughts and write their memoirs and do whatever they wanted to do without a lot of press people and whatever bothering them," Lunar Receiving Laboratory manager Richard Johnston believed. In quarantine, the men of Apollo 11 prepared for debriefing and mission report-writing; played Ping-Pong, gin rummy, and snooker; jogged the halls and rode a stationary bicycle; watched *Ice Station Zebra*, *If It's Tuesday, This Must Be Belgium*, and *Goodbye, Columbus* (twice); and read their massive outpouring of mail, telegrams, and press clippings, one of which, from the *Philadelphia Inquirer*, asked: "Will this magnificent accomplishment serve as inspiration . . . or will the inspiration be abandoned before the veiled censure of those who seem to suggest the solution of all human dilemmas lies in turning away from space to other priorities?"

Mike Collins: "Duke Ellington was playing his new composition 'Moon Maiden' in the Rainbow Room and we are welcome there. Ditto the Steel Pier in Atlantic City, which offers the three of us $100,000 for a one-week stint. Another offers to name a hybrid orchid after me, and I sign a release authorizing a race horse to be called Michael Collins. May he orbit the track at unheard-of velocities, even in the mud. There are congratulatory messages from the Montgomery Police Department, the Catholic Daughters of America, the American Fighter Pilots Association, the Pope, the Peace Corps . . . on and on it goes. There are honorary memberships in a host of organizations, my favorite being the Camel Drivers Radio Club of Kabul, Afghanistan."

When Aldrin learned of the size of the audience that had followed every

step of their mission, he told Armstrong: "Neil, we missed the whole thing."
Eventually, in the tedium of quarantine, it became common for the crew to
stand by the windows, hoping someone would pass by, so they could be
waved at.

Buzz Aldrin was surprised that no one at the debriefings asked about the
mysterious "flicker flashes" of light he'd reported seeing in both spacecraft.
Aldrin: "Then I found it wasn't under 'anomalies' [in the NASA mission re-
port] but under 'meals' or something. It took about three months for it to get
out." When he then asked Armstrong if he saw the same thing, the commander said, "Yeah. Saw about a hundred of them."

Aldrin: "What do you suppose they were?"

Armstrong: "I don't know."

Eventually it was shown that these were radiation particles hitting the
eyeball's internal liquid at zero-g.

There was much for the men to recover from in quarantine, both physically
and emotionally. "As soon as the human body goes into weightlessness, the
blood that we're used to pooling in our lower extremities . . . is redistrib-
uted throughout your body because there's no gravity pull," biomedical
technician Pamela Donaldson said. "Your body senses then that it's got too
much blood because the sensors are in your neck and great veins of your
chest. The brain says, 'we've got to unload some of this fluid,' and the body
begins diuresis. In the process, not only do we get rid of the plasma volume
portion of the blood, but we also get rid of our blood cells. . . . And it occurs
pretty fast after you get into space. What else happens? When you lose water,
you also lose salt. And when the body loses salt . . . we keep pumping up the
hormone aldosterone that controls sodium. The hormones don't seem to
work up there like they do on the ground, and why? We don't know. [Then
when] you land back on Earth, just the opposite happens. All of a sudden
blood pools in your lower extremities, you feel faint. You sit down or drink
something to make up the volume difference [and] your heart, great veins,
neck sensors are all saying, 'hey! where's all the blood?' So, you retain fluids,
you retain salt, and you build back up the blood volume over a two-week
period."

It went unreported in the press of the time that the jet carrying NASA administrator Thomas Paine and the moon rocks left Hawaii before the crew.
LRL manager Richard Johnston:

Gilruth called me one day and said, "That lab is in terrible trouble. Go over and straighten it out."

I've never seen so many problems. . . . I'm home and I get a call, and they said, "You'd better come out here. You're not going to believe what we did." They dissolved the film. I couldn't believe it. It was just one problem after another.

When I went over there we really hadn't thought through what we had to do. Just had not thought through what needed to be done. So I got the guys from recovery and the people who built the suits that the recovery people were going to use, and I took over as a spokesman with a thing called the Interagency Committee for Back Contamination. We wrote a couple of reports that we sent to them, and we met with them down in Atlanta, Center for Communicable Disease.

Chuck Berry was still there. Chuck is a nice guy, but, boy, he can be a pain in the neck. I remember me flying down there one day and he was telling me what he was going to talk to the people about. I said, "Chuck, we're not ready to talk about that and I don't want you to bring it up." What do you think he did? He brought it up.

We had a group of physical scientists and medical scientists, you know, life scientists. They didn't even talk to each other. They wore different color suits. But I went over. I got everybody together in a big conference room they had over there, and I said, "Look. I didn't ask to come over here, but I have been sent over here to get this place straightened out. I'm going to do it, and you're going to help me."

One day I went down to the guy who was handling where they were going to store the lunar material. I said, "Would you show me where you're going to put that?" He was going to put it in his lab. He was talking about taking this material that we'd spent millions of dollars going to the Moon, and he was just going to have it in his lab. I said, "You've got to be kidding me." He said, "No."

So I went to Gilruth and I said, "If it's okay with you, I'm going to get the shop and we're going to build a place to put the lunar material. That idiot's got this lab and he's going to take the material down there and leave it."

He said, "Dick, whatever you need to do, go do it."

So we built an interior—really like a vault, had wiring all around the thing, and fire alarms and everything. We did it in a couple of days.

Security remained a problem. "There were some interns that stole a safe full of lunar samples," science and applications technical assistant Fred Pearce said. "When they did that, they dumped all the data. They stole the

safe, took the samples, and dumped all of the records, which was a tremendously damaging thing to lose all the records off of the samples in the safe."

Still, for selenologists and geologists, NASA became paradise found.
"When you go to a college, the ugliest, dirtiest building is usually where
you'll find the geologists," geochemist Robin Brett said, "but here [at NASA]
we work in expensive laboratories with rubber gloves set in gleaming
cabinets. . . . Normally, we spend a whole year doing research and maybe
writing a couple of papers. Here, we discover something new every day, we
barely have time to write it up, and before we're finished we have to give a
press conference about it." The Apollo program would ultimately collect
2,200 samples—842 pounds of lunar rock—which is today safeguarded in
nitrogen-sealed boxes at Houston's Johnson Space Center. Contamination
control officer Mike Reynolds:

> We had a series of rubber gloves that were used to work in the vacuum
> system, and these rubber gloves were the only thing that we had in the Lunar
> Receiving Laboratory that was top secret. We got them from the Air Force.
>
> Every time we had a break in the system—a piece of glove opened up like
> that [imitates drum sound], it'd hit the drum. The doors would lock, lock you
> inside there, and whatever you were doing, if you were doing something in the
> cabinet and the glove caught, you got out, and you sat down and you waited
> till the [imitates drum sound] that was clear. So they completely closed that
> building off.
>
> It had its own air-conditioning system, its own air handling system, own
> exhaust system. All the exhaust went through burners, and they didn't push
> the air outside, they pushed the air through an exhaust system, and they used
> that air to burn it up so that they had nothing to the outside. They had done a
> lot of unique things in the design of the Lunar Receiving Laboratory.
>
> When you go to a meeting, there are ten, fifteen people at a meeting, and
> they discuss how they want to do something; they've got about fifteen differ-
> ent ideas of how to do it. Well, can you imagine fifteen guys or twenty guys,
> Ph.D.'s from Caltech, Chicago, MIT, Harvard. It was unbelievable.

Again, to prevent contagion, the lunar samples were raked in ultraviolet
light, washed in peracetic acid, and then aerated with nitrogen, before entering a vacuum chamber. They were then brushed and dusted, and set on an
elevator that raised them to an observation port with a binocular telescope
and a camera. At the latter stages, technicians manipulated the rocks and
dirt with space suit gloves, which was such difficult and tedious work that
the techs had to be replaced every hour on the hour.

"When the first rock was lifted up inside the vacuum chamber, just one glance was enough to show that it was igneous," Harvard geologist Clifford Frondel said. "One glimpse answered ages of speculation, and you could see theories falling all around. Finding titanium in it was another turning point, because it presumably indicated very high internal moon temperatures. Another was the extreme age of the rocks—over three and a half billion years! Then there were all those glass balls in the dust. That was a thriller. They're very pretty—they come in reds and browns and yellows—but they're very hard to interpret."

Lunar Receiving Laboratory director of processing John Annexstad:

It was like living in a fishbowl. We were paid basically forty hours a week and we were all government employees, but we were all young, and this whole system was so new, and it was so exciting that I think the average person was working fifty, sixty, seventy hours a week. We'd come into the lab and we would work extra hours. We would work in the evenings. We would come in on weekends. Nobody ever worried about time or anything like that. We were all mission-oriented, very definitely . . . the overall feeling was one of incredible pride, almost sitting there and just saying, "My goodness sakes. Look at what we're doing. Look at what is going on."

Nobody knew how to do anything. Just how do you take apart a lunar sample? I remember being asked that, and I said, "The only experience I've ever had in taking apart rocks was I used to work in a stone quarry once." They said, "Well, maybe you've got some ideas." Everything was new. Nobody could tell you, "Don't do that," or, "You can't do this," because we didn't know if it could be done or not. So it was just a wonderful, wonderful experience.

One day I was involved with the cutting of a large piece of Apollo 11, a 300-gram piece. The man who was the head of the McDonnell Center for Space Science was a fellow by the name of Bob Walker. Now, Bob Walker is a member of the National Academy of Sciences, who was an honored member of the Analysis Planning Team, was very, very well known and has good friends all over the system. He's a top-flight scientist. He came in that day to see us cutting this sample. . . . I looked over at him, and he's shaking, literally shaking. His face is gray, and the sweat is just pouring off his forehead. Here's a man who has devoted his life to cosmic materials, meteorites and everything else. He's looking at his sample from the Moon, that he's going to be looking at, a beautiful piece, and some work that he's going to be doing with it, and he's almost beside himself. I've never forgotten this, because at that instant it was probably one of the clearest messages that I had ever gotten that told me of the incredible importance of our work and where we were in relationship to science. It's something that you'll never forget.

The Department of Agriculture's Dr. Charles H. Walkinshaw sprinkled moon dust on thirty-three different biological specimens, from algae to tobacco. On the simpler plants, it produced lush, thriving results, worthy of high-grade fertilizer, and similar to what Walkinshaw could also trigger with a terrestrial source—volcanic ash.

The forty-four pounds of rock Apollo 11 carried from the Moon turned out to be primarily igneous basalt—similar to that found on the Earth's seabeds—but with high levels of titanium, which meant that the Moon was, at one time, volcanic. This was initially a contentious finding; when an early researcher reported that "the evidence for lava flows was 'overwhelming,'" another replied that that depended on who, exactly, was being overwhelmed. Dating set their age at an approximate 3.65 billion years. The other major category of rock samples was regolith breccias, formed by the fusing of soil from meteoric impact. Geologists described the dust, meanwhile, as being like the "bottom of a coalbin," with a sparkle from a "high percentage of glassy beads," and noted that it clumped and could be cut like cake flour. Since it had come not from the Moon, but from outer space, the dust was as old as the Milky Way. There was no evidence that there had ever been any water or any kind of life on the Moon, but one scientist noted that, "If we could find out why the moon died, we might discover what made the earth live." One previously unknown mineral was named armalcolite, after Armstrong, Aldrin, and Collins.

"It was a horrendous shock to most of us, the Apollo 11 basalts coming up to be 3.7 to 3.8 billion years old, older than anything we'd dated here on Earth. Holy Toledo!" University of Texas Department of Geological Science chairman William Muehlberger exclaimed. "Right off the bat. And that's some of the youngest stuff on the Moon, because it's sitting there in very big sheets, not very much pounded up by meteorites, just the top few meters are all smashed up by them, but the rest of them are visibly—they've got folds in them and you can see the edges of the lava flows and all that kind of good stuff." Lunar sample principal investigator Michael Duke:

Whenever you see a rock being sawed in the laboratory or in a quarry, they're always dumping huge amounts of water on it to cool the saw blade as it cuts through the rock. But we couldn't do that because the lunar samples would react with the water, so you didn't want to expose them. So we developed some saws. We developed a band saw that was actually a converted meat cutter saw, and used a diamond blade to cut the rocks.

Throughout the Apollo Program and in most NASA programs, the engineers are in charge. The engineers are the ones that make the systems work, and they really have to look carefully at every little angle that is important to the mission. The scientists typically recognized that, but still chafed a lot at the restrictions and maybe what looked like lack of respect sometimes to the scientists, that they couldn't get their views heard by the engineers, and when they could get listened to, there was this long, complex system of reviews and reports and discussions that had to go on before even the smallest thing was changed.

So during the Apollo Program, there got to be quite a conflict between the scientists and the engineers. At least on the surface there appeared to be. As we progressed from the Apollo 11 mission to the later Apollo missions and the engineers got a little bit more comfortable with the missions, the scientists were able to play a larger part. By the end of the program, there was quite a lot of respect among the scientific community for the way in which the engineers pulled this all off.

Project Apollo and the first Moon landing would have a profound effect on another aspect of science, in a very unexpected way. The speaker at one NASA scientific banquet was British astrophysicist Fred Hoyle, who had predicted in 1948 that, once a photograph of the Earth had been taken from space, a whole new way of thinking about the planet would result. As he told the attendees: "You have noticed how, quite suddenly, everybody has become seriously concerned to protect the natural environment. It happened almost overnight, and one can understand how one can ask the question, 'Where did this idea come from?' You could say, of course, from biologists, from conservationists, from ecologists, but after all, they've really been saying these things for many years past, and previously they've never even got on base. Something new has happened to create a worldwide awareness of our planet as a unique and precious place. It seems to me more than a coincidence that this awareness should have happened at exactly the moment man took his first step into space."

21

Through You, We Touched the Moon

R eleased from quarantine on August 10, Collins was surprised that he could smell a "warm and moist and inviting and reassuring" planet. "The weather was foul, but I smelled Earth, unspeakably sweet and intoxicating. And wind," rhapsodized Cosmonaut Andriyan Nikolayev on his own first day home. "How utterly delightful; wind after long days in space."

One of the first questions Buzz Aldrin was asked was, how long it would take to get his life back to normal, and he replied, "I wish I knew the answer. . . ." NASA had negotiated a gentlemen's agreement with the press to leave the men and their families alone for six weeks of R&R, but it wasn't universally respected. A pack of insistent Italian paparazzi forced Aldrin and Collins to detour through the grounds of Ellington AFB in order to go shopping, while Armstrong one afternoon found three Japanese photographers lounging in his backyard.

On Wednesday, September 13, at 5:00 a.m., the men, their families, and a NASA retinue boarded Air Force II to begin the domestic leg of the "Giant Step" tour. They were met at La Guardia airport by New York City mayor John Lindsay, and helicoptered to Wall Street, where they took their places in open-air convertible limousines for a ticker-tape parade before a record-breaking four million cheering citizens through the Canyon of Heroes. Neil Armstrong: "Sometimes they threw a whole stack of [computer] punch cards from the eighty-seventh floor of a building, and, when they didn't come apart, it made like a brick. We had a couple of dents in our car from cards that didn't quite open." The reception was thrilling, heartfelt, overwhelming, and in some ways, frightening. "We were advised not to reach out to shake hands because we could be pulled from cars and couldn't be rescued easily," Aldrin remembered.

They flew to Chicago, where the reception was even more extreme. Aldrin remembered ending that parade covered in sweat, with confetti and streamers stuck to his face and his neck, and his jaws in pain from so much smiling. That evening in Los Angeles was a state dinner for 1,440 at the Century Plaza Hotel, including President and Mrs. Nixon, Mamie Eisenhower, Chief Justice Warren Burger, forty-four state governors, Jimmy Doolittle, Rudy Vallee, Bob Hope, Red Skelton, Rosalind Russell, and Billy Graham. Vice President Spiro Agnew gave each astronaut, as well as flight controller Steve Bales (who symbolized the 400,000 "unknowns" who'd been a part of the Apollo Program), a Medal of Freedom, the nation's highest civilian honor. When Neil Armstrong rose to describe how the astronauts felt at their nation's reception, the notorious Ice Commander's voice choked, and his eyes were wet with emotion: "I was struck this morning in New York by a proudly waved but uncarefully scribbled sign. it said: 'Through you, we touched the Moon.' It was our privilege today to touch America. I suspect that perhaps the most warm, genuine feeling that all of us could receive came through the cheers and shouts and, most of all, the smiles of our fellow Americans. We hope and think that those people shared our belief that this is the beginning of a new era—the beginning of an era when man understands the universe around him, and the beginning of the era when man understands himself."

After making appearances on *Meet the Press* and *Face the Nation* (where Collins announced he was leaving NASA, Aldrin said he hoped to get slotted into another mission, and Armstrong said, "I am available to serve in any capacity that they feel I can contribute best"), by Saturday the entourage was back in Houston for another parade, a barbecue at the Astrodome, entertainment by Flip Wilson and Bill Dana, and dinner with Frank Sinatra at the Beefeater-themed Royal Coach Inn. While the Armstrongs then took a secret vacation in Colorado, the Aldrins spent a week at Acapulco's Las Brisas resort under the care of manager Frank Brandstetter, who'd made it his business to guarantee that vacationing astronauts were happy in Mexico. Immediately after, each returned to his hometown—Armstrong to Wapakoneta, Ohio; Aldrin to Montclair, New Jersey; and Collins to his "adopted" New Orleans, Louisiana. At Aldrin's homecoming celebration banquet, the featured speaker was retired U.S. senator Albert Hawkes, whose comments staggered the astronaut: "In all my years as a senator, in all the many votes and suggestions I have made, I shall remember that to me, the most significant decision I made was to nominate a young man from Montclair, New Jersey, as a cadet at West Point. His accomplishments exceeded my wildest dreams."

On Tuesday, September 16, 1969, the three flew to Washington to speak to a Joint Session of Congress. Before a thunderous ovation from an audience of nearly every major American leader and a number of foreign dignitaries, Armstrong made prefatory remarks. Then Aldrin (who privately said he was more frightened of speaking before Congress than he had been in landing on the Moon) took the podium: "Charles Lindbergh said, 'Scientific accomplishment is a path, not an end; a path leading to and disappearing in mystery.' The first step on the moon was a step toward our sister planets and ultimately toward the stars. 'A small step for a man,' was a statement of fact; 'a giant leap for mankind,' is a hope for the future. What this country does with the lessons of Apollo applies to domestic problems, and what we do in further space exploration programs will determine just how giant a leap we have taken."

"During the flight of Apollo 11, in the constant sunlight between the earth and the moon, it was necessary for us to control the temperature of our spacecraft by a slow rotation not unlike that of a chicken on a barbecue spit," Mike Collins said. "As we turned, the earth and the moon alternately appeared in our windows. We had our choice. We could look toward the Moon, toward Mars, toward our future in space . . . or we could look back toward the Earth, our home, with its problems spawned over more than a millennium of human occupancy. We looked both ways. We saw both, and I think that is what our nation must do."

Neil Armstrong:

The plaque on the *Eagle* which summarized our hopes bears this message: "Here men from the planet Earth first set foot upon the Moon. July 1969 A.D. We came in peace for all mankind."

Those nineteen hundred and sixty-nine years had constituted the majority of the Age of Pisces, a twelfth of the great year that is measured by the thousand generations the precession of the Earth's axis requires to scribe a giant circle in the heavens.

In the next twenty centuries, the Age of Aquarius of the great year, the age for which our young people have such high hopes, humanity may begin to understand its most baffling mystery: Where are we going?

The Earth is, in fact, traveling many thousands of miles per hour in the direction of the constellation Hercules—to some unknown destination in the cosmos. Man must understand his universe in order to understand his destiny.

Mystery, however, is a very necessary ingredient in our lives. Mystery creates wonder, and wonder is the basis for man's desire to understand. Who knows what mysteries will be solved in our lifetime, and what new riddles will become the challenge of the new generations?

Science has not mastered prophecy. We predict too much for the next year yet far too little for the next ten. Responding to challenge is one of democracy's great strengths. Our successes in space lead us to hope that this strength can be used in the next decade in the solution of many of our planet's problems.

Several weeks ago I enjoyed the warmth of reflection on the true meanings of the spirit of Apollo. I stood in the highlands of this nation, near the Continental Divide, introducing to my sons the wonders of nature, and pleasures of looking for deer and for elk. In their enthusiasm for the view, they frequently stumbled on the rocky trails, but when they looked only to their footing, they did not see the elk.

To those of you who have advocated looking high, we owe our sincere gratitude, for you have granted us the opportunity to see some of the grandest views of the Creator. To those of you who have been our honest critics, we also thank, for you have reminded us that we dare not forget to watch the trail.

We carried on Apollo 11 two flags of this Union that had flown over the Capitol, one over the House of Representatives, one over the Senate. It is our privilege to return them now in these halls which exemplify man's highest purpose—to serve one's fellow man.

We thank you on behalf of all the men of Apollo, for giving us the privilege of joining you in serving—for all mankind.

Believing that they were finished, the men headed for their cars, but instead were first sent to be photographed for a Japanese sculptor who was going to carve their likenesses, and then dispatched to meet with congressional wives and children for a two-hour personal talk on Apollo 11—none of which NASA had informed them about beforehand. While using the men's room, they were discovered by Armed Services Committee chairman Mendel Rivers of South Carolina, who insisted on getting his autographs then and there.

Accompanied by six NASA PR officers, four men from the Voice of America, two men from the United States Information Agency, two security guards, two secretaries, a photographer, a doctor, a porter, and a representative from the White House, the Apollo 11 men and their wives began the next leg of "Giant Step"—a goodwill tour of twenty-three countries in thirty-eight days, including state appearances before Marshal Tito, England's Queen Elizabeth, the pope, Japan's Emperor Hirohito, Iran's Shah Pahlavi, and Spain's Generalissimo Franco. In a single day, they lunched with the king and queen of Belgium, and then dined that night with the king and queen of the Netherlands. In Rome, a party at Gina Lollobrigida's lasted until dawn; in Berlin, the reception for one thousand was so poorly managed that the astronauts had to be spirited away; in Zaire, President Mobutu Sese Seko gave each crewman one of his signature leopard hats. Australia's celebratory dinner, hosted by the newly elected prime minister, had the previous PM in attendance, bitterly insisting how the ceremony would have been much, much better if he were running it (and the country). Like all canny service wives, Joan, Jan, and Pat made sure to get all their holiday shopping done overseas. Administrative assistant Geneva Barnes:

> By the Canary Islands, we were already getting tired. People were starting to become ill. It just started like one or two people with flu-like symptoms, and as we got on into Europe, everybody had been ill. . . . When we got to London, Dr. Carpentier went on national television to deny that the astronauts had brought back a lunar sickness and that all the staff were exposed to this.
>
> In one day we flew from Paris in the morning, to Amsterdam for a lunch that the astronauts spoke at. Then we landed in Brussels in the evening, and I believe the astronauts had an event after we arrived, after having been on the road all day. I don't know how they kept up, frankly. They were the ones that were on the front lines, they were the ones making the appropriate remarks and the speeches to the heads of state. We were doing our best to keep up with them. Also, with that much closeness, you'd get on each other's nerves. Somebody very cleverly designed sort of a silly thing to break the ice, keep everything on the proper perspective, the "personality of the day." If you were caught being unkind to somebody or having a temper tantrum, you could assure that the next day you were going to appear in one of these write-ups.
>
> After the trip was over and Neil Armstrong was assigned to headquarters as the deputy associate administrator for aeronautics, I went to work for him as a public affairs assistant and answered all of his public mail in his role as an astronaut. . . . I don't know that you ever really get to know him. He was very supportive of me while I was doing that job. There was a lot of mail. Anything that came in that I thought deserved his special attention and people would want him to sign certain things that they'd send in, or they'd ask him for special inscriptions on photographs. I would line these things up on a table in my office. At the end of his work day, he would come in and sign all those things. . . . He was very conscientious about doing that. I enjoyed working for him. I remember we got one request from a group in his home state, in Ohio, who wanted him to send them the toenail clippings from his left foot. That was the first human foot to step on the moon. They wanted to auction them off.

For the astronauts, the "Giant Step" tour, like all of their PR and statesmanship duties, was at once thrilling and grueling, exhilarating and oppressive.

The accompanying public realations men, operating incognito, talked to local people about their reaction. "[In Turkey] they said the thing that they liked about Americans was the fact that nothing seems impossible . . . everywhere that we went—that same feeling, nothing seems to be impossible," MSC public affairs officer William Der Bing remembered, a little wistfully. Mike Collins: "People, instead of saying, 'Well, you Americans did it!,' everywhere, they said 'We did it!' We, humankind, we the human race, we, people, did it! The warmth of shared experience was remarkable given the origins of the space race in an atmosphere of fear and belligerence." The overseas opinion was unanimous: the United States of America was the most admired country in the world. The PR motivation for NASA's birth had been achieved beyond anyone's wildest dreams. Apollo 11 marked the global apogee of American power, prestige, and influence.

The men returned to the United States, first stopping at the White House, where Pat Nixon gave them a tour of her husband's collection of historic gavels, and the president commented that night at dinner that Romania's President Ceauşescu had finally agreed to a state meeting in the wake of Apollo 11, a diplomatic breakthrough that Nixon believed was worth the cost of the entire space program. When the president then asked about their future plans, Collins mentioned he'd like to continue doing diplomatic work like they had done with "Giant Step," and Nixon immediately called the secretary of state to get him an appointment.

At the AFL/CIO National Convention in Atlantic City, New Jersey, Buzz gave the keynote address. "Imagine these big stevedore types . . . all sitting on the edges of their seats," Der Bing said. "The fact that Buzz had a union card didn't hurt him any. . . . Golda Meir was sitting there just as smug and like a Cheshire cat, and all of a sudden, for no apparent reason, she punched big George Meany, this boxer type, in the ribs with her elbow, and looked him straight in the eye and said, 'Not bad for a Jewish girl from Milwaukee, huh?' So that brought down the house."

Armstrong next accompanied Bob Hope on his Christmas USO tour of Vietnam, Guam, Taiwan, Turkey, Italy, and Germany. The American tabloids used this generous effort on behalf of American soldiers to report on a scandalous affair between Armstrong and fellow USO volunteer Connie Stevens, an affair that both publicly denied. The following May, Armstrong toured the Soviet Union, giving Premier Aleksei Kosygin a moon rock and a flag of the USSR that had traveled to the lunar surface.

During the foreign travel for "Giant Step," Buzz Aldrin, usually a reasonably outgoing man, had begun to lapse into periods of brutal silence. After the Los Angeles Century Plaza dinner, Joan Aldrin wrote in her diary: "It is a sparkly, shiny new adventure and it was a glorious day. I loved all of those people." Three months later, she noted: "The tinsel is tarnished. Buzz, who was never comfortable with all this, pushes loyally on. I cooperate, but I am tired and unhappy." In Sweden, the couple had a terrible fight. "We fell into an uneasy silence which I ended by saying I felt all six of us were fakes and fools for allowing ourselves to be convinced by some strange concept of duty to be sent though all of these countries for the sake of propaganda, nothing more, nothing less," Buzz Aldrin remembered. "We proceeded to get drunk and we both cried. That night we slept like two frightened children, hanging onto each other."

Afterward, Collins said, "Fame has not worn well on Buzz. I think he resents not being the first man on the Moon more than he appreciates being the second." Aldrin conceded, "I knew that anyone who was on the first lunar landing was certainly going to be propelled into the public view in an enormous way. That awareness was troublesome and interfered during the mission. But it's nothing like what happens after the mission, and for the rest of your life, you are the person now, not just the average fighter pilot, who did this and that pretty well, but this guy walked on the moon! I've got to uphold that image for the rest of my life. What do I do?" The answer would be long in coming, leaving Aldrin lost and in misery for a number of years.

Neil Armstrong had now spent his entire adult life as a military pilot, a test pilot, and an astronaut. Had he been aware that, after becoming the first man on the Moon, no one in the United States would ever want to take the chance of letting him fly anything that posed even the slightest risk again? Like John Glenn before him, who Kennedy had personally decided was too much of a public relations treasure to be allowed on any further missions, Armstrong as a symbol of American heroism, ingenuity, technology, and the whole of NASA was regarded too highly to be sent back to outer space or even on another test flight. "I never asked the question about returning to spaceflight, but I began to believe that I wouldn't have another chance, although that never was explicitly stated," he said. Armstrong claimed to be perfectly content with this dramatic reversal in his life, but "he was a pilot, and he was always happier when he was flying," Jan Armstrong later admitted. Instead of getting back into Deke Slayton's Apollo rotation, he was made deputy associate administrator for aeronautics in the Office of Advanced Research and Technology, at a salary of $36,000. Considering Neil Armstrong's keen interest in state-of-the-art aeronautical engineering, this would seem a position tailor-made for him.

He would barely last a year.

22

When All Those Curves Lined Up

Just after the Apollo 11 landing, CBS commentator Eric Sevareid said of Neil Armstrong and Buzz Aldrin, "We're always going to feel, somehow, strangers to these men. They will, in effect, be a bit stranger, even to their own wives and children. Disappeared into another life that we can't follow. I wonder what their life will be like, now. The Moon has treated them well, so far. How people on Earth will treat these men, the rest of their lives, that gives me more foreboding, I think, than anything else."

In his new position at the Office of Advanced Research and Technology, Armstrong sought to promote digital fly-by-wire, a system that replaced the ropes, pulleys, and hydromechanics of the stick-and-rudder era with a computer and signal transducers that could respond simultaneously to pilot control, autopilot, and the ship's instruments. MIT Instrumentation Lab director of space guidance and analysis Richard Battin:

> The F-8 aircraft was going to be an experimental plane so that a digital computer would fly the airplane. [When the pilot] moved the stick he wasn't controlling any of the flight control systems directly, he was merely sending digital signals through a wire, that's the fly-by wire, to control the ailerons and elevators and what have you. But he had no direct control, so if the computer failed, the plane crashed.
>
> The computer was going to be developed by IBM, and it was going to be a triplex system, three computers. The idea was to test [these computers] in this F-8 system, because they would be part of the shuttle. But the IBM computers were late. They wanted to fly the F-8, and so they asked Neil, "What shall we do?"
>
> Neil said, "Well, the most reliable computer I know is the computer that took me to the Moon. Couldn't we use one of those computers to fly the F-8?"
>
> Well, everybody thought he was crazy. They said, "How are you going to do that?"

He said, "Well, get a hold of MIT."
So we actually did exactly that.

Though he made some progress with his proposals at OART, Armstrong was continually sidelined by a flood of speaking engagements, dinner invitations, and other PR requests. To place the demands he faced in perspective, consider that, after John Glenn returned from being the first American in orbit, he immediately received half a million letters and telegrams, and tried to answer them all. NASA, however, seemed incapable of making any attempts to relieve Armstrong's pressure. In August of 1971, when Armstrong resigned, the agency (now dramatically reduced from its 1960s peak, with all future lunar missions cancelled) did little to try to keep him.

Instead, having recently been awarded his master's from the University of Southern California (alongside an eventual nineteen honorary doctorates), Armstrong became a professor of aerospace engineering at the University of Cincinnati, and moved his family to Rivendell, a 300-acre dairy farm in Lebanon, Ohio, named for J.R.R. Tolkien's elf haven. Was he disappointed by this turn of events? "I know I could make a million dollars in personal appearances on the outside, but I just want to be a university professor and be permitted to do my research," he said. Armstrong created two new courses for the UC engineering curricula on experimental flight mechanics and aircraft design but, once again, was hounded by reporters, photographers, and in 1974, Gina Lollobrigida, on assignment for *Ladies' Home Journal*. When UC became a state school, Armstrong's unusual status became a bureaucratic problem, and in 1979, he resigned to become a national spokesman for Chrysler automobiles, as well as serving on various corporate boards of directors, including that of the Utah aerospace firm Thiokol, creator of both the X-15 and the Space Shuttle rocket engines. He helped fund-raise for Purdue, the YMCA, and the Cincinnati Museum of Natural History. In April 1985, he went to the North Pole with Sir Edmund Hillary, and in 1986, served as vice chairman for the presidential commission on the *Challenger* disaster.

In 1990 Armstrong's parents died, and his wife Janet left him. In February 1991, he suffered a heart attack. Between his speaking fees, stock options, and investing, however, he had finally made some money, and was worth over $2 million. In 1994, Neil and Janet Armstrong divorced after thirty-eight years of marriage. "The man needed help," Janet explained. "I couldn't help him. He really didn't want me helping him. . . . It wasn't an easy thing to do—I cried for three years before I left. I just couldn't live with the personality anymore." In 1992, mutual friends introduced Neil to the recently widowed Carol Knight, and they decided to marry in 1994.

During this period, the interminable Shuttle/Space Station era, when NASA seemed to fall into a holding pattern of somnolent torpor, its most historic and galvanizing figure, Neil Armstrong, was nowhere in evidence. Considering the stellar quality of his speeches and the excitement accompanying his rare public appearances, Armstrong's withdrawal was a loss for more than just the American space agency. Jim Lovell talked with the first man about his notorious reclusiveness: "Sometimes I chastise Neil for being too Lindbergh-like. I tell him, 'Neil, Charles Lindbergh flew across the Atlantic on private funds and had a private group build his airplane and everything else, so he had all the right to be as reclusive as he wanted. But you went to the Moon on public funds. The public taxpayers paid for your trip and gave you all that opportunity and fame, and there is a certain amount of return that is due them.' And Neil's answer to that is, 'I'd be harassed all the time if I weren't reclusive.' And he's probably right."

Others, however, have a different opinion of Armstrong's epic withdrawal, including Gene Cernan: "Over the years I think [many] have questioned Neil's devotion to privacy, even criticized the fact that he wasn't more open on a daily basis with the press like some other people are. Maybe he wasn't comfortable with that. He had his own reasons. I'd just like to say, it could have been anyone who walked [first] on the moon: it could have been Neil, it could have been Buzz, it could have been Wally, it could have been any one of our colleagues. But I don't think any one of us—any one of us—who would have had that opportunity, could have handled it with as great and as honorable dignity as Neil Armstrong has handled the responsibility of being the first human being to step foot on the surface of the moon."

Neil Armstrong: "I recognize that I'm portrayed as staying out of the public eye, but from my perspective it doesn't seem that way, because I do so many things, I go so many places, I give so many talks, I write so many papers that, from my point of view, it seems like I don't know how I could do more. But I recognize that from another perspective, outside, I'm only able to accept less than one percent of all the requests that come in, so to them it seems like I'm not doing anything. But I can't change that."

Neil Armstrong, especially, would be struck by the paradox of modern celebrity, of being both idolized as a hero and degraded by tabloid leers, both worshipped as a god and owned like a slave. That he heard the call to prayer on the lunar surface and converted to Islam is a myth. That his neighbor, Mr. Gorsky, was told by his wife, "You'll get oral sex when the kid next door walks on the Moon!" is a joke invented by Buddy Hackett, though it is true that in 2005, he fought with his Lebanon barber after the man sold his hair for $3,000. In his hometown of Wapakoneta, the State of Ohio then built a Neil Armstrong Air and Space Museum, and the town changed the name of its airport to Armstrong, but neither had the grace to ask his permission beforehand. When Hallmark cards then used Armstrong's name and words on a Christmas ornament, he sued and reached a settlement, which he donated to Purdue.

At one point, he tried to sum up his experience: "The future is not something I know a great deal about. But I did live in Washington for a time and learned that lack of knowledge about a subject is no impediment to talking about it. . . . It would be presumptuous of me to pick out a single thing that history will identify as a result of this mission. But I would say that it will enlighten the human race and help us all to comprehend that we are an important part of a much bigger universe than we can normally see from the front porch. I would hope that it will help individuals, the world over, to think in a proper perspective about the various endeavors of mankind as a whole. Perhaps going to the Moon and back in itself isn't all that important. But it is a big enough step to give people a new dimension in their thinking—a sort of enlightenment.

"After all, the Earth itself is a spacecraft. It's an odd kind of spacecraft, since it carries its crew on the outside instead of the inside. But it's pretty small. . . . From our position on the Earth it is difficult to observe where the Earth is and where it's going, or what its future course might be. Hopefully, by getting a little farther away, both in the real sense and the figurative sense, we'll be able to make some people step back and reconsider their mission in the Universe, to think of themselves as a group of people who constitute the crew of a spaceship going through the universe. If you're going to run a spaceship, you've got to be pretty cautious about how you use your resources, how you use your crew, and how you treat your spacecraft."

Of his life after Apollo 11, Armstrong said, "I was pleased doing the things I was doing. That's the sum and substance of it." And then he laughed. "If they offered me command of a Mars mission, I'd jump at it." Since, in January 1998, the seventy-seven-year-old John Glenn returned to spaceflight, perhaps there's still hope.

After returning from the "Giant Step" tour, Buzz Aldrin tried to get his life in order and back to normal. Instead, he essentially found himself in the emotional position of a famous rock star just coming off an enormous world tour, trying to wind down from the adrenaline highs of appearing night after night, alongside the backstage offers of everything and anything, to returning home to pay bills, do the laundry, and empty the cat box. "I don't think either my mother or my father were really prepared for what happened," Buzz's son Andrew said. "In fact, they were exquisitely underprepared. . . . The whole thing was very tough on my dad. Everybody wanted him to be a hero and they created a role for him he didn't want. He'd lived his entire life as a scientist and an engineer and nothing in his background prepared him for what happened. To be all of a sudden in the spotlight was hell for him. There were so many demands on his time and everyone wanted answers he couldn't give. They'd ask over and over: 'How did it feel?' The guy was a scientist; he could tell you how they made it happen but didn't have the words to describe how it felt. My dad was trained to be a pilot, an engineer and an astronaut, not a public figure or a philosopher. From his point of view, it was time to move on to the next mission, something which never happened."

Goddard liaison William Easter: "After the flight, [Buzz] lived in Nassau Bay, and I was at a cocktail party at his house, and his wife was really upset. She was bending my ear because she said nobody would offer a Buzz a job. I didn't know anything about that. I said, 'He probably wants to be president of the company. I'm sure he wouldn't take a job as an engineer. He'd want something big.' But as far as I know, Buzz never did go to work for anybody in a big job. Really, neither did Armstrong. We've talked about that a lot in the past, he could have been a millionaire, done anything he wanted to do. I remember he never had any money. My wife sold his house down in Seabrook and he needed the money out of it. I thought, 'Jeez, here's a guy that landed on the Moon, somebody should be giving him some money.' It didn't work that way."

Unable to right himself in the enormous wake of Apollo 11, the second man fell into a cataclysm of depression, becoming debilitated. Buzz Aldrin: "My life was highly structured and there had always existed a major goal of one sort or another; finally there had been the most important goal of all, and it had been realized—I had gone to the moon. What to do next? What possible goal could I add now? There simply wasn't one, and without a goal I was like an inert Ping-Pong ball being batted about by the whims and motivations of others. I was suffering from what poets have described as the melancholy of all things done. . . .

"The correct term for what now happened to me is a quasi-medical term, 'dysfunction.' What it means is that I stopped. Stopped everything. I'd go to my office in the morning, determined to work a full day and then go home to more work. I'd sit down at my desk and stare out the window. A few hours would go by and I'd drive to the beach in Galveston and walk. Then I'd go home for dinner, turn on the television, and get a bottle of Scotch. Or I'd not go home at all until everyone was in bed. . . . Every time I decided to get help I began to cry; the only help available was official Air Force treatment and the matter would go into my record. . . . [Then I fell] even deeper into despair. I could see no hope, no possibility of controlling anything. I began staying up nearly all night every night with some vague fear of sleeping in the darkness. . . . [I told one doctor that] 'I don't believe I could kill myself because I don't think I'd be able to decide how to do it.'"

After his return to the USAF, Buzz had to pay his psychiatrist personally, since there is no insurance for the destruction of a service career when an officer undergoes mental-health treatment. When, four years after Apollo 11, journalist Paul Hendrickson interviewed Aldrin, the writer went back to his car and sobbed, saying that Buzz "is the Disintegrated American Hero . . . the one who didn't make it back all in one piece," a man who at times looks at the Moon and thinks, "You son of a bitch . . . you're the one that got me in all this trouble." Gene Cernan: "One result of space travel was that I had become much more philosophical, at times unable even to focus on minor problems back on Earth because they just seemed so small in comparison to what I had experienced and the places I had been. My fellow astronauts who went to the moon encountered varying degrees of the same disease: We broke the familiar matrix of life, and couldn't repair it."

Eventually, one of America's most admired men needed to be secretly hospitalized. Buzz Aldrin: "Apollo 11 may have been a small step for Neil but it was a beginning of a tremendous hurdle for me. And that hurdle eventually led to the disease of alcoholism. And I'd like you to know that I'm certainly a changed person now . . . I find myself able to cope with many things that used to baffle me [and] have a sense of comfort that far surpasses anything I've had before. The space program was kindergarten in comparison to coping with the culminating effects of alcoholism. You get into a pattern of just gradually spiraling down."

In time, Buzz became one of the bravest of all of America's astronauts by publicly revealing his mental illness, to the extent of writing a memoir on his troubles and his recovery, *Men from Earth*, and serving as chair for the National Mental Health Association. After seventeen years, though, his marriage fell apart, and he and Joan divorced. Aldrin moved to Los Angeles, where he met and married Lois Driggs Cannon. (Their license plates say: "MARS GUY" and "MOON GAL.") He has since written a number of novels, helped produce the computer game *Buzz Aldrin's Race into Space,* and actively promotes the future of manned spaceflight. His nonprofit organization, ShareSpace, is dedicated to finding methods so that people besides astronauts, cosmonauts, and the rich can experience outer space travel.

If the costs of rocket payloads continue to drop precipitously as they have over the past decades, extraterrestrial tourism will become a reality in two generations or so. Besides the spectacular views and roller-coaster thrills, there is another sensation that space travelers will be able to experience, one that has been enjoyed by only a very select few human beings in history: Because of zero gravity, space tourists will be able to fly.

When asked about his dad's current life, Andrew Aldrin said, "It's a game called, 'spin the Buzz on the globe.' Put a pin on there, he'll be anywhere." About Apollo, Buzz notes, "Many [now say that going to the Moon] was not only a drain on the nation's financial resources but also useless. Something is useless only if we do not know how to use it." And about his life today: "I wasn't close to my father. He came from the era of aviation pioneers and they would get together and talk about the past. I always thought that was kind of sad. I'd think: 'Why aren't they talking about the future?'"

It is ironic that the happiest post–Apollo 11 life seems to have been lived by the one man who didn't walk on the Moon. After a career with the State Department in the wake of Nixon's White House phone call, Michael Collins became the first director of the world's most popular museum, Smithsonian's National Air and Space. He has, however, admitted, "I share with [Buzz] a mild melancholy about future possibilities, for it seems to me that the list of exciting things to do here on earth has diminished greatly in the wake of the lunar landings. I just can't get excited about things the way I could before Apollo 11; I seem gripped by an earthly ennui which I don't relish, but which I seem powerless to prevent. . . . I get very tired, not so much anymore because—well, for a whole variety of reasons, but in those days I got tired of being asked the same things over and over and over and over and over and over and over again."

Today, Mike is still married to Pat, and in his spare time, paints watercolors—not of outer space, but of Florida landscapes and wildlife. Michael Collins: "I've always wondered why it is that, because we're crew members of a particular mission, that we all of a sudden became experts in trying to figure out what the future ought to be . . . I ran across something the other day that

I thought made the point that I'm trying to make, that I feel a little uncomfortable in trying to take the role of judging where we are and what we're doing. This is a little article that was printed in *Test Pilots Quarterly*, taken from the *Illustrated World*, May 1914, and titled 'Folly to Cross the Atlantic in Air,' written by Orville Wright, aviator. And I'll read it rather quickly, parts of it: 'It's a fair possibility that a one-man machine without a float and favored by wind of, say, 15 miles an hour, might succeed in setting across the Atlantic. But such an attempt would be the height of folly. When one comes to increase the size of the craft, the possibility rapidly fades away. This is because of the difficulties of carrying sufficient fuel.' . . .

"Apollo 11 was perceived by most Americans as being an end, rather than a beginning, and I think that is a dreadful mistake. Frequently, NASA's PR department is blamed for this, but I don't think NASA could have prevented it. It's simply the American way, to view a televised spectacular and think of it as the Super Bowl. Then followed confusion and a trace of irritation. Why was the Super Bowl being played over and over again? . . . The magic was gone, despite all the talk of more sophisticated scientific instruments and more extensive explorations using the lunar rover. The only thing that could have titillated the public and gotten the momentum back was a manned expedition to Mars, and that seemed impractical even to the program's most ardent supporters. So the focus returned from moon to earth, and the orbiting Skylab was loaded with cameras to record in as much detail as possible the ravages to our planet, as a first step in repairing the damage. So that's where we are. . . .

"I think it is premature to make a judgment on the manned space program and its possible value to mankind. We simply don't know yet what it may mean to us."

If the history of the Apollo 11 crewmen, forty years after touchdown, has not been one of unalloyed triumph neither has the legacy of the mission itself. At the time, even Britain's august and reserved *Economist* proclaimed that, "The human race's way of sublimating its highest aspirations has been to build the greatest and grandest artifact that the technology of the time can achieve. Through the pyramids, the Parthenons and the temples, built as they were on blood and bones . . . the line runs unbroken to the launch pad of Apollo 11. . . . Unlike the pyramids and the cathedrals, the exploration of space will have so many practical justifications that our descendants will think us mad that we ever doubted its value."

Historians, however, have since raised the question: Was Apollo 11 the crowning glory of the Space Race, or one of its many wins, losses, and ties? Another very different ending to that contest began in November 1961, when the astronaut slotted to follow John Glenn in an orbit was the cigarillo-smoking, blue-eyed, steel-haired and -nerved, onetime dairy farmer and Edwards test pilot from Sparta, Wisconsin, Deke Slayton. While Slayton was in the middle of his training, however, USAF physicians detected a very slight heart murmur, and recommended he be pulled from rotation.

Deke wanted to quit the program, and made overtures to return to the air force. Chief of Staff LeMay, however, decided that, if a flier wasn't good enough for NASA, he certainly wasn't good enough for the USAF. Instead, the Mercury fliers unanimously made Deke Slayton their chief, and he became head of the Astronaut Office.

Soon after Apollo 11, and encouraged by Nixon's passion for détente, NASA administrator Thomas Paine began corresponding with Soviet Academy of Sciences chief Mstislav Keldysh—just as Webb and Gilruth had done before him under Kennedy—to see if there was any possibility of the superpowers' working together in space. In January 1971, a delegation led by George Low met with Keldysh in Moscow, where the two men discussed how competition, not cooperation, had spurred on their respective agencies' progress. That observation highlights one hope: that American manned space exploration in the future, either through NASA, the Pentagon, or commercial interests, may fully reignite once again in the face of competition from other nations.

On May 24, 1972, Nixon and Premier Aleksei Kosygin jointly signed an "Agreement Concerning Cooperation in the Exploration and Use of Outer Space for Peaceful Purposes," and on July 15, 1975, Valeri Kubasov and Aleksei Leonov launched from Tyuratam aboard Soyuz 19. Leonov was the first human to walk in space and the most famous cosmonaut in the Soviet Union after Gagarin; their launch was the first in the twenty-year history of the Baikonur Cosmodrome to be broadcast live on television. Seven and one-half hours later, on July 15, 1975, the forty-eight-year-old Deke Slayton—finally cleared by agency physicians to fly—alongside Tom Stafford and Vance Brand, lifted off on the last Apollo mission. After two days of orbiting, while flying five miles a second, Stafford docked with Soyuz over the city of Metz, France, fulfilling President Kennedy's dream of cooperating with the USSR on space exploration, a dozen years after his assassination. After an exchange of flags and bear hugs and an assortment of scientific experiments, the American and Soviet ships parted, the Apollo mission becoming NASA's last manned spacecraft splashdown. Some feel that Apollo-Soyuz was the true ending of the Space Race, making it, like the Missile Race, a tie.

———

With the Apollo program having come to on end, nearly all of its astronauts would spend the rest of their lives trying to answer one question: After you've been to the Moon, what do you do next? The same could be said for NASA. TRW's Simon Ramo: "Kennedy's Science Advisor, Jerry Wiesner, and those advising him—the PSAC, the President's Science Advisory Committee—they were very much against the psychological PR program, what they called it. They correctly predicted that we would land some astronauts on the moon and we would put the Russians to shame, and once we had done that, there would be lack of interest and the moon program would be stopped and never go anywhere, and that was what happened."

By examining Apollo 11's history in the context of the Space and Missile Race, however, it becomes immediately clear why the agency's public mandate has faltered. If NASA can no longer find a way to present itself as a vital part of American identity, that may be only partly its own fault. On the one hand, the agency has, for its entire life, suffered in being a creature born, in part, of public relations and propaganda, at the some time that its employees do not consider public relations a significant and worthwhile endeavor. "When [JSC director George W. S.] Abbey came up here, we had a meeting not too long ago," Guenter Wendt said, "and I asked Abbey, 'You know, that's a problem. You guys don't do publicity.' I said, 'The people don't even know. I called the Public Affairs Office and I said, "Who is the present manager of the Space Station?" They didn't know. 'Now,' I said, 'How come I don't see him in Congress? How come I don't see him on TV? How come I don't see him on talk shows? How in the heck am I going to convince an average factory worker that he should support the Space Station if nobody knows about it? They don't even know who it is or what it is or what it does. You do a lousy job of publicizing things!'"

Without its status as a theater of the Cold War—in both meanings of that term—of serving as a noble proxy for nuclear Armageddon, there is little reason for maintaining a dramatically expensive, state-of-the-art federal space program. And there is a crucial third factor: the lack of an American public determined that their nation be a global leader in science and technology. It seems that, of late, the United States has relinquished this vanguard position as well. For one prominent example, the world's most powerful collider, where key questions of particle physics may be solved, is no longer the

American Fermilab's Tevatron, but the Large Hadron Collider in Geneva, Switzerland. The United States was going to build a Tevatron successor in Waxahachie, Texas, but Congress canceled it in 1993. Fermilab and Stanford's colliders will both be shut down within a few years. As *Newsweek* explained, building a state-of-the-art collider "requires the political will to lavish money on a project that has no predictable practical return, other than prestige and leadership in the branch of science that delivered just about every major technology of the past hundred years."

Yet, at the same time, what American doesn't thrill to the extraordinary photographs of the Hubble Telescope, the shots of Earth storms from the International Space Station, or the discoveries made by various NASA robots? The truth is that the citizens of the United States have spent fifty years deeply conflicted about space and technology, as fearful of Sputnik and Gagarin as they are proud of Glenn, Armstrong, and GPS, alternately convinced that exploring the cosmos is an honorable pursuit or an absurd waste of taxpayers' money. This ambivalence is even reflected in the nation's popular culture. When Stanley Kubrick's *2001: A Space Odyssey* was released the year before Apollo 11 in 1968, after all, there was so much optimism about the future of space exploration that it was plausible to imagine that the movie's images of lunar towns, elegant space stations, computers with a consciousness, and missions to Jupiter would all be reality in thirty-three years' time. Yet, during the following year of 1969, while Armstrong and Aldrin were touching the Moon, the number-one song in America was a lament on the horrors of future technology:

> In the year 6565
> Ain't gonna need no husband, won't need no wife.
> You'll pick your son, pick your daughter too.
> From the bottom of a long glass tube.

For most of the twentieth century, inventing new technologies, being yearly the nation with the most Nobel prizes, going into outer space, and being first on another planet were considered profoundly American achievements. Immediately after the triumph of Apollo 11, however, the public seemed to reverse its thinking, now collectively judging the United States no longer powerful enough, smart enough, or wealthy enough to embark on such monumental undertakings. The country would lose its War on Poverty, its War on Drugs, its War in Vietnam, and though the Space Race had been won, it, too, would be abandoned. NASA researcher Catherine Harwood: "I remember Chris Kraft telling me once that he knew it was all over when he looked at his little monitor in Mission Control and we were driving on the Moon for the first time. And he looked over at the networks, and they were all still showing soap operas." Bob Gilruth: "The last landing we made on the Moon, the networks wouldn't pay for putting it on television. We had to pay for that ourselves. Did you know that? We had been so successful that people thought they knew all about it and they did not think that they wanted to watch it."

In January 1970, six months after Apollo 11 returned to Earth, Wernher von Braun wrote a new magazine article on the future of planetary exploration for his old *Collier's* editor, Cornelius Ryan, now working at *Reader's Digest*. Ryan's boss, the magazine's editor in chief, rejected the story, saying that his tens of millions of readers weren't interested in outer space. Agency funding under Nixon was so curtailed that, in Huntsville, the era became known as "The Great Massacre." By 1976, of von Braun's original 118-member team, only eight were still NASA employees. The year before, von Braun himself had been diagnosed with intestinal cancer; he died at the age of sixty-five on June 16, 1977.

In September 1969, a White House Space Task Force headed by Vice President Agnew had reported that the United States could choose from three future options for NASA: a Space Shuttle and space station for $4 to $5.7 billion a year; a manned mission to Mars for around $8 billion a year; or Mars, the Space Shuttle, and both Earth and lunar space stations, for $8 to $10 billion annually. Nixon hadn't attended the launch of Apollo 11 so that, if it were a failure, it wouldn't color his new administration; his vice president's main interest in the agency was, as Tom Paine reported, "in playing golf with the astronauts." Nixon chose a fourth option: the Shuttle, with no space station, save one made from Apollo leftovers to be called Skylab. The president was talked out of shutting down NASA almost entirely by Office of Management and Budget director Caspar Weinberger, who told the president that the administration shouldn't take actions that made it look as though "our best years are behind us."

Flight dynamics chief Jerry Bostick: "The last couple of years of the Apollo program and throughout Skylab there was a lot of personal questioning going on among the people that worked at JSC about what we want to do next, what does the country want to do, what does NASA want to do, what did I want to do. I personally was terribly bored with Skylab. I argued that there was no need to have the trajectory guys even go to the control center.

That was a very radical idea at the time, probably still is, but there's not anything that we had to do that we couldn't do from our office. So why go over there and sit at the console and twiddle your thumbs for twenty-four hours? So there was a lot of that kind of discussion. Of course I lost that debate. They wanted warm fuzzies. The flight director wanted to see people right there. He wanted to go and talk to them, and even some of the younger people who hadn't spent all those years in the control center, they wanted to go over there because, hey, that's where the excitement is.

"Well, guess what? The excitement's gone."

Simulator technician Richard Williams: "We had all been so hyped on this thing of going to the Moon. And then, to all of a sudden wake up one day with the realization of 'there's no more' . . . Why didn't we plan for something further on? . . . I was just devastated. Of course, this whole area, with layoffs was just very [hard hit]. . . . There was no diversification for these guys that had just finished launching the Apollo launch vehicle, which was probably one of the greatest engineering marvels of its time. They would [end] up on the streets, out of work, with no place to go. I knew a couple of engineers that were actually at the gas station pumping gas."

Max Faget: "We don't have any bold people in NASA anymore, absolutely none of that. Everything is done very conservatively. We'll never get back to the moon unless we change NASA. . . . the government has moved to such a safety-conscious attitude, no risk-taking approach to life, that I just don't think we'll do it. We'll talk about it a lot. We might send unmanned vehicles to the moon. I hope we do. But to send a man to the moon and the planets is going to take an earthquake in management to do that, an earthquake in our culture to do that. The culture is moving absolutely very strongly in the wrong direction."

Buzz Aldrin's son Andrew (who is today an aerospace industry executive): "Now that I understand more about what really goes into developing space vehicles and what really did not go into sending us to the Moon, the risks that we were taking, I'm appalled. No, not appalled, that's not the right word . . . in retrospect it's very frightening. And the flip side of that is that if you talk to the old engineers, they're just like, 'Jeez, we're not willing to take those risks anymore.' Well, we're not, and maybe that's a good thing, and maybe it's a bad thing—I don't know."

What NASA has done, post-Apollo, is to dramatically expand its work in science, while returning to its origins. "The sciences right now [in 2007] are about thirty-two percent of our overall budget, and in Apollo days I think it was like, I don't know, seventeen percent," NASA deputy administrator Shana Dale pointed out. "We have a healthy, robust science program that is accomplishing absolutely wonderful science for the United States, and for the world at this point. I mean if you look at what we're doing in earth science, $1.5 billion a year is taking 3-D images of solar flares on the sun. Which is important in and of itself for science, but also important for telecommunications, satellites and astronauts on the International Space Station. Planetary science, for example Mars Rovers, or New Horizons going to Pluto, and the space shuttle . . .

"We [also] want to be really, really supportive as the [commercial space service] community takes off and they're able to provide those space transportation capabilities for us. That frees up NASA funding to be able to push out the frontier in terms of other areas of exploration that we want to engage in. . . . You know, NASA's predecessor agency, NACA, one of its critical functions was to sort of midwife, if you will, a new industry. The aviation industry. And we still have that responsibility, but now it's the space industry."

Others, though, have criticized NASA for creating make-work projects to keep its various centers staffed and the country's aerospace corporations supported. Northwestern University engineering professor Peter Voorhees: "Since 1990, NASA has spent literally billions of dollars [with the International Space Station] building up a world-class microgravity program that has been basically squandered. There's a perfect example of snatching defeat from the jaws of victory." "It would be ironic," George Washington University's Space Policy Institute director John Logsdon commented, "if research breakthroughs came through the work of our partners, rather than research that we've chosen to forgo."

A 1999 Gallup Poll found 89 percent of Americans believing that NASA did indeed send men to the Moon, while 6 percent believed the lunar landings were faked, and 5 percent could not make up their minds. After a 2001 Fox television special, *Conspiracy Theory: Did We Land on the Moon?*, a 2006 Dittmar Associates study revealed that among eighteen- to twenty-six-year-olds, "twenty-seven percent expressed some doubt that NASA went to the Moon," with 10 percent indicating that it was "'highly unlikely' that a Moon landing had ever taken place." About this curious and widely spreading conviction, administrative aide Vivian White reported: "Mr. Armstrong believes that the only thing more difficult to achieve than the lunar flights would be to successfully fake them."

Meanwhile, many of those who do believe that the Mercury, Gemini, and

Apollo programs actually existed now look back at them as a dead end, a co-
lossal folly, a piece of history as distant and forgotten as bra-burning demon-
strators wreathed in patchouli. Geochemist Robert Brett: "Can you imagine
the people [in the year 2200] looking back at us and saying, 'That was the
time when people were taking trips to the moon and sending unmanned
spacecraft to the planets—and then it all stopped'? I can't see that happening, except for one of two reasons: either in the interim we will have passed
through more dark ages, or we will have investigated the moon and the planets and found that they weren't worth any further effort." GlobalSecurity.org
director John E. Pike: "Maybe we'll find out that spaceflight turned out to be
a historical aberration, like zeppelins. That would be the end of the frontier. I
just don't want to go there."

Who among us hasn't fallen in love and, walking home in weightless ecstasy,
looked up to see an enormous moon, fulgent and blinding, perched against a
dark and starless sky, or a harvest moon, rusted and looming, or a perfect,
silvered crescent, a view that shocks us with its beauty and its power? It might
seem that the men who went up to the Moon and back would have forever
lost this feeling, but it's gratifying to learn that they haven't, that the sweet
white light of the Moon still can be just as beautiful and romantic for them
as it is for us gravity-hindered mortals. "[Today when I look at the Moon,] I
try to feel like everybody thinks I should," Frank Borman laughed. "Which
is, in awe: 'I can't believe I was really there.' And sometimes I do. But most
often I find I just revel in the beautiful Moon."

Just like the child who, after being swept up in an adult's arms, raised over
his head, and then returned to the floor, after feeling the rush of danger and
disorientation and then the safety of both feet on the ground, says, "Do it
again! Do it again! Again! Again!" the Apollo astronauts yearn for that thrill,
but for them, there is no "again." It may be easy for civilians to point out how
privileged they were to have that one shot in the sky, that voyage like no voyage
before, but civilians can never appreciate the feeling of knowing that there
will never be an "again."

Gene and Marta Kranz, their six children, and their many grandchildren
still live in Texas, as does Chris Kraft, who remained in Houston, explaining
that the region is an excellent place to golf. John Hodge and Glynn Lunney
both continued at NASA with the Shuttle; Gerry Griffin became Johnson
Space Center director and president of the Houston Chamber of Commerce;
Steve Bales became Johnson's deputy director of operations; Guenter Wendt

retired to a Kennedy Space Center suburb. Charlie Duke, after his own trip to the Moon, became a beverage distributor and a Christian lay witness in New Braunfels, Texas. George Mueller left NASA when the Apollo program was terminated to become CEO of Burroughs, and then Kistler Aerospace. After serving as secretary of the air force, Bob Seamans became dean of MIT's Engineeering School; he died in 2008. George Low stayed with NASA until 1976 and died in 1984; Max Faget died at his Houston home in 2004 at the age of eight-three. Chimpanzee #61, Ham, died in 1983 at the age of twenty-seven; he is buried at the New Mexico Museum of Space History in Alamogordo.

In 2013, the mighty Hubble space telescope—which can look back in time as far as eleven billion years (just shy of the thirteen-billion-years-ago origins of the universe), and has so far uncovered forty billion never-before-known galaxies—will be replaced by a new orbiting eye, featuring a twenty-one-foot-diameter mirror and a tennis-court-sized sunshield. It will be named in honor of NASA administrator James Webb, who died on March 27, 1992, and is buried at Arlington.

When in 1981 the U.S. government's National Archives released its documents from the 1947 Dora-Nordhausen War Crimes Trials, it was revealed that one von Braun team member, Arthur Rudolph, had been an overseer of Mittelwerk, with responsibility for the paucity of the slaves' meals, and for helping the SS identify, punish, and execute prisoners accused of sabotage. The Justice Department's Office of Special Investigations met with Rudolph at his San Jose, California, retirement home in 1983, and told him he could either renounce his American citizenship and return to Germany, or remain in the United States, be tried for war crimes, and lose his NASA pension. The seventy-six-year-old Rudolph returned to Germany.

After the fall of the Soviet Union, Russia signed a $115 million a year, twenty-year rental agreement with newly independent Kazakhstan for its Cosmodrome in March of 1996. Leninsk, the neighboring suburb built for space workers, was officially renamed Baikonur, while Kaliningrad, home to the Soviet Mission Control, is now known as Korolyov.

NACA, the father of NASA, lives on today through NACA car ducts and airplane cowling and airfoils.

Of the many corporations that built Apollo 11–Saturn V, all have been acquired by Boeing, save IBM, and Grumman (which merged with Northrop).

Columbia, the Apollo 11 Command Module, is currently exhibited in the main hall of the Smithsonian's National Air and Space Museum, alongside the *Spirit of St. Louis* and the X-15.

Over the course of its four decades, the Cold War cost America almost 100,000 lives (the great majority in Korea and Vietnam), and nearly $8 trillion in defense. As I write this, the entire cost of going to the Moon is spent, in 2008 dollars adjusted for 600 percent inflation, every 540 days in the Iraq War.

Aerial reconnaissance technology has continued on its wayward development. By 1977, the National Reconnaissance Office had engineered a whole new species of surveillance, variously known as Keyhole-11, 5501, "Kennan," or "Crystal." Descended from the USAF's Samos, Crystal beamed its data to intermediary satellites, which then beamed them home to the CIA, providing so much data that it overwhelmed the Office of Imagery Analysis, a problem that has continued for the past three decades: humans who can't keep up with their robots. On Christmas Eve, 1997, Earlybird I, the first civilian recon satellite, was launched, and by that year, seven nations and the European Union each had their own surveillance moons in orbit, monitoring weather, wildlife, pollution, civil engineering, damage assessment, spying, and arms control. From 2001 to 2005, the NRO spent $4 billion developing what it called Future Imagery Architecture, which was canceled before a single rocket was launched. In 2008, the Bush administration asked for a $1.7 billion replacement program, Broad Area Space-Based Imagery Collector, even though this would duplicate GeoEye, an already-orbiting and taxpayer-assisted commercial satellite that sells half of its photographs to the government. At the same time, recent U.S. successes in Iraq and Pakistan have been attributed to a new "sense-through-the-wall" radar technology employed by satellites and flying drones.

What is, then, the future of manned spaceflight? A number of scientists believe that there is little in the stars for humans, but a bounty of opportunity for robots. NASA has had success after success, after all, with its Pioneers, Mariners, Vikings, Magellans, Rangers, Galileos, Cassinis, Sojourners, Voyagers, Pathfinders, Spirits, Opportunities, Clementines, Hubbles, and Huygens—a history that has many agency-watchers urging it to focus entirely on its pathbreaking drones. If NASA gives up manned spaceflight, however, under the terms of its lease, it will have to relinquish Houston's Johnson Spacecraft Center back to Rice University.

Today, it seems that America's course in space has almost fully diverged from that of the rest of the world. Since 1981, the Pentagon's annual space budget has been bigger than NASA's, with many now estimating it is over

twice the size of the civilian agency's. The U.S. Department of Defense currently controls over half of the world's military satellites and 95 percent of
the globe's military space budget, with spending expected to rise from $17.5
billion in 2003 to $25 billion in 2010 (with NASA, respectively, at $15 and
$18 billion).

There is something of a rationale for this. Historically, the United States
has reaped the greatest benefits of any nation from space. Currently, its corporations amass 75 percent of the profits from all space businesses in what
has become a significant commercial arena. Between 1998 and 2007, 421
telephone, Internet, radio, TV, reconnaissance, GPS, and defense satellites
were launched. Today, there are so many artificial moons in orbit that they,
combined with four million pounds of orbiting refuse, form a veritable exo-
skeleton enrobing the Earth.

America's preeminence in this area means that the United States has the
most to lose from any attempts to sabotage or destroy these artificial moons.
When one telecommunications relayer, Galaxy 4, malfunctioned on May 19,
1998, it stopped the transmission of Newark bomb squad radios, Chicago
elevator repair alarms, an assortment of online financial data feeds, the *All
Things Considered* radio program, and forty million pagers. Pentagon officials
have insisted that, with so many satellites forming a galaxy so rich in commercial and military applications, the system will need to be defended, and
will of necessity become a new battlefield. "It doesn't take much imagination
to realize how badly war in space could unfold," Steven Lee Myers said in
The New York Times. "An enemy could knock out the American satellite system and a barrage of anti-satellite weapons, instantly paralyzing American
troops, planes and ships around the world. Space itself could be polluted for
decades to come, rendered unusable. The global economic system would
probably collapse, along with air travel and communications."

The consequences of a war in space are in fact so cataclysmic that arms-
control advocates regularly attempt to propose a global ban on all weapons
beyond the Earth's atmosphere. Instead, the Pentagon's lofting ambitions
were given a profound boost in 2002, when President Bush withdrew the nation from the Antiballistic Missile Treaty, which had successfully banned
space weapons for the previous thirty years. American officials have since
made it clear that they are committed to enabling the waging of war in space,
a commitment that has historically prompted competitive nations to parry
and thrust. General Kevin P. Chilton, commander, U.S. Strategic Command,
explained that this was entirely a defensive posture: "Our adversaries understand our dependence upon space-based capabilities and we must be ready

to detect, track, characterize, attribute, predict and respond to any threat to our space infrastructure." Others have not sounded quite so defensive. "We haven't reached the point of strafing and bombing from space," said Air Force Secretary Pete Teets in 2003. "Nonetheless, we are thinking about those possibilities."

"After the next country introduces space weaponry, then what do we do?" Center for Defense Information analyst Dan Smith asked. "Live with a new, unpredictable threat orbiting right above us? Or commit an act of war by preemptively removing their weapons from space?"

Will this posture trigger a new (and integrated) Space and Missile Race? The USAF's Strategic Master Plan calls for a future "expeditionary aerospace force" to achieve "Full Spectrum Dominance." Though little concrete is known about the air force's efforts at very high altitudes, one of the service's assistant secretaries, Keith Hall, noted: "With regard to space dominance, we have it, we like it, and we're going to keep it." One goal of the air force's Space Command, begun on September 1, 1982 (ten months before the navy started its own competitive Space Command), is to keep track of every foreign satellite so that, in the event of a war, they can be targeted and disabled or destroyed. After Desert Storm (which the DoD admitted was the first war to employ military satellites), in 1994 the air force additionally began operating a Space Warfare Center in Colorado Springs to develop such technologies as satellite reconnaissance video that could be broadcast directly into pilots' visors. The center's motto? "In Your Face from Outer Space."

In 2005, the USAF launched the XSS-11, a satellite that can interfere with the transmissions of satellites of other nations, and plans to install in orbit Rods from God, a weapon that drops flechettes of uranium, titanium, or tungsten which, by the time they strike their Earth-based targets, reach a speed of 7,200 miles an hour. The service also is developing microwave guns to destroy the electronics of enemy satellites, Global Strike spaceships armed with one-half-ton warhead ballistic missiles, space-fired laser cannons, and swarmbots. There is a problem with these grand notions, however, which happens to be the same one that afflicted the whole of Apollo. A Tomahawk missile costs $600,000 to fire; a space-based laser, $100,000,000.

Perhaps something the Defense Department might consider defending against is the possibility of a catastrophic asteroid or comet strike. In 1908 Russia was hit by one such collision, which had the force of a hydrogen bomb; the Gulf of Carpentaria, north of Australia, was bombarded in A.D. 536; and the Indian Ocean was struck 4,800 years ago with such power that it is believed the resulting tsunamis inspired the Bible's story of Noah and the flood. After tracking 5,388 asteroids and comets that fly in near-Earth trajectories, NASA has labeled 835 of them "potentially hazardous" and 186 as "impact risks," but others believe a more accurate, and more ominous, number is 20,000. Rusty Schweickart: "There is a slim but real possibility—about 1 in 45,000—that an 850-foot-long asteroid called Apophis could strike Earth with catastrophic consequences on April 13, 2036. What few probably realize is that there are thousands of other space objects that could hit us in the next century that could cause severe damage, if not total destruction. . . . Scientists have a good grasp of the risks of a cosmic fender-bender, and have several ideas that could potentially stave off disaster. Unfortunately, the government doesn't seem to have any clear plan to put this expertise into action."

Over the past forty years, the United States government has assigned a future de facto preeminence in space to the military, rather than NASA. "We're giving up our civilian space leadership, which many of us think will have huge strategic implications," national security expert Joan Johnson-Freese commented. "Other nations are falling over each other to work together in space; they want to share the costs and the risks. Because of the dual-use [civilian and military] issue, we really don't want to globalize." The governments of Russia, China, Japan, India, and the European Community, in fact, are very interested in the civil prospects up there. While India set a new record by launching ten satellites from a single rocket, the European Space Agency built a spaceport in French Guiana where it launches both scientific and commercial satellites, including a competitor to the American Global Positioning System called Galileo. As science fiction author Bruce Sterling proposed, "Imagine this scenario: It's 2029, and a lunar mission lands at Tranquility Base. A crew of heroic young Indians—or Chinese—quietly folds and puts away America's sixty-year-old flag. If the world saw that on television, wouldn't the gesture be worth tens of billions of rupees or yuan?"

The rest of the world also has more prosaic reasons for its extraterrestrial R&D. Thirty years from now the global population of around ten billion people will need five to ten times today's energy; John Young, for one, predicts that outer space could be a source of massive solar power: "Some people are starting to get very concerned about the increase in the number of people on Earth that are gonna be using larger amounts of fossil fuels. If the Chinese get to the stage where they want a car in every garage, boy it's gonna be a big deal." Moon bases can generate solar power; there's plenty of land, and no cloud cover. The University of Houston's David Criswell believes microwave beams transmitting solar energy "could provide the clean, safe, low-cost commercial electric energy needed on Earth."

Another significant source of power, helium-3, is rare on our planet but common on the Moon. When mated with deuterium, helium-3 becomes an excellent source for clean and safe nuclear fusion reactors. "Helium-3 is a product of the solar wind that is embedded in the surface materials of the Moon," Jack Schmitt said. "And over four billion years and turnover in the regolith or the debris layer that covers the Moon, it has been mixed in; and there is a certain constant steady-state concentration of helium-3— particularly in areas that are rich in titanium. . . . Because it's so valuable— its energy equivalent value today, relative to coal, would be about three billion dollars a metric ton! And so you only need about sixty to one hundred kilograms to operate a large power plant (that's a—say a gigawatt power plant—a thousand megawatt power plant) for a year. . . . helium-3 is chucky jam full of energy." One shuttle load of twenty-five tons could supply American energy needs for a year, and it's estimated that the lunar soil holds ten times the energy of all of Earth's current fossil fuels combined.

With such promise, the Chinese have decided to map the Moon inch by inch to explore its geological bounty. Beijing's announcement of a lunar base by 2024 triggered an equivalent proclamation by President Bush that the United States would have its own lunar base by 2021, and Russia (which now has both a private space corporation, RSC Energia, and a state agency, Roskosmos) announced its own outpost, to be operational by 2015. As of 2008, though, while China's space program had 200,000 employees, NASA had only 80,000, 17,000 of whom will retire when the Space Shuttle program ends.

In that same 2004 Moon base announcement, Bush and NASA proposed the Constellation program to send Americans to Mars, using a spaceship three times the size of Apollo to be assembled in Earth orbit—von Braun's original *Collier's* method. A Mars mission, though, would require a voyage of six to seven months and then a stay on the red planet of eighteen months while astronauts waited for Mars and Earth to realign in their orbits for the journey home—in total, a 900-day mission, to be preceded by two Mars shots, one with supplies, and the other with living quarters. Additionally, long-term voyages will have to overcome the severe psychological problems that have revealed themselves aboard the Russian space station, Mir. Soviet psychologists discovered that, for the first two months in outer space, cosmonauts were busy and happy; then, their attitudes changed, and their work became meaningless and boring; eventually, the isolation of space turned the Mir spacefarers anxious, hallucinatory, irritable, and morose. Cosmonaut Valeri Ryumin: "All the conditions necessary for murder are met if you shut two men in a cabin measuring eighteen feet by twenty and leave them together for two months."

Little mentioned at the time of President Bush's announcement was the fact that his father, when president, had made nearly word-for-word the same pronouncement—that NASA should return "back to the moon, back to the future, and this time to stay," establishing a lunar base in preparation for sailing to Mars—in a speech at the National Air and Space Museum to mark Apollo 11's twentieth anniversary on July 20, 1989. His comments were based on "Pioneering the Space Frontier," a report from a commission of, among others, Neil Armstrong, Thomas Paine, and Chuck Yeager. Funding levels, however, did not match dreaming levels, or rhetoric. Lunar and Planetary Institute scientist Paul Spudis: "The hot rumor floated around the grapevine that NASA was talking about thirty years and $600 billion. When everybody heard those numbers, they kind of choked." It seems as though history is repeating itself: while NASA administrator Michael Griffin called Constellation "Apollo on steroids," one agency-watcher, commenting on the lack of support from Congress and the White House, called it "Apollo on food stamps."

Charlie Duke: "I think the wind was taken out of the sails of manned deep space missions because we didn't discover any life out there. If we'd turned the camera on Mars and there'd been little green men looking back at us, we'd have been there by now."

Buzz Aldrin: "When we came back from the moon, there was a Space Task Group doing a study of our future . . . even the most relaxed and slower-paced future that was envisioned then had us reaching Mars in the 1990s. So we certainly haven't lived up to those expectations. . . . Those of us in the space program do realize what we're capable of doing. And hate to see that, because of the stretched out [amount of time] big projects take and with the expectancy of the general public, there's an impatience to move to something else when U.S. leadership is being contested by many other nations today. It may be difficult for us to regain that leadership if we let it slip away because of inattention and [inadequate] allocation of resources."

"For the long-term mastery of spaceflight by our nation as a strategic capability, Apollo did more damage than good," Michael Griffin would come to believe:

> We built up an industrial base, we built up a set of expectations, we accomplished one of the most marvelous things that's ever been done, and then we dismantled it all. It brings to mind the fable of the tortoise and the hare. . . .

I am convinced that if NASA were to disappear tomorrow, if we never put up another Hubble Space Telescope, never put another human being in space, people in this country would be profoundly distraught. Americans would feel that we had lost something that matters, that our best days were behind us, and they would feel themselves somehow diminished. Yet I think most would be unable to say why. . . .

When Charles Lindbergh was asked why he crossed the Atlantic, he never once answered that he wanted to win the $25,000 that New York City hotel owner Raymond Orteig offered for the first nonstop aircraft flight between New York and Paris. Burt Rutan and his backer, Paul Allen, certainly didn't develop a private spacecraft to win the Ansari X-Prize for the $10 million in prize money. They spent twice as much as they made. Sergei Korolyov and the team that launched Sputnik were not tasked by their government to be the first to launch an artificial satellite; they had to fight for the honor and the resources to do it.

I think we all know why people strive to accomplish such things. They do so for reasons that are intuitive and compelling to all of us but that are not necessarily logical. They're exactly the opposite of acceptable reasons, which are eminently logical but neither intuitive nor emotionally compelling. Most of us want to be, both as individuals and as societies, the first or the best in some activity. We want to stand out. This behavior is rooted in our genes. . . .

In that sense, the value of space exploration really is in its spinoffs, as many have argued. But it's not in spinoffs like Teflon and Tang and Velcro, as the public is so often told—and which in fact did not come from the space program [what did come are dialysis machines, CAT scans, MRI machines, space suit technology used for firefighters and oil derrick workers, and great strides forward in solid-state electronics, plastics, metals, lubricants, coatings, insulation, packaging, and water purification].

The real spinoffs are, just as they were for cathedral builders, more fundamental. Anyone who wants to build spacecraft, who wants to be a subcontractor, or who even wants to supply bolts and screws to the space industry must work to a higher level of precision than human beings had to do before the space industry came along. And that standard has influenced our entire industrial base, and therefore our economy.

As for national security, what is the value to the United States of being involved in enterprises which lift up human hearts everywhere? What is the value to the United States of being a leader in such efforts, in projects in which every technologically capable nation wants to take part? The greatest strategy for national security, more effective than having better guns and bombs than everyone else, is being a nation that does the kinds of things that make others want to do them with us.

Before Apollo 11, travel to other planets was a subject for science fiction. Now, it is a reality; we know how to do it, and we have done it successfully. We went to the Moon because the universe is our home, and the universe is beautiful and mysterious. The force that drops ripe fruit from the orchard into your hands is the same force that holds the Moon to the Earth. And, to this day, we don't fully understand something as elemental to life as gravity, the law of attraction, or how the cousins electricity and magnetism act as two variations of the same thing, such as space and time.

Regardless of anyone's opinions on the history and value of manned space-flight, however, sooner or later, human beings will be forced to leave the Earth. In five billion years, the sun will run out of fuel, expanding and reddening, swallowing Mercury and Venus, evaporating all of our planet's water, and then melting it. Novelist Ray Bradbury: "If the Sun dies, if the Earth dies, if our race dies, then so will everything die that we have done up to that moment. Homer will die, Michelangelo will die, Galileo, Leonardo, Shakespeare, Einstein will die, all those will die who now are not dead because we are alive, we are thinking of them, we are carrying them within us. And then every single thing, every memory, will hurtle down into the void with us. So let us save them, let us save ourselves. Let us prepare ourselves to escape, to continue life and rebuild our cities on other planets: we shall not long be of this Earth! And . . . all that matters is that somehow life should continue, and the knowledge of what we were and what we did and learned: the knowledge of Homer and Michelangelo, of Galileo, Leonardo, Shakespeare, of Einstein! And the gift of life will continue."

A mere twenty-five years from guided missile to man on the Moon, and then . . . nothing. Those dismayed by this history can find solace in the fact that, initially, the Wright brothers' creation was widely derided as worthless; few saw much benefit in Lindbergh's crossing the Atlantic; and it took 115 years to get from Columbus to Jamestown. When China's Zhou Enlai was asked by Henry Kissinger for his thoughts on the consequences of the French Revolution, he legendarily said, nearly 180 years after Robespierre, "It's too early to say." So maybe Apollo 11 wasn't the end, but just the beginning. Even with NASA adrift with its questionable space station erector set and its Space Shuttle bus-and-truck operation, shouldn't a nation as rich, as energetic, as brilliant, and as ingenious as the United States—a nation founded by explorers—always be reaching for new frontiers? "We need art as we need dreams," Wally Schirra said. "Without our dreams we wouldn't be where we are: dreaming of going to other planets, to other solar systems, and finding other earths, our earth, among billions of stars."

Korolyov. Sputnik. ICBMs. U-2. Johnson. Cuba. Gagarin. Kennedy. The race to the Moon began with a confluence of historic forces. Will such an astounding achievement ever come our way again? Flight director Gerry Griffin:

I wonder if twenty years from now, or maybe fifty—let's say fifty years from now, people are going to look back and—we did all of that roughly in the mid-1900s, say. Just to make it for the sake of argument. They're going to look back and say, "You know, those people created that great hardware. They did all of those neat missions to the Moon back in the mid-1900s or so! And then they did—they stopped." And they'll wonder why.

I think that's going to be an interesting question to ask. There was this little snippet of time when we went to another place, left this planet. Went to another place. It took place over a little short period of time, and then it just goes on and on. "Why did they do that? I mean, was it just a Cold War thing? Was it—?" I don't think so. . . .

When we finished the Apollo Program, Jack Schmitt had a fellowship at Caltech. He had a little money left in it that he had actually spent most of it. But he had a little left. And he pulled together about, as I recall, twenty-five or thirty people that had all worked on Apollo. We went out to Caltech and spent about three days . . . to discuss what we had done, why we had done it, how did we do it, how were we able to have this little point in history where we could do—this was right after the flight of Apollo 17, so it was all fresh in our minds. But none of us had thought about anything majestic like that. We'd all been—had our heads down.

And Kraft was there. And Armstrong, and Conrad, and Schmitt. . . . And it was a great experience, because we—you know, none of us had been into anything touchy-feely yet. We'd been too busy with—and this was kind of one of those touchy-feely kind of things.

What did we do, and why did we do it, and how did we do it?

And we all had our ideas, and the Cold War was obviously a big piece of that. But Armstrong did something very interesting.

By that time he was up in Cincinnati, I think, teaching engineering. And he got up at a blackboard and he drew four curves. They looked kind of like mountain peaks. And he had them out all like this. And he had one of them titled "Leadership." He had one titled "Threat." He had one of them titled "Good Economy." And he had one of them, I think was the last one, was "Peace" or "World Peace," something like that. He said, "My theory is that when you get all of those curves in conjunction, when they all line up together, you can do something like Apollo. Apollo, or something like it, will happen. And we happened to be ready for that when all of those curves lined up." And, he kind of stole the show of this whole three-day get-together. And he was right on.

I've used that several times with younger people in NASA who sometimes get a little discouraged. You know, "How long are we going to stay in low-Earth orbit?" And, "When are we going to break out?" And I use that story to tell them that, "You've got to be ready so that when those curves do line up again, that we as a nation can take it on and do it."

So to me, that's what Apollo was—it's kind of becoming more meaningful to me now, that's what allowed us to have that little point in history. It didn't last too long, if you stop and think about it. It's kind of almost a blip. And—but what a great program and what a great result in the American and even world-wide spirit that it caused. I know I sound like I'm preaching a little bit here but I really do believe that was the—that was the worth of Apollo.

When old dreams die, new ones come to take their place.
God pity a one-dream man.

—Robert Goddard

Acknowledgments

NASA's History Division is home to a dazzling array of archivists and librarians who were immensely helpful, notably John Hargenrader, Colin Fries, and Liz Suckow at headquarters in Washington; Shelly Henley Kelly and Jean Grant at Johnson; and Elaine Liston and Barbara Green at Kennedy. Additionally, Press Librarian Kay Grinter and Public Affairs Specialist Manny Virata, both of Kennedy, went beyond the call of duty in a difficult post-9/11 environment.

The greatly talented and ever generous National Air and Space Museum's Roger Launius is one of our profession's undersung heroes.

Michael Collins, Chris Kraft, Guenter Wendt, and Andrew Aldrin were gracious with their time and their insights in details missing from the historic record.

The unrelenting support and encouragement of Viking's Rick Kot, Wendy Wolf, and Laura Tisdel have consistently transformed one author's despair into hope.

Stuart Krichevsky, Shana Cohen, and Kathryne Wick of the Stuart Krichevsky Agency are infinitely tireless and meticulous and what any writer needs on his or her side.

In London, Sarah Lutyens of Lutyens and Rubinstein and Roland Philipps and Helen Hawksfield of John Murray couldn't have been more thoughtful or professional.

My task of creating a fresh look at the history of Apollo 11 and the Space and Missile Race would have been impossible without the outstanding contributors of NASA's Oral History Project: Summer Chick Bergen, Michelle Buchanan, Carol Butler, Catherine Harwood, Sandra Johnson, Michelle Kelly, Doyle McDonald, Roy Neal, Jennifer Ross-Nazzal, Kevin Rusnak, Jim Slade, Steven Spencer, Ron Stone, Glen Swanson, Doug Ward, and Rebecca Wright.

While the bibliography covers in full the book's primary and secondary sources, I want to note the outstanding writers, journalists, and historians to whom I'm especially indebted: Piers Bizony, William E. Burrows, Matthew Brzezinski, Andrew Chaikin, Henry S.F. Cooper, Edgar Cortright, Dwayne Day, Oriana Fallaci, Sylvia Fries, John Gaddis, Seth Goddard, James Hansen, David Harland, John Logsdon, Jack Hitt, Norman Mailer, Walter McDougall, Charles Murray and Catherine Bly Cox, Frederick Ordway, Dennis Piszkiewicz, Asif Siddiqi, Andrew Smith, Tom Wolfe, Andrew Young, Brian Silcock, and Peter Dunn.

I'd finally like to honor Eric M. Jones, whose magnificent *Apollo 11 Lunar Surface Journal* is a keystone of scholarship, and Robert Sherrod, whose archival trove, love for NASA, and journalistic humanism were a profound inspiration.

Notes

ix   "In 1969, a few months after Apollo 11 landed on the Moon": Kolbert.

1. Behemoth

3   "Somebody in our shop came up with the idea": Cortright.
5   "By the time of Apollo 11": Ibid.
5   "an ardent Nazi": Brzezinski, 261.
5–6   "One wonder to me was that no Saturn V": Kelly, Michael Collins interview.
6   "I heard Neil Armstrong one time say": Slade.
6   "It is a monster, that rocket": McDonald, Guenter Wendt interview.
6   "Standing up at night and the lights are on it": Smith, 305.
6   "You remained on the pad while the LOX prechilled": Fries, 24.
7   "There was a Juno that went up and turned ninety degrees": Ibid.
7   "The principal reason for not landing on land": Bergen, George Mueller interview.
8   "You don't get shot up": Butler, Ernie Reyes interview.
8   "The slide wire": Collins, *Carrying the Fire*, 113.
8   "One day I got a call": author interview; McDonald, Guenter Wendt interview.

2. The General's Command

11   "First, I'd like to hear from Lee James": Sherrod, "Untitled Manuscript," 317
13   "Neil used to come home with his face drawn white": Mailer, 249.
13   "opened my eyes to the immensity of time" . . . "I hate geology": Harland, 33.
14   "The toughest job we had to train for": Armstrong, Collins, and Aldrin, 210.
14   "A lot of people thought about the kind of people we were": Dale and Howard.
15   "April 3, 1969. I have a list": Low, Special Notes.

3. Anything but What He Is

17   "I suspected that it was highly unlikely that": Makara.
18   "our marriage wouldn't be the same": Goddard, "A Giant Leap."
18   "Broke out in blotches last night": Aldrin, *Return*, 202.
19   "the flight plan as now drawn calls for Aldrin": Snider.
19   "I thought back to the intense and private discussions": Kraft, 322.
19   "Goddamn it, Aldrin, you got a reputation": Aldrin, *Men*, 199.
19   "came flapping into my office": Cernan and Davis, 231.
20   "Neil, who can be enigmatic if he wishes": Aldrin, *Return*, 207.
20   "I thought about it": Kraft, 322.
20   "We had procedures guys": Sherrod, Deke Slayton interview.
20   "had to crawl all over": Sherrod, Colonel Collins interview.
20   "Originally, some of the early checklists were written to show a copilot first exit": Collins, *Carrying the Fire*, 348.

21 **"The story told to me was"**: Ross-Nazzal, Michael Reynolds interview.
21 **"When you're getting ready to go to the moon"**: Kelly, Alan Bean interview.
23 **"I've got a MiG at zero"**: Wolfe, 31.
23 **"We reporters would watch from the beach"**: Dale and Howard.
24 **"Super! Really enjoyed it"**: Smith, *Moondust*, 304.
24 **"Being a military test pilot was the best background"**: Sington.
25 **"The tension between the astronauts and the public affairs office"**: Powers.
25 **"a curious mixture of high competence and near imbecility"**: Mailer, 39–42.
25 **"What [the press] really wanted to know was"**: *In the Shadow of the Moon.*
26 **"I'm going 99.9 percent of the way there"**: Mailer, 39–42.
26 **"Well . . . that's an unpleasant thing to think about"**: Ibid.
28 **"Owen [Maynard] and I got together one morning and we said"**: Murray and Cox, 112.
29 **"I'll tell you, it would have been damn easy"**: Halvorson.
29 **"Annie, we cannot guarantee you safe return of John"**: McDonald, Guenther Wendt interview.
29 **"Going to the moon is a picnic, a trifle, a party trick"**: Fallaci, 220.
29 **"What is the purpose of a newborn baby"**: Young, Silcock, and Dunn, 211.
29 **"I always felt that the risks that we had in the space side of the program"**: Ambrose and Brinkley.
30 **"I got a phone call from an Associated Press reporter"**: Butler, Charles Berry interview.
31 **"If you get into trouble up there"**: Harland, 129.

4. The Sons of Galileo

32 **"It ruled out the matadors"**: Sherrod, "Untitled Manuscript," Chapter: Men for the Moon.
32 **"I like fighter pilots"**: Goddard, "A Giant Leap."
33 **"Fighter pilots can afford to be irresponsible and impetuous"**: Goddard, "A Giant Leap."
33 **"You're flying machines"**: Dale and Howard.
33–34 **"When Deke was a test pilot, I was surrounded by widows"; "Being a test pilot is more dangerous"; "People are always fascinated by anything new"; "For heaven's sake, I loathe danger"; "All shy, [and] shy people"; "When NASA began looking for astronauts"**: Fallaci, 116, 138, 43, 100, 297.
35 **"Am I motivated"**: Sherrod, "Untitled Manuscript," Astronaut notes from Brian O'Leary's book, 65.
35 **"Most of them fall into the top two percent of the population intellectually," "With a few exceptions, they're extremely disciplined people"**: Young, Silcock, and Dunn, 153.
36 **"We knew the world would not be the same"**: AJ Software & Multimedia, *Atomic Archives,* http://www.atomicarchive.com/Movies/Movie8.shtml.
37 **"Man is a Tool-using Animal"**: cited in Shapin.
38 **"I knew I didn't want to be any of the human skill-oriented people"**: Fries, 22.
38 **"Just by my nature, I can't stand to be around anything that I don't know how it works"**: Rusnak, John Aaron interview.
38 **"We used to subscribe to *Doc Savage,* and that had some quite far-out scientific things"; "We made gunpowder"**: Fries, 22.
38 **"Any intelligent fool can make things bigger"**: Harris.
39 **"If I am an engineer"**: Logsdon, "Managing the Moon Problem."
39 **"After spending a lot of time with different subcultures that I intuitively knew were nerdy"**: Nugent.

39   "You could identify the houses of the NASA guys": Kraft 296.
40   "We had gotten to the place we had decided": Caldwell Johnson oral history.
40   "We live in a society exquisitely dependent on science and technology": cited in Shapin.
41   "I am, and ever will be, a white-socks, pocket-protector, nerdy engineer": Home-Douglas.

5. Mr. Cool Stone

42   "Theirs is not to reason why": Walsh.
42   "During the summer, I'd catch crabs for bait": Armstrong, Collins, and Aldrin, 160.
42   "Marry him": Ibid., 166.
43   "Like most of the early astronauts": Kelly, Michael Collins interview.
43   "always the easy-going guy who brought levity into things": In the Shadow of the Moon.
43   "never transmits anything but the surface layer": Smith, Moondust, 95.
44   "But that didn't mean that Neil was all pragmatist": Collins, Carrying the Fire, 314.
44   "that I was too competitive, too insensitive to others": Ibid., 133.
45   "One of the guys who was representing the Air Force in meetings [with NASA] was Buzz Aldrin": Butler, Melvin Brooks interview.
46   "[His] doctoral thesis on space rendezvous": Kraft, 199.
46   "You wouldn't want to sit near him in a party": In the Shadow of the Moon.
46   "If Buzz were a trash man and collected trash": Mailer, 322.
46   "Buzz would always say, 'Well, now I'm coming up on terminal phase'": Butler, Jerry Bostick interview.
46   "Buzz knew more than anybody here about rendezvous": Kelly, Alan Bean interview.
47   "never been great on human relations or public relations, either one": Sherrod, Deke Slayton interview.
47   "extraordinarily remote": Mailer, 27.
47   "If you tell Neil that black is white": Goddard, "A Giant Leap."
47   "If I would marry him and come along in the car": Sawyer.
47   "Pilots take no special joy in walking": Ibid.
48   "That can't be pertinent to Apollo history": Sherrod Archives, miscellaenous notes.
48   "People are a third-rank category of things interesting to talk about": Makara.
48   "There is an obvious kind of order about the physical world that has to be impressive": Greene.
48   "If something is in me which can be called religious": Harris.
48   "Instead of screaming and running for a doctor": Sawyer.
48   "Neil Armstrong was probably the coolest under pressure of anyone": In the Shadow of the Moon.
48   "John always acts as if he were being watched by an army of Boy Scouts": Fallaci, 97.
48   "All through the preparation for the mission, I was absolutely amazed at how quiet, how calm he was": Wright, Gene Kranz interview.
49   "I was disappointed by the wrinkle in history that had brought me along one generation late": Hansen, 53.
50   "a seven-year program: Two years of [university], then go to the navy": Ambrose and Brinkley.
50   "The main problem is overconfidence": Mallick, 26.
50   "I'd be lying if I said they'd done me any good": Fallaci, 297.

51   "They had five pilots, and, if memory serves, seventeen aircraft": Ambrose and Brinkley.

51   "The general pattern was: first, by classroom lectures": Kelly, Michael Collins interview.

52–53 "There's very little time for wondering"; "I can remember several different system problems in the flights": Ambrose and Brinkley.

54   "We were very fortunate. It could have turned ugly": Hansen, 136.

54   "It was a terrible time": Ibid., 166.

54   "Soaring is something that's very easy to do": Ambrose and Brinkley.

54   "I could, by holding my breath, hover over the ground": cited in Brinkley.

54   "Joining NASA wasn't an easy decision": Ambrose and Brinkley.

6. Don't Eat Toads

57   "training was [only] about one-third of our time and effort": Ambrose and Brinkley.

58   "I considered divorce but I wondered if any man could love me the way I wanted to be loved": charlieduke.net.

58   "We work, like slaves we work": Fallaci, 82.

58   "I missed the Vietnam War": In the Shadow of the Moon.

58–59 "You rotated faster and faster until you were spinning violently"; "We didn't know how to train an astronaut in those days": Armstrong, Collins, and Aldrin, 208, citing Virgil I. Grissom, We Seven, New York: Simon and Schuster, 149.

59   "I don't think it's gotten nearly enough credit": Butler, Jack Schmitt interview.

60   "The thing that surprises people on their initial flights": Harland, 68.

60   "Probably in 1968, my boss in the program office": Rusnak, Charles Haines interview.

61   "All the pilots, to my knowledge": Ambrose and Brinkley.

62   "If you don't know what to do, don't do anything"; "To err is human, but to do so more than once is contrary": Murray and Cox, 274.

62   "simulators really are good because they create a sense of confidence in oneself": Neal, Alan Shepard interview.

62   "The simulation organization in those days": Butler, Carl Shelley interview.

63   "We had to develop the flight rules": Johnson, Jay Greene interview.

63   "What we did [in Panama] was pair of them off": Wright, Morgan Smith interview.

64   "I was having to try to organize a brand new training program": Butler, Jack Schmitt interview.

65   "Until Eagle was about ten thousand feet high": Kraft, 320.

66   "What got me, when I got into that lunar thing": Rusnak, John Llewellyn interview.

66   "Neil—hit abort!"; "Apollo 11, we recommend you abort"; "You guys are making too much damn noise out here": Chaikin, 173.

66   "They had less than seven months to train for the most ambitious space mission in history": Chaikin.

66   "Neil was quoted as saying and it was": Butler, Jack Schmitt interview.

67   "Being an astronaut was the most interesting job I ever expect to have": Apollo 11 Thirtieth Anniversary Press Conference.

67   "Training for the lunar mission was probably the most difficult time of my entire life": Wright, Gene Kranz interview.

69   "Fate has ordained that the men who went to the moon to explore in peace will stay on the moon to rest in peace": cited in "Apollo 11," Modern Marvels.

7. A Way to Talk to God

71 **"I remember that we did not go to the Cape to watch the launch":** "Race to the Moon."

71 **"as strong as horseradish":** Hansen, 126.

72 **"Will you let the children stay up and watch the moon walk?":** Mailer, 115.

72 **"We're here in front of the trim, modest suburban home of Squarely Stable":** Wolfe, 340.

73 **"broiled sirloin":** Apollo News Center, KSC press release, February 24, 1969.

73 **"I'm sick of steak!":** Harland, 106.

74 **"I went up to [Reverend Abernathy] and I said":** Butler, Charles Berry interview.

75 **"I succumbed to the awe-inspiring launch":** Harland, 132.

75 **"On Apollo 11 I carried prayers, poems, medallions, coins, flags":** Collins, *Carrying the Fire*, 311.

76 **"We ran a detailed rendezvous meeting of some sort":** Rusnak, Kenneth Young interview.

76 **"I was very tempted to sneak a piece of limestone up there":** Harland, 65.

76 **"We suit technicians had been working in the suit room since 3:30":** Buchanan and Spencer, Joe Schmitt interview.

77 **"They have to be heavy, insulated, bulky":** Kelly, Michael Collins interview.

77 **"Nylon comfort gloves followed by the suit gloves":** Buchanan and Spencer, Joe Schmitt interview.

78 **"My gut feeling was that we had a 90 percent chance":** Apollo 11 Post-Flight Press Conference.

78 **"When you get to the base of this gigantic gantry, it's empty":** *In the Shadow of the Moon.*

78 **"We walked through a sealed compartment painted gray":** Buchanan and Spencer, Joe Schmitt interview.

79 **"Here on Earth usually":** "Race to the Moon."

79 **"It's a hostile environment, and it's trying to kill you":** Goldstein.

80 **"These things were canceled more often than they were launched":** Ambrose and Brinkley.

80 **"A lot of people ask the question":** McDonald, Guenter Wendt interview.

81 **"We were our nation's envoys":** Collins, *Carrying the Fire*, 350.

81 **"I was certainly aware that this was a culmination of the work":** Ambrose and Brinkley.

83 **"The reality is, a lot of times you get up and get in the cockpit":** Ibid.

84 **"The hold-down mechanism would release the rocket only after all five engines":** Cortright.

84 **"There was this enormous light":** Cooper.

85 **"You give me ten billion dollars":** Ross-Nazzal, John Creighton interview.

85 **"At liftoff, I cried for the first time in twenty years"; "Everything else that has happened in our time is going to be an asterisk":** Harland, 132.

85 **"I think it is equal in importance to that moment in evolution when":** Mailer, 39.

85 **"Shake, rattle, and roll!":** Collins, *Carrying the Fire*, 370.

85 **"like being a rat in the jaws of a giant terrier":** Smith, *Moondust,* 38.

85 **"[After crew insertion] we would go to a very forward position":** author interview; McDonald, Guenter Wendt interview.

8. How the Pyramids Were Built

91 **"The Soviet government always was supportive":** Burrows, 460, citing "Venus Unveiled," *Nova.*

91   "The Apollo program was a non-military one": Lowman.
92   "It seemed likely that if the Germans had succeeded in perfecting these new weapons": Ordway, 248.
93   "At night I would stand spellbound looking at the moon": Fallaci, 208.
93   "It never occurred to me that [innocent bystanders]": Neufeld, 46.
94   "On the afternoon of October 19, 1899": Goddard, Diaries.
95   "seems to lack the knowledge ladled out daily in high schools": "A Correction," *New York Times.*
95   "valuable negative information": Goddard, Diaries.
95   "Further investigation and experimentation have confirmed the findings of Isaac Newton": "A Correction," *New York Times.*
96   "will free man from his remaining chains": "Soviet Satellites," *Life.*
96   "if Oberth wants to drill a hole, first he invents the drill press": Ordway, 12.
96   "One weekend in 1930 we started this run": Young, Silcock, and Dunn, 138.
97   "The early history of rocketry reads like an account of the burning of witches": Mailer, 168.
97   "I had no illusions whatsoever as to the tremendous amount of money necessary": Neufeld, 128.
98   "My refusal to join the [Nazi] party": Piszkiewicz, 45.
99   "short of stature, heavily built": Vladimirov, 121.
99   "taught us much, not only that people suffered": McDougall, 24, citing VI Lenin, *Polnoe sobrainie sochinenii,* 5th ed., Moscow, 1958–65, vol. 26, p. 116.
101   "Today, the spaceship was born": "Wernher von-Braun," *Man Moment Machine.*
101   "I have dreamed that the rocket will never be operational against England"; "The bird will carry a ton of amatol in her nose"; "It seems to me that the sole consequence of that high impact velocity is that"; "The A4 is a measure that can decide the war": Ordway, 42–45.
101–2   "could guarantee secrecy"; "The steady stream of convoys from Buchenwald unloaded their human cargo"; "It was at Dora that I realized how the pyramids were built": Michel, 60–62; 160.
102   "a cold, ruthless schemer": Ibid., 160.
103   "You proposed to me that we use the good technical education of detainees available to you at Buchenwald": Piszkiewicz, 202.
103   "I do not claim that von Braun, Reidel": Michel, 97.
104   "I have been [to Mittelwerk] 15 to 20 times": Piszkiewicz, 52.
104   "The problem of amorality is very, very old": Young, Silcock, and Dunn, 25.
104   "Gigantic effort and expense went into developing": Ordway, 248.
105   "Germany has lost the war": Fallaci, 210.
106   "Hey, I've got a nut here": Ordway, 10.
106   "The thinking of the scientific directors of this group is": Brzezinski, 9.
106   "This is absolutely intolerable": McDougall, 44.
107   "to kidnap our leading lights from us": Ordway, 260.
108   "a prisoner of peace": "Wernher von Braun," *Man Moment Machine.*
108   "We can dream about rockets and the Moon until Hell freezes over": Ordway, 361.
109   "The cost of one modern heavy bomber": McDougall, 114.
110   "We can lick gravity": Fries, 162.
110   "In the end, the amnesty given to him by the United States": Dyson.

9. Total Cold War
111   "Huntsville was a [beautiful] southern town when we came here": "Wernher von Braun," *Man Moment Machine.*

111   "We knew that people here ran around without shoes": Dewan.

111   "He had the instinct and intuition of an animal": Bizony, 40.

111   "Von Braun could sell ice to Eskimos": John Johnson, Jr.

113   "Do you realize the tremendous strategic importance of machines of this sort": Young, Silcock, and Dunn, 92.

115   "We gawked at what he showed us": Piszkiewicz, 111.

117   "I suspect we will have to pass Russian customs when we finally reach the moon": Ibid., 388.

118   "You'll never get court-martialed for saying [the Soviets] have a new type of weapon": Brzezinski, 58.

118–19   "Our only solid Central Intelligence Agency information came from the Corona satellite photography": Seamans, Oral History.

119   "every nook and cranny available to it in the basin of world power": McDougall, 88.

119   "You have a row of dominos set up,"; "I have come to the conclusion that some of our traditional ideas of international sportsmanship": Gaddis, 117, 164.

119   "We are facing an implacable enemy whose avowed objective is world domination": "Report on Covert Activities."

120   "At the appointed hour, our neighbors came to our yard to help me watch it": Hickam.

121   "Thanks to Comrade Korolyov and his associates": Piszkiewicz, 111.

121   "Mr. President, Russia has launched an Earth satellite"; "The Russians, under a dictatorial society": Brzezinski, 179.

121   "Now the Communists have established a foothold in outer space": Wolfe, 49.

121   "a National Week of Shame and Danger"; "If the intercontinental missile is, indeed": Brzezinski, 172.

122   "In the open West, you learn to live closely with the sky": McDougall, 141.

122   "The issue [of Sputnik] is one which, if properly handled, would blast the Republicans out of the water": Ibid., 148.

122   "I'll be damned if I sleep by the light of a Red Moon": Brzezinski, 175.

122   "It was as if, overnight, his nation had been vaulted to a preeminent position": Ibid., 179.

123   "enough weight allowance to put a powerful atomic bomb on the moon": Ibid., 210.

123   "I wonder if our plans for the next great breakthrough are adequate": Ibid., 185.

123   "Let us not pretend that Sputnik is anything but a defeat for the United States": McDougall, 145, citing *Life*, October 21, 1957, pp 19–35.

124   "like sausages"; "The fact that the Soviet Union was the first to launch an artificial earth satellite": McDougall, 238.

125–26   "Why doesn't somebody go out there, find it, and kill it?"; "When you get back to Washington"; "Do you hear her?": Piszkiewicz, 116–20.

125   "Almost every reference to Army-developed hardware": Brzezinski, 256.

126   "a grapefruit": Logsdon, "Legislative Origins."

126   "aimed at the stars, but often hit London": Piszkiewicz, 111.

126   "Don't say that he's hypocritical": Neufeld, 118.

128   "Finally, we've got a chance to build something that doesn't have any guns on it": Bizony, 68.

129   "Both from consideration of our prestige as a nation as well as military necessity": Bilstein.

130   "We had tracking stations all around the world in foreign countries": Ward, Jim McDivitt interview.

131   "The history of the Soviet space program is also the history of the milita-
       rization of space": Lucas.
131   "The space program was a paramilitary operation in the Cold War, no
       matter who ran it": McDougall, 174.
132   "Without [Lyndon] Johnson, the launching of Sputnik could very well
       have led": Ibid., 149.

10. The Bluff at Nobleman's Grave

135–36 "There was absolutely no deliberate attempt to violate Soviet airspace";
       "I must tell you a secret": Wright, "Legal Aspects."
136   "The Soviet headstart in space was so disturbing": Bissell, 135.
137   "The very day that Francis Gary Powers was standing in the dock": Ibid.,
       138.
138   "During the Cold War's space race": Krepon and Katz-Hyman.
139   "I wouldn't want to be quoted on this": Burrows, 159fn.
139   "was not about to hock his jewels": Logsdon, 35.
139   "If we let scientists explore the Moon": Seamans, "Project Apollo."
139   "to establish whether there are any valid scientific reasons": Eisen-
       hower.
140   "Empowered by U-2 intelligence": Bissell, 131.
141   "That's so I don't forget to order them to spare the city when the rockets
       fly": Gaddis, 54.
141   "Our missiles were still imperfect in performance and insignificant in
       number": McDougall, 255.
141   "Control of space means control of the world": Brzezinski, 252.
142   "The people of the world respect achievement": Kennedy, John F., Speech
       at Portland, Oregon, September 7, 1960.
142   "One state acts to make itself safer, but in doing so diminishes the secu-
       rity": Gaddis 27.

11. The Fluid Front

146   "the equivalent of trying to remove a man's appendix": Sherrod, "Untitled
       Manuscript," chapter 4, 31.
146   "trying to get information out of a fire hydrant"; "That was one of the
       best lunches we've had together": Bizony, 19.
146   "NACA was a group of engineers": Neal, Alan Shepard interview.
146   "Most of us came in from aircraft flight test": Dale and Howard.
147   "We were going to launch a pig": Ross-Nazzal, Alan Kehlet interview.
147   "One day [at McDonnell in St. Louis]": McDonald, Guenter Wendt inter-
       view.
147   "Computing at that time consisted of row after row of women": Fries, 73.
148   "The site had been selected and announced about September 6, 1961":
       Powers.
148   "Moving to Houston was magnificent": Murray and Cox, 111.
149   "Sometime later, when he was shown the spacecraft": Bizony, 23.
149   "Some of the medical members [of the committee under Jerome Wiesner]
       caucused": Low.
150   "We were in Hangar S at the time where the Air Force": McDonald,
       Guenter Wendt interview.
150   "that the Russians will, for the next five to ten years, beat us to every
       spectacular exploratory flight": Sherrod, "Untitled Manuscript," chapter 2, 10.
150   "In the late fifties, early sixties, we had sometimes twenty, thirty launches
       a week here": McDonald, Guenter Wendt interview.

151 **"We did a lot of things that had never been done before":** Rusnak, Grady McCright and Rob Tillett interview.

151 **"Here we go"; "The earth was gay with a lavish palette of colors":** Burrows, 312.

151 **"At 10:55, Cosmonaut Gagarin safely returned to the sacred soil of our Motherland"; "You have made yourself immortal"; "Now, let the other countries":** "Cruise of the Vostok," *Time.*

152 **"What is this! We're all asleep down here!":** Bizony, 25.

152 **"national extinction"; "The Roman Empire controlled the world because it could build roads":** Smith, *Moondust,* 130.

152 **"create a red dust and turn the whole moon red":** Aldrin, *Men,* 68.

152 **"while no one is more tired than I am":** News Conference, April 12, 1961, Public Papers of the Presidents, John F. Kennedy 1961, 262–63.

152 **"If this is a race, do we want to catch up?":** "Soviet Satellite Sends U.S. Into a Tizzy," *Life.*

153 **"symbolized the nation's lack of initiative, ingenuity":** Sorensen, *Kennedy,* 523–24.

153 **"Is there anyplace we can catch them?":** Sherrod, "Untitled Manuscript," chapter 2, 11.

153–54 **"immediately sensed that the possibility of putting a man on the moon could galvanize public support":** Sorensen, *Counselor,* 336–38.

154 **"There are a number of paradoxes in the story of the Bay of Pigs":** Ibid., 314–20.

155 **"There are limits to the number of defeats I can defend in one twelve-month period":** Bissell, 151.

155 **"I don't think anyone can measure it"; "if a determined national effort is made":** Launius, "Apollo."

156 **"We need a major national space program for prestige purposes":** Sherrod, "Untitled Manuscript," chapter 2, 20.

156 **"We do not have a good chance of beating the Soviets":** von Braun.

156 **"Would you rather have us be a second-rate nation":** Launius, "Apollo."

157 **"light this candle":** McDougall, 211.

157 **"brush the monkey shit":** Aldrin, *Men,* 70.

157 **"The mission only lasted twenty minutes":** Dale and Howard.

157 **"There's a picture of me sitting on the sofa":** Neal, Alan Shepard interview.

158 **"The president was impressed with the world's reaction to the Shepard flight"** . . . **"I didn't want to sound negative":** Kranz, *Failure,* 36.

158 **"found his suggestion horrifying":** Seamans, "Project Apollo."

158 **"this took away all argument against the space program":** McDougall, 322.

159 **"All over the world we're judged by how well we do in space":** Wolfe, 227.

159 **"I believe that this nation should commit itself":** Kennedy, "Special Message to Congress."

160 **"We choose to go to the moon in this decade":** Kennedy, "Address at Rice University."

160 **"embodied everything [Kennedy] had said for a year and longer about striving to get this country moving again":** Sorensen, *Counselor,* 335.

160 **"our efforts in space from low to high gear":** Sorensen, *Kennedy,* 524.

160 **"There's been much conjecture about President Kennedy's motivation when he addressed":** Seamans, "Project Apollo."

161 **"Can't you fellows invent some other race here on earth":** "John F. Kennedy and the Space Race," The White House Historical Association, http://www.whitehousehistory.org/04/subs/04_a03_b02.html.

161 **"The cost":** Young, Silcock, and Dunn, 88.

161   "This is, whether we like it or not, in a sense a race": Bizony, 82.
161   "It's that challenge that best explains the emotional hold of the Kennedys": Herbert.
161–62 "It certainly does not make sense to me"; "just nuts": Smith, *Moondust*, 134, 339.
162   "I always noticed that when we became NASA": Johnson, Milton Silveira interview.
162   "I could hardly believe my ears": Sherrod, "Untitled Manuscript," chapter 4, 1.
162   "Now I'm going to lie down and sleep": Aldrin, *Men*, 88.
162   "more than 30,000 people, in fact the best and most qualified people": Gaddis, 113.
163   "After we had the initial press conference in April 1959": Powers.
163   "I insist on only two conditions": Wolfe, 116.
164   "If I thought about the odds at all": Kraft, 156.
164   "the personal background of each man, his family": letter from Glenn to Sherrod, June 23, 1969, in Sherrod, "Untitled Manuscript," chapter 4, "The Astronauts Peddle Their Story," 15.
165   "were burdened with expenses they would not incur were they not": Callaghan.
165   "whatever property rights there may be in the stories of the astronauts": Webb, "Memorandum for Mr. Callaghan."
165   "If a society editor called up and said": Sherrod, "Untitled Manuscript," chapter 4, "The Astronauts Peddle Their Story," 33.
165   "As soon as we got selected, we were told that": Ward, Jim McDivitt interview.
166   "*Life* was the one thing I trusted": Smith, *Moondust*, 250.
166   "All of the Mercury flights had trouble": Mailer, 170.
166   "In November of '59, I was working in the labor and delivery room": Wright, Dee O'Hara interview.
168   "Its 1.5 million pounds of thrust was an order of magnitude": Seamans, NASA Oral History.
170   "always the epitome of politeness": John Disher, oral history.
170   "Some of the people at headquarters referred to Marshall": Ibid.
171   "By the time you had to do all of the work necessary to fly a single stage": Bergen, George Mueller interview.
171   "To the conservative breed of old rocketeers who had learned the hard way"; "It sounded reckless": Cortright.
172   "throw a hedgehog at Uncle Sam's pants": Thrall.
172   "The fate of Cuba and the maintenance of Soviet prestige": Gaddis, 76.
172   "Total shock": Klein.
172–73 "mop up Cuba in seventy-two hours"; "almost as bad as the appeasement at Munich"; "You'll have to invade . . . as quick as possible"; "All I know is that when you were walking along a Texas road": Sorensen, *Counselor*, 288.
173   "If the American president gives into pressure": Klein.
173   "We're eyeball to eyeball and I think": Piszkiewicz, 144.
174   "persuaded everyone who was involved in it—with the possible exception of Castro": Gaddis, 76.
174   "Are you in sufficient control to prevent my being undercut by NASA": Bizony, 99.
175   "Webb called me yesterday to comment on three interconnected aspects": Bundy.
176   "It was with great attention that we studied President Kennedy's proposal for a joint moon project": Welsh.

176   "participating in a manned lunar landing"; "moondoggle": Loory.
176   "I would like you to assume personally the initiative and central respon-
       sibility": Kennedy, "National Security Action Memorandum No. 271."
176   "My father decided that maybe he should accept": "Krushchev Accepted
       Joint Moon Mission Offer."
177   "The nations of the world, seeking a basis for their own futures"; "It is
       not too much to say that in many ways the viability of representative
       government": Bizony, 381.
177   "You could never get President Kennedy to think beyond what he had to
       do at nine o'clock tomorrow morning": McDougall, 398.
178   "We're the richest country in the world, the most powerful": Mackenzie
       and Weisbrot.
178   "Gemini was an unsung hero in terms of the readiness": Dale and Howard.
178   "There was a very bitter feeling among a lot of us": Ross-Nazzal, Alan Keh-
       let interview.
179   "I've got everything I want": Seamans, "Project Apollo."

12. The Transfiguration

180   "I remember one of the first times I went out": Harwood, Frank Borman
       interview.
181   "There are a lot of things wrong with this spacecraft": Neal, Alan Shepard
       interview.
181   "The first flight in a program is very intense": Wright, Gene Kranz interview.
181   "It isn't that we don't trust you, Joe"; "How can we get to the moon if we
       can't talk between three buildings": Murray and Cox, 180.
182   "We've got a fire in the cockpit": Kranz, *Failure*, 200.
182   "It was like when you were a kid and you put a firecracker": Murray and
       Cox, 188.
182   "It all happened much faster than I can tell it": *Apollo One*.
182   "The technicians, once they knew they could not put the fire out on the
       capsule and open the door": Butler, Ernie Reyes interview.
183   "I'd better not describe what I see": *Apollo One*.
183   "I've never seen a facility or a group of people, a group of men, so
       shaken": Wright, Gene Kranz interview.
184–85 "I would like to observe that we have some materials in spacecraft
       012" . . . "You are not going to fool around with wire bundles after we
       test it down at the Cape": Young, Silcock, and Dunn, 200.
185   "We think that what happened, there was probably an electrical short":
       Apollo One.
185   "Hindsight is wonderful": Slade, Max Faget interview.
185   "I'd known Gus for a long time": Ambrose and Brinkley.
185   "Each flight is but one of the many milestones we must pass": Sherrod,
       "Untitled Manuscript," chapter 4, 3.
185   "If there is a serious accident in the space program"; "If we die, we want
       people to accept it": Bizony, 105.
186   "You can't believe the impact": *Apollo One*.
186   "We met in one of our big conference rooms at the Manned Spacecraft
       Center": Kraft, *Flight*, 110.
186   "It's one thing you learned as an engineer": Wright, John Hodge interview.
187   "How do you explain to the public at large": "Washington Goes to the Moon."
187   "strongly of the view that they would prefer to have a company like
       North American": Bizony, 73.
187   "North American was five times as large as Martin": McDougall, 374.

188 **"[North American] made the first good flight of an X-15 in November"; "I walked out behind Gus Grissom talking to some guys":** Caldwell Johnson oral history.

188 **"continual failure by North American to achieve the progress required":** Young, Silcock, and Dunn, 207.

188 **"a terrible blow to him":** "Washington Goes to the Moon."

189 **"My wife is a wonderful, wonderful person who was a complete support system":** Harwood, Frank Borman interview.

189 **"I was on the beach with Jo Schirra for the last Atlas test firing"; "We all knew that if we weren't there":** Smith, *Moondust,* 250, 253.

190 **"The night I received this news, Joan and I crossed the backyard":** Aldrin, *Men,* 153.

190 **"Press on!"; "If we die, we want people to accept it":** Sherrod, "Untitled Manuscript," chapter 4, 9.

190 **"Apollo 1 was a greater contributor to the entire Apollo program than had it flown and been a success":** *Apollo One.*

190 **"All of us, every single one of us [was] part of a group that had gone through Mercury, had gone through Gemini":** Neal, Alan Shepard interview.

191 **"We could not compete with you Americans":** Burrows, 404.

192 **"If I don't make this flight"; "They knew they had problems for about two hours before Komarov died"; "If I ever find out [Leonid Brezhnev] knew about the situation"; "without realizing it because of the terrible weather conditions":** Bizony, 140, citing Alexei Leonov, *Two Sides of the Moon.*

193 **"I came home so many nights thinking":** Ward, Jim McDivitt interview.

193 **"Indeed a blue Monday":** Low, Special Notes.

193 **"What kind of two-bit garbage are you running up in Bethpage?":** Bizony, 172.

194 **"In the hallway, Deke was almost bouncing with anticipation":** Kraft, *Flight,* 284.

194 **"A better question is: 'What *couldn't* have gone wrong?'":** *Secrets of the Moon Landings.*

194 **"What can we accomplish in it?":** Sherrod, "Untitled Manuscript," chapter 12, "Apollo 8 Certified and Launched," 3.

194 **"would involve risks of great magnitude":** Collins, *Carrying the Fire,* 304.

195 **"I do not favor a manned flight of Saturn V":** Sherrod, "Untitled Manuscript," chapter 12, 8.

195 **"Are you out of your mind?"** Murray and Cox, 322.

195 **"I'm going to walk out the day you do":** Bizony, 177.

195 **"It is very important that we are there first":** "Has U.S. Settled."

197 **"My husband [Frank] came home and, as best he could":** "Race to the Moon," *American Experience.*

197 **"Just how do we tell Susan Borman":** Sherrod, "Untitled Manuscript," Chapter 12, 5.

197 **"My legs felt peculiar, as if they didn't belong 100 percent to me":** Goddard, "A Giant Leap."

198 **"We were very excited about [Apollo 8]":** Ambrose and Brinkley.

198 **"An important Marshall facility was the Dynamic Test Tower":** Cortright.

199 **"It was all there in our emotions as they took off"; "We had our own paparazzi"; "I remember *Life* magazine taking our family pictures":** "Race to the Moon."

199 **"At the time I didn't know it was motion sickness":** Harwood, Frank Borman interview.

199   "The one nice thing about being on Earth, if someone gets sick": "Race to the Moon."

200   "Are you going to take communion every thirty seconds": Aldrin, *Men*, 220.

200   "Stop looking out the window and get back to work!": "Race to the Moon."

200   "[On Apollo 8] I had two great concerns": Harwood, Frank Borman interview.

200   "On Apollo 8, when we did the translunar injection burn": Ward, Gerry Griffin interview.

201   "was like being on the inside of a submarine": Aldrin, *Men*, 223.

201   "We were like three schoolkids looking through": "Race to the Moon."

201   "It was the most awe-inspiring moment of the flight"; "It was almost as if we were discovering the earth for the first time": "Race to the Moon."

201   "Hey, don't take that [picture], it's not scheduled": Sherrod, "Untitled Manuscript," chapter 14, "The Squares Ascendant," 4.

201   "After all of the preparation, after the split-second planning of everything on the mission": *Secrets of the Moon Landings*.

202   "I think that was probably the most magical Christmas Eve I've experienced in my life": Wright, Gene Kranz interview.

202   "When we opened up the dinner for Christmas": Harwood, Frank Borman interview.

203   "We hit the water with a real bang": Ibid.

203   "By now the spacecraft was a real mess, you know": "Race to the Moon."

203   "The crew was well taken care of": Sherrod, "Untitled Manuscript," chapter 14, 7.

13. The Great Black Sea

207   "If I'd had to reach a switch with all that vibration going on": *In the Shadow of the Moon*.

208   "There's a big change, it's from 4 g's to a minus one and a half": Butler, Jack Schmitt interview.

208   "[On Apollo 8,] we went from plus six to minus a tenth g": "Race to the Moon."

208   "This big fireball comes roaring up the length of that booster": *In the Shadow of the Moon*.

208   "We've got skirts up": this and all ensuing shipboard and mission control commentary can be found online. Apollo 11 Technical Air-to-Ground, PAO Spacecraft Commentary, and Command Module On-Board Transcripts, http://history.nasa.gov/alsj/a11/a11trans.html.

210   "We were horrified at the lunar rendezvous approach the first time we saw it": Low.

210   "When we became convinced that [LOR] was the only way": Caldwell Johnson oral history.

210   "had all the money, Wiesner had only me": Sherrod, "Untitled Manuscript," chapter 4, 30.

210   "[Finally] von Braun got up and he said": Caldwell Johnson oral history.

211   "Barfing anywhere is no fun, but barfing in space is different": Wright, Rusty Schweickart interview.

211   "Up there you go around every hour and a half, time after time after time": The Sustainability Institute, http://www.sustainer.org/dhm_archive/index.php?display_article=vn252astronautsed.

212   "Sitting on God's front porch": Cernan and Davis.

212   "The atmosphere on edge presents a striking sight": Pettit.

213   "driving an aircraft carrier with an outboard motor": Collins, *Carrying the Fire*, 311.

214   "I was going in to see Bob [Gilruth] about some of the stuff": Wright, John Hodge interview.

14. The Birth of the Moon

216   "A scuba diver uses a tank of air in sixty minutes": Cortright.

216   "You don't pass anyplace on the way"; "a blackness that is almost beyond conception": *For All Mankind*.

216   "just as beautiful and strange as anything conjured by a child's imagination": Smith, *Moondust*, 46.

217   "you produce a most amazing array of tiny jeweled spheres": Pettit.

217   "we could have shut down our altitude-control thrusters": Aldrin, *Return*, 220.

218   "Nothing goes to the bottom of the bag in zero gravity": *For All Mankind*.

218   "urine dump at sunset": "Race to the Moon."

219   "Feeling weightless . . . it's . . . a feeling of pride": Fallaci, 151.

219   "Contrary to [weightlessness] being a problem": Mailer, 141.

219   "With no gravity pulling down on the loose fatty tissue": Collins, *Carrying the Fire*, 386.

220   "to volunteer for service as a crew member on the Apollo mission to the moon": Brooks, *Chariots for Apollo*, 22.

220   "If you had to single out one subsystem as being most important": Cortright.

221   "It's like a horizon in an airplane": Wright, Chuck Deiterich interview.

221   "As it was to flash crazy lights in Buzz's face all the way to the lunar surface": Collins, *Carrying the Fire*, 342.

222   "The computer interface in those days was a device called the DSKY": Rusnak, Jack Garman interview.

222   "He's growing a mustache!": Mailer, 284.

223   "It reminds me of a wake": Armstrong, Collins, and Aldrin, 410.

223   "In the community I lived in when I was a kid": author interview with Caroline Scott.

223   "I didn't know much about my father's job": "Race to the Moon."

223   "When my eldest son Rick was eight or nine years old in 1965": Smith, *Moondust*, 85.

223   "One Saturday morning we were having breakfast at a long table we had": Wright, Jim McDivitt interview.

224   "Fine. . . . What is history, anyway?" Collins, *Carrying the Fire*, 390.

224   "I remember one day picking up a copy of *Life* magazine": Muson.

225   *Don't you dare come around. Don't you dare:* Harland, 384.

227   "It was a totally different moon than any moon I'd ever seen before": Light.

227   "I was sure that it would be": Armstrong, Collins, and Aldrin, 324.

227   "I feel that all of us are aware that the honeymoon is over": Collins, *Carrying the Fire*, 390.

228   "So that's what he was doing with the *World Book* in his study": Armstrong, Collins, and Aldrin, 207.

229   "It's hard to wrap your mind around a place where nothing ever happens": Gugliotta.

230   "quite a lot of the darned thing is still quite mysterious": Gugliotta.

15. The *Eagle* Has Wings

233   "an ugly and unearthly bug": Armstrong, Collins, and Aldrin, 250.

233   "The complexity of [the other lunar landing designs]": Slade, Max Faget interview.

235   "The tense period of the separation of the LM from Columbia": Hansen, 455.

235   "It's like a dramatic television show": Young, Silcock, and Dunn, 238.

236   "We didn't know this until after the mission, but the crew had not fully depressed": Wright, Gene Kranz interview.

236   "On Apollo 11, we wound up five miles off target": Ross-Nazzal, Floyd Bennett interview.

16. One of Those Sad Days When You Lose a Machine

237 **"Everything we needed to go to the Moon with, we had to create":** Smith, *Moondust*, 129.

237 **"I've got probably thirty or forty records":** Wright, Gene Kranz interview.

237 **"I am Flight and Flight is God":** Kraft, *Flight*, 2.

237 **"The day that Cliff Charlesworth came into the office":** Wright, Gene Kranz interview.

239 **"We didn't have computers on the console":** Johnson, Jay Green interview.

239 **"[The men and women of Mission Control] were in their twenties":** Butler, Jack Schmitt interview.

240 **"I think I am back in the trenches again":** Kranz, *Failure*, 142.

241 **"You take on such a literal involvement that you picture yourself as":** Cooper.

241 **"were a team to do trajectory operations":** Wright, Chuck Deiterich interview.

242 **"The Apollo 11 mission, like many missions":** Wright, Gene Kranz interview.

242 **"I've never never seen things so tense around here"; "Oh God, I can't stand it":** Harland, 214, 231.

243 **"Today is our day"; "The next thing I do is I have the doors of Mission Control locked":** Wright, Gene Kranz interview.

243 **"Gemini and Apollo were computerized":** Smith, *Moondust*, 116.

244 **"It was my twenty-first flight in the Lunar Landing Research Vehicle":** Armstrong, Collins, and Aldrin, 214.

245 **"The frightening films show that he escaped death by just two-fifths of a second":** Kraft, 313.

245 **"Offhand, I can't think of another person"; "It's one of those sad days when you lose a machine":** Hansen, 332.

245 **"Immediately, as soon as we acquire telemetry":** Wright, Gene Kranz interview.

246 **"Scooting in at 3,000 mph at 47,000 feet":** Dale and Howard.

246 **"When [the LM] pitches over and you get your first look":** *For All Mankind*.

246 **"Here around the Earth you get used to an orbit":** "To the Moon," *Nova*.

247 **"We now get to the point where it's time to start engines":** Wright, Gene Kranz interview.

248 **"When I heard Neil say '12 02' for the first time":** Ward, Charlie Duke interview.

248 **"[On] the Apollo guidance computer":** Wright, Richard Battin interview.

248 **"During the descent, when we started having problems":** Sherrod, Buzz Aldrin interview.

248 **"[The simulation trainers] would think up the problems":** Rusnak, Jack Garman interview.

249 **"[During lunar module simulator runs for Apollo 11]":** Wright, Gene Kranz interview.

250 **"Gene Kranz, who was the real hero of that whole episode":** Rusnak, Jack Garman interview.

251 **"In the Control Center":** Harland, 227.

251 **"The concern here was not with the landing area":** "Apollo 11 Technical Crew Debriefing."

251 **"We were concerned, very concerned, at the time":** Armtrong, Collins, and Aldrin, 287.

253 **"We see Neil take over manual control":** Wright, Gene Kranz interview.

253 **"He slowed our descent from twenty feet per second":** Aldrin, *Men*, 267.

253 **"As we dropped below a thousand feet":** "Apollo 11 Technical Crew Debriefing."

253 **"We kept watching his forward velocity":** Ross-Nazzal, Floyd Bennett interview.

253 **"I see the vehicle going across the surface of the moon":** "To the Moon."

254   "At the end, all we knew was that the LM was descending": Wright, Don Lind interview.
254   "I was in the room off the Mission Control Center, the Sun Room": Butler, Wilmot Hess interview.
254   "Deke Slayton is sitting next to me": Harland, 231.
254   "Carlton calls out in hushed tones": Wright, Gene Kranz interview.
254   "For some reason I'm not sure of": "Apollo 11 Technical Crew Debriefing."
255   "Not only did you have these thrusters": Rusnak, Bob Carlton interview.
256   "Yes, it was touch-and-go there at the end": Ross-Nazzal, Floyd Bennett interview.
256   "In simulations, someone's training you to give a certain response": Jones, "First Lunar Landing."
256   "I never dreamed we would still be flying this close to empty": Wright, Gene Kranz interview.
257   "Just before [Apollo 11]": Johnson, Milton Silveira interview.
257   "was packed. You could have heard a pin drop": Rusnak, Bill Easter interview.
257   "I changed my mind several times, looking for a parking place": Harland, 235.
258   "We'd watched hundreds of landings in simulation": Rusnak, Jack Garman interview.
258   "Carlton was just ready to say, 'Fifteen seconds'": Wright, Gene Kranz interview.
259   "I was absolutely dumbfounded": Jones, "First Lunar Landing."
259   "The only thing that was out of normal": Wright, Gene Kranz interview.
260   "We were so excited I couldn't even pronounce": Secrets of the Moon Landings.
260   "If there was any emotional reaction to the lunar landing": Aldrin, Return, 228.
260   "a real high in terms of elation": Sawyer.
260   "The first thing that happens when you land is you experience": "To the Moon."
260   "As I went out of the Control Center, the Moon was up": Wright, Wayne Koons interview.
260   "I can remember during that time, though": Johnson, Robert Heselmeyer interview.
261   "The first thing you have to remember": Rusnak, Ed Fendell interview.
262   "I was with Joan Aldrin that night": Wright, Dee O'Hara interview.
262   "My mind couldn't take it all in": Harland, 242.
262   "The speculation by the TV commentators": Hansen, 477.
263   "Listen! Aren't you all excited": Chaikin, 203.

17. "Mr. President, the *Eagle* Has Landed"

264   "Right after they landed on the Moon": Ross-Nazzal, Ray Melton interview.
265   "If it got up about four hundred degrees": Rusnak, Thomas Kelly interview.
265   "The night that Apollo 11 landed": Ross-Nazzal, Ray Melton interview.
265   "That [West Crater] was a big dude!": Jones, "First Lunar Landing."
267   "We had thought, even before launch, that if everything went perfectly": Harland, 251.
268   "We felt like two fullbacks trying to change positions inside a Cub Scout pup tent": Aldrin, Men, 289.
268   "The biggest problem is that the gloves are balloons": Butler, Jack Schmitt interview.
269   "We tried to pull the door open and it wouldn't come open": "Apollo 11," Modern Marvels.
269   "I think it's dangerous": Korspeter.
269   "It's taken them so long because Neil's trying to decide": Armstrong, Collins, and Aldrin, 309.

270 **"The a was intended":** Apollo 11 Thirtieth Anniversary Press Conference.
271 **"Be descriptive!":** Harland, 260.
272 **"Armstrong surprised everybody":** Walsh.
272 **"When we ask him about it later, he'll say":** Hansen, 480.
272 **"When I was in orbit around the Moon":** Kelly, Alan Bean interview.
273 **"The surface of the moon was like fine talcum powder":** Aldrin, "What It Feels Like to Walk on the Moon."
273 **"The horizon seems quite close to you":** Bradley.
273 **"The light was sometimes annoying,"; "Stepping out of the LM's shadow was a shock":** Nash.
274 **"serenity . . . peacefulness . . . unreal clarity"; "unbelievably beautiful naked charcoal ball"; "until you've been there":** Reinert.
274 **"I think the feelings I had the whole time was the feeling of awe":** Ward, Charlie Duke interview.
274 **"The sun itself was brighter than any sun that I had ever seen":** Butler, Jack Schmitt interview.
274 **"No, it made me feel really, really small":** "Neil Armstrong," *Sixty Minutes.*
274 **"[The Earth is] very delicate"; "It's the abject smallness of the Earth"; "Everything that I know":** Light.
274 **"From space there is no hint of ruggedness to [the Earth]":** "Race to the Moon."
275 **"When you're on the moon":** Aldrin, "What It Feels Like to Walk on the Moon."

18. To Rediscover Childhood

277 **"There was a committee formed to determine":** Rollins, Jack Kinzler interview.
277 **"Peter Flanigan [from the White House] called":** Scheer.
278 **"As I recall the story, Al Shepard bought the first Hasselblad":** Butler, Jack Schmitt interview.
278 **"It wasn't until we were back on earth":** Aldrin, *Return,* 230.
279 **"On Apollo 11, nobody wondered why we never released":** Bergen, Richard Underwood interview.
280 **"We wanted the flag to be able to suspend itself nicely":** Rollins, Jack Kinzler interview.
280 **"We don't want another Tang":** Platoff.
281 **"He doesn't know what's going on. Poor Mike!":** "Apollo 11," *Modern Marvels.*
281 **"[At that moment] I didn't have any great feeling of":** *In the Shadow of the Moon.*
282 **"Oh, hurry up and get the samples":** Cooper.
282 **"To fall on the moon is to rediscover childhood":** Light.
282 **"You can actually just fall over on your face like a dead man":** Young, Silcock, and Dunn, 349
282 **"Every time I threw something":** Light.
282 **"a unique, almost mystical environment":** Mailer, 395.
282 **"We got excellent sampling":** Butler, Jack Schmitt interview.
283 **"One component of the scientific community":** Wright, Don Lind interview.
284 **"The one thing that gave us more trouble than we expected":** "Apollo 11 Technical Crew Debriefing."
285 **"When I watched my dad bouncing about on the moon":** author interview, Scott.
285 **"You know how it is—you have this gung-ho, can-do attitude":** Ivins.
285 **"something symbolic of mankind":** Sherrod, "Untitled Manuscript," unnumbered chapter, "A Matter of Relics," 2.

286   **"The LEC was a great attractor of lunar dust":** "Apollo 11 Technical Crew Debriefing."

286   **"The awe and wonder is pushed into the background":** Goddard, "A Giant Leap."

287   **"I thought a little, and then I stopped thinking altogether":** Fallaci, 117.

287   **"When you're part of the pioneering effort":** Aldrin, "What It Feels Like to Walk on the Moon."

287–88   **"I don't know just what the temperatures were outside"; "We cleaned up the cockpit and got things pretty well in shape"; "We wouldn't be breathing all that dust"; "It was very chilly in there"; "One is that it's noisy"; "I was on the engine cover":** "Apollo 11 Technical Crew Debriefing."

288   **"When Neil Armstrong and what's-his-name landed on the moon":** Butler, Ernie Reyes interview.

289   **"I am alone now, truly alone":** Collins, *Carrying the Fire*, 388.

19. A Tenuous Grasp

291   **"The critical thing was the takeoff":** Rusnak, Joseph Gavin interview.

291   **"one of the best-tested engines in the universe":** Armstrong, Collins, and Aldrin, 408.

291   **"What if that burner would not ignite":** Hansen, 536.

292   **"I have skimmed the Greenland ice cap in December":** Collins, *Carrying the Fire*, 411.

293   **"As long as that thing is lit,"; "They may not be much on show biz"; "Astronauts get along so well because they don't talk"; "Walter Crankcase"; "always likes to get you psyched up for tragedy":** Armstrong, Collins, and Aldrin, 361, 409.

293   **"In space, say you wanted to pitch up":** Ambrose and Brinkley.

294   **"all hell [broke] loose. . . . The docking process begins":** Collins, *Carrying the Fire*, 372.

295   **"just like the old fighter pilot's life":** Armstrong, Collins, and Aldrin, 382.

295   **"I allowed the platform to go into gimbal lock":** Ambrose and Brinkley.

295–96   **"A test pilot's job is identifying problems and getting the answers"; "The sun flashed through the window about once a second"; "It was a great disappointment to us":** Ambrose and Brinkley.

297   **"Your body simply had to be anchored":** Aldrin, *Men*, 162.

297   **"I grabbed the docking collar":** Kelly, Michael Collins interview.

297   **"I instantly took action to correct the angle"; "Me, who couldn't repair the latch on my screen door" "grabbed Buzz by both ears and I was gonna kiss him on the forehead":** Collins, *Carrying the Fire*, 413.

298   **"The biggest joy was on the way home":** *In the Shadow of the Moon.*

301   **"You are literally on fire":** Ibid.

301   **"You have to understand that parachutes in a packed condition":** Slade, Max Faget interview.

301–2   **"On the Apollo missions that I was out on"; "They were worried about contaminating the seas":** Rusnak, John Stonesifer interview.

302   **"I can remember this young frogman, Navy Seal, pulling the hatch back":** "Race to the Moon."

303   **"I closed [the hatch], but it wouldn't lock":** Armstrong, Collins, and Aldrin, 418.

303   **"The command module lands in the Pacific Ocean":** Kelly, Michael Collins interview.

303   **"At one time the capsule was going to be lifted up":** Kelly, Jerome Hammack interview.

304   "To me, the marvel is that it all worked like clockwork": *In the Shadow of the Moon.*

20. We Missed the Whole Thing

306   "I was really impressed at some of the folks' chugging ability": Johnson, Robert Heselmeyer interview.

306   "[The splashdown party for] Apollo 11 was the big blowout": Johnson, Jay Greene interview.

307   "We now stand at what is undoubtedly the greatest decision point": Young, Silcock, and Dunn, 282.

308   "They have something, a sort of wild look": Fallaci, 60.

308   "Even if it isn't harmful to the mice or the quail": Cooper.

308   "The LRL had a colony of I don't know how many white mice": Kelly, Michael Collins interview.

309   "We were worried about there being something toxic": Wright, Fred Pearce interview.

309   "Most of the crews who landed on the Moon": Bergen, Richard Johnston interview.

309   "Duke Ellington was playing his new composition": Collins, *Carrying the Fire,* 443.

310   "Neil, we missed the whole thing": Epstein.

310   "Then I found it wasn't": Sherrod, Buzz Aldrin interview.

310   "As soon as the human body goes into weightlessness": Fries, 107.

311   "Gilruth called me one day and said": Bergen, Richard Johnston interview.

311   "There were some interns that stole a safe full of lunar samples": Wright, Fred Pearce interview.

312   "When you go to a college, the ugliest, dirtiest building": Cooper.

312   "We had a series of rubber gloves that were used": Ross-Nazzal, Mike Reynolds interview.

313   "When the first rock was lifted up inside the vacuum chamber": Cooper.

313   "It was like living in a fishbowl": Butler, John Annexstad interview.

314   "the evidence for lava flows was 'overwhelming'"; "bottom of a coalbin"; "high percentage of glassy beads"; "If we could find out why the moon died": Cooper.

314   "It was a horrendous shock to most of us, the Apollo 11 basalts": Butler, William Muehlberger interview.

314   "Whenever you see a rock being sawed in the laboratory": Butler, Michael Duke interview.

315   "You have noticed how, quite suddenly": Cooper.

21. Through You, We Touched the Moon

316   "warm and moist and inviting and reassuring": Collins, *Carrying the Fire,* 454.

316   "The weather was foul, but I smelled Earth": The Sustainability Institute, http://www.sustainer.org/dhm_archive/index.php?display_article=vn252astronautsed.

316   "I wish I knew": Walsh.

316   "Sometimes they threw a whole stack of [computer] punch cards": Hansen, 566.

317   "I was struck this morning in New York by a proudly waved": "Homage," *Time.*

317   "In all my years as a senator, in all the many votes": Aldrin, *Return,* 38.

320   "By the Canary Islands, we were already getting tired": Swanson.

321   "[In Turkey] they said the thing that they liked about Americans": Der Bing.

321   "People, instead of saying, 'Well, you Americans did it'": *In the Shadow of the Moon.*

321 "Imagine these big stevedore types": Der Bing.

322 "It is a sparkly, shiny new adventure and it was a glorious day"; "The tinsel is tarnished": Aldrin, *Return*, 28.

322 "We fell into an uneasy silence which I ended": Aldrin, *Return*, 67.

322 "Fame has not worn well on Buzz": Collins, *Carrying the Fire*, 95.

322 "I knew that anyone who was on the first lunar landing": *In the Shadow of the Moon*.

322 "I never asked the question about returning to spaceflight"; "he was a pilot, and he was always happier when he was flying": Hansen, 587.

22. When All Those Curves Lined Up

324 "We're always going to feel, somehow, strangers to these men": Hansen, 422.

324 "The F-8 aircraft was going to be an experimental plane": Wright, Richard Battin interview.

325 "I know I could make a million dollars in personal appearances on the outside": Collins, *Carrying the Fire*, 461.

325 "The man needed help, I couldn't help him. He really didn't want me helping him": Hansen, 640.

326 "Sometimes I chastise Neil for being too Lindbergh-like": Ibid., 606.

326 "Over the years I think [many] have questioned Neil's devotion to privacy": Apollo 11 Thirtieth Anniversary Press Conference.

326 "I recognize that I'm portrayed as staying out of the public eye": Ambrose and Brinkley.

327 "The future is not something I know a great deal about": "Neil Armstrong," *Chicago Daily News*.

327 "I was pleased doing the things I was doing"; "If they offered me command of a Mars mission, I'd jump at it": Apollo 11 Thirtieth Anniversary Press Conference.

328 "The whole thing was very tough on my dad": author interview, Scott.

328 "After the flight, [Buzz] lived in Nassau Bay": Rusnak, William Easter interview.

328 "My life was highly structured": Greene.

329 "is the Disintegrated American Hero": Ibid.

329 "One result of space travel was that I had become much more philosophical": Cernan and Davis.

329 "Apollo 11 may have been a small step for Neil but it was a beginning": Newman.

330 "It's a game called, 'spin the Buzz on the globe"; "Many [now say that going to the Moon]"; "I wasn't close to my father": Scott.

330 "I share with [Buzz] a mild melancholy": Collins, *Carrying the Fire*, 461.

330 "I've always wondered why it is that": Newman.

331 "The human race's way of sublimating its highest aspirations": *The Economist*, July 26, 1969.

333 "Kennedy's Science Advisor, Jerry Wiesner, and those advising him": Butler, Sam Ramo interview.

333 "When [JSC Director George W. S.] Abbey came up here": McDonald, Guenter Wendt interview.

334 "requires the political will to lavish money on a project": Guterl.

335 "I remember Chris Kraft telling me once": Harwood, Don Fuqua interview.

335 "in playing golf with the astronauts"; "our best years are behind us": Piszkiewicz, 182.

335 "The last couple of years of the Apollo program": Butler, Jerry Bostick interview.

336   **"We had all been so hyped on this thing of going to the Moon":** Fries, 116.

336   **"We don't have any bold people in NASA anymore":** Slade, Max Faget interview.

336   **"Now that I understand more about what really goes":** Smith, *Moondust*, 244.

336   **"The sciences right now [in 2007] are about thirty-two percent":** Rogers.

337   **"Since 1990, NASA has spent literally billions"; "It would be ironic":** Schwartz, "Destination."

338   **"Can you imagine the people [in the year 2200]":** Cooper.

338   **Maybe we'll find out that spaceflight turned out to be a historical aberration":** Schwartz, "Destination."

338   **"[Today when I look at the Moon,] I try to feel":** Harwood, Frank Borman interview.

341   **"It doesn't take much imagination to realize how badly war in space":** Myers.

341   **"Our adversaries understand our dependence upon space-based capabilities":** Myers.

342   **"We haven't reached the point of strafing and bombing from space":** Weiner, "Air Force."

342   **"After the next country introduces space weaponry":** Hitt.

342   **"With regard to space dominance, we have it, we like it, and we're going to keep it":** Hall, Speech to the National Space Club, 1997.

343   **"There is a slim but real possibility":** Schweickart.

343   **"We're giving up our civilian space leadership":** Kaufman.

343   **"Imagine this scenario":** Sterling.

343   **"Some people are starting to get very concerned":** Smith, *Moondust*, 210.

343   **"could provide the clean, safe, low-cost":** Ferris, "New Pathway to the Stars."

344   **"Helium-3 is a product of the solar wind":** Butler, Jack Schmitt interview.

344   **"All the conditions necessary for murder":** Burrows, 513.

345   **"The hot rumor floated around the grapevine":** Chang, "Allure of an Outpost."

345   **"Apollo on steroids":** Leary.

345   **"Apollo on food stamps":** "Do We Have the 'Right Stuff,'" *Sixty Minutes*.

345   **"I think the wind was taken out of the sails":** Ward, Charlie Duke interview.

345   **"When we came back from the moon":** Buckman.

345   **"For the long-term mastery of spaceflight":** Berger.

347   **"If the Sun dies, if the Earth dies":** Fallaci, 14.

347   **"It's too early to say":** recounted in Pearlstein.

347   **"We need art as we need dreams":** Fallaci, 145.

348   **"I wonder if twenty years from now":** Ward, Gerry Griffin interview.

Sources

Akens, David S. *Saturn Illustrated Chronology.* Historical Office and Management Services Office, NASA George C. Marshall Space Flight Center, January 20, 1971.

Ambrose, Stephen E., and Douglas Brinkley. "Neil A. Armstrong." NASA Johnson Space Center Oral History Project, September 19, 2001.

Aldrin, Buzz (as told to Mike Sager). "What It Feels Like to Walk on the Moon." *Esquire,* June 2001.

Aldrin, Buzz, and Malcom McConnell. *Men from Earth.* New York: Bantam Books, 1989.

Aldrin, Jr., Colonel Edwin E. "Buzz," with Wayne Warga. *Return to Earth.* New York: Random House, 1973.

"Apollo 11," *Modern Marvels,* season 10, episode 25.

Apollo 11: The Eagle Has Landed. Robert Garofalo, director. 1997.

"Apollo 11, The First Landing." BBC, September 1, 2000, http://www.bbc.co.uk/dna/h2g2/A429086.

"Apollo 11 First Post-Landing Press Conference." Apollo 11 Files, NASA Historical Reference Collection, NASA Headquarters, Washington, D.C.

Apollo 11 Post-Flight Press Conference, 10:00 A.M. CST, August 12, 1969, Manned Spacecraft Center, Houston, Texas, http://history.nasa.gov/ap11ann/FirstLunarLanding/ch-1.html.

"Apollo 11 Technical Air-to-Ground Voice Transcription." Apollo 11 Files, NASA Historical Reference Collection, NASA Headquarters, Washington, D.C.

"The Apollo 11 Technical Crew Debriefing." NASA Mission Operations Branch Flight Crew Support Division, July 31, 1969.

Apollo 11 Thirtieth Anniversary Press Conference, Kennedy Space Center, July 16, 1999, http://history.nasa.gov/ap11ann/pressconf.htm.

"Apollo Mission Simulator Instructor Handbook." Downey: NAA, Inc., July 1, 1965.

Apollo News Center, Kennedy Space Center press release, February 24, 1969.

Apollo One: Tragedy to Triumph. WLIW, producer. September 19, 2003.

"Apollo Operations Handbook, Block II Spacecraft." NAS 9-150, April 15 1969, revised October 15, 1969.

"Apollo Operations Handbook, Extravehicular Mobility Unit." CSD A 789 (1), Houston: Manned Spacecraft Center, August 1968.

Apollo Program Qualification Test Summary Report, Z65-10615. Apollo Support Department, General Electric Company, Daytona Beach, Florida, February 15, 1964.

"Apollo Program Summary Report." JSC-09423, Lyndon B. Johnson Space Center, Houston, Texas, April, 1975.

Arabian, Donald D., Chief, Apollo Test Division, NASA. "Apollo 11 Mission Evaluation Plan." Houston: Manned Spacecraft Center, June, 1969.

Armstrong, Neil. "The Engineered Century." National Press Club, February 22, 2000.

———. "Remarks at the Dedication of Grissom and Chaffee Halls, Purdue University." May 2, 1968.

Armstrong, Neil, Michael Collins, and Edwin E. Aldrin Jr. *First on the Moon: A Voyage with Neil Armstrong, Michael Collins and Edwin E. Aldrin, Jr.* Written with Gene Farmer and Dora Jane Hamblin. Epilogue by Arthur C. Clarke. Boston: Little, Brown, 1970.

Arnett, Bill. "The Eight Planets." http://seds.lpl.arizona.edu/nineplanets/luna.html.

"AstroSpies." *Nova,* February 12, 2008.

Atkinson, Nancy. "A Cold War Meeting in Space 33 Years Ago Today." *Universe Today,* July 17, 2008.

Barkun, Michael. *A Culture of Conspiracy: Apocalyptic Visions in Contemporary America.* Berkeley: University of California Press, 2006.

Barr, Stephen. "The Idea Factory That Spawned the Internet Turns 50." *The Washington Post,* April 7, 2008.

"The Beginner's Guide to Aeronautics," http://www.grc.nasa.gov/WWW/K-12/airplane/index .html

"Benefits from Apollo: Giant Leaps in Technology." *NASA Facts,* Lyndon B. Johnson Space Center, July 2004.

Benson, Charles D., and William Barnaby Faherty. *Moonport: A History of Apollo Launch Facilities and Operations.* Washington: NASA Scientific and Technical Information Office, 1978.

Bergen, Summer Chick, interviewer. Interview with John R. Brinkman. NASA Johnson Space Center Oral History Project, Houston, Texas, March 16, 2001.

———. Interview with Robert G. Chilton. NASA Johnson Space Center Oral History Project, Houston, Texas, April 5, 1999.

———. Interview with Charles H. Feltz. NASA Johnson Space Center Oral History Project, Temecula, California, March 9, 1999.

———. Interview with Richard S. Johnston. NASA Johnson Space Center Oral History Project, Houston, Texas, August 11–December 2, 1998.

———. Interview with Owen G. Morris. NASA Johnson Space Center Oral History Project, Houston, Texas, May 20, 1999.

———. Interview with George E. Mueller. NASA Johnson Space Center Oral History Project, Kirkland, Washington, August 27, 1998 and January 20, 1999.

———. Interview with Henry O. Pohl. NASA Johnson Space Center Oral History Project, Houston, Texas, February 9, 1999.

———. Interview with Alan M. Rochford. NASA Johnson Space Center Oral History Project, Houston, Texas, September 15, 1998.

———. Interview with Richard W. Underwood. NASA Johnson Space Center Oral History Project, Houston, Texas, October 17, 2000.

———. Interview with Chester A. Vaughan. NASA Johnson Space Center Oral History Project, Houston, Texas, December 16, 1998.

Berger, Eric. "A Look at Future Beyond Shuttle." *The Houston Chronicle,* March 16, 2008.

Bigham, James P. Southwest Texas State University Oral History Project, Horseshoe Bay, Texas, May 26, 1999.

Bilstein, Roger E. *Orders of Magnitude: A History of the NACA and NASA, 1915–1990.* Washington: National Aeronautics and Space Administration Office of Management Scientific and Technical Information Division, 1989.

Bissell, Richard M., Jr. *Reflections of a Cold Warrior: From Yalta to the Bay of Pigs.* New Haven: Yale, 1996.

Bizony, Piers. *The Man Who Ran the Moon: James E. Webb and the Secret History of Project Apollo.* New York: Thunder's Mouth Press, 2006.

Blumenthal, Ralph. "Astronauts at the Mall, and Rocket Scientists at the Bar." *The New York Times,* April 25, 2004.

———. "For Residents of Space City, New Mission Provides a Reason to Walk a Little Taller." *The New York Times,* January 10, 2004.

———. "It's Lonesome in This Old Town, Until You Go Underground." *The New York Times,* August 21, 2007.

Boone, W.F. "Study on National Needs and NASA Capabilities—NASA-DOD Relationships," April 8, 1969. Archives, NASA Headquarters, Washington, D.C.

Brinkley, David. "A Walk on the Bright Side." *The New York Times,* November 6, 2005.

Broad, William J. "From the Start, the Space Race Was an Arms Race." *The New York Times,* September 25, 2007.

———. "NASA Forced to Steer Clear of Junk in Cluttered Space." *The New York Times,* July 31, 2007.

Brooks, Courtney G., James M. Grimwood, and Lloyd S. Swenson, Jr. *Chariots for Apollo: A History of Manned Lunar Spacecraft.* Washington: NASA Scientific and Technical Information Branch, 1979.

Brooks, David. "The Alpha Geeks." *The New York Times,* May 23, 2008.

Brown, Kenneth A. *Inventors at Work.* Redmond: Microsoft Press, 1988.

Brown, Patricia Leigh. "A Cult of Backyard Rocketeers Keeps the Solid Fuel Burning." *The New York Times,* October 14, 2006.

Bruning, William. NASA Johnson Space Center Oral History Project. December 8. 1971.

Brzezinski Matthew. *Red Moon Rising.* New York: Times Books/Henry Holt, 2007.

Buchanan, Michelle T., with Steven C. Spencer, interviewers. Interview with Joe W. Schmitt. NASA Johnson Space Center Oral History Project, Friendswood, Texas, July 1997.

Buckman, Adam. "Lost in Space: Buzz Aldrin Laments Lack of U.S. Exploration." *The New York Press,* June 20, 2008.

Buenneke, Richard H., Richard DalBello, R. Cargill Hall, and Roger Launius. "National Space Policy: Does it Matter?" Washington: The George Marshall Institute, May 12, 2006.

Bumpus-Hooper, Lynne. "Men on the Moon Look at Future and See Mars." *The Orlando Sentinel,* July 17, 1999.

Bundy, McGeorge. "Memorandum for the President, Subject: Your 11 a.m. appointment with Jim Webb." September 18, 1963. Archives, NASA Headquarters, Washington, D.C.

Burrows, William E. *This New Ocean.* New York: Random House, 1998.

Bush, George. "Remarks by the President at 20th Anniversary of Apollo Moon Landing." National Air and Space Museum, July 20, 1989.

Butler, Carol, interviewer. Interview with John O. Annexstad. NASA Johnson Space Center Oral History Project, Houston, Texas, March 15, 2001.

———. Interview with Charles A. Berry. NASA Johnson Space Center Oral History Project, Houston, Texas, April 29, 1999.

———. Interview with Jerry C. Bostick. NASA Johnson Space Center Oral History Project, Marble Falls, Texas, February 23, 2000.

———. Interview with Melvin F. Brooks. NASA Johnson Space Center Oral History Project, Glendale, Arizona, March 25, 2000.

———. Interview with Anthony J. Calio. NASA Johnson Space Center Oral History Project, McLean, Virginia, April 12, 2000.

———. Interview with Michael B. Duke. NASA Johnson Space Center Oral History Project, Houston, Texas, October 13, 1999.

———. Interview with R. Bryan Erb. NASA Johnson Space Center Oral History Project, Houston, Texas, October 14, 1999.

———. Interview with Donald T. Gregory. NASA Johnson Space Center Oral History Project, Houston, Texas, October 20, 2000.

———. Interview with Wilmot N. Hess. NASA Johnson Space Center Oral History Project, Berkeley, California, April 22, 2002.

———. Interview with Jack R. Lousma. NASA Johnson Space Center Oral History Project, Houston, Texas, March 7, 2001.

———. Interview with Glynn S. Lunney. NASA Johnson Space Center Oral History Project, Houston, Texas, January 28, 1999.

———. Interview with William R. Muehlberger. NASA Johnson Space Center Oral History Project, Austin, Texas, November 9, 1999.

————. Interview with Dr. Simon Ramo. NASA Johnson Space Center Oral History Project, Redondo Beach, California, April 6, 1999.

————. Interview with Reyes, Raul E. "Ernie." NASA Johnson Space Center Oral History Project, Titusville, Florida, September 1, 1998.

————. Interview with Harrison "Jack" Schmitt. NASA Johnson Space Center Oral History Program, Houston, Texas, May 30, 1984, July 15, 1999.

————. Interview with Philip C. Shaffer. NASA Johnson Space Center Oral History Program, Houston, Texas, January 25, 2000.

————. Interview with Carl B. Shelley. NASA Johnson Space Center Oral History Program, Houston, Texas, April 17, 2001.

————. Interview with Leon "Lee" Theodore Silver. NASA Johnson Space Center Oral History Program, Houston, Texas, May 5, 2002.

————. Interview with Troy M. Stewart. NASA Johnson Space Center Oral History Program, Houston, Texas, September 21, 1998.

Cabbage, Michael. "Is Love Affair with the Moon Over?" *Orlando Sentinel,* July 16, 1999.

Callaghan, Richard L. Memorandum for Mr. James E. Webb. Subject: Meeting with President Kennedy on Astronaut Affairs. August 30, 1962. Archives, NASA Headquarters, Washington, D.C.

Cernan, Eugene, and Don Davis. *Last Man on the Moon.* New York: St. Martin's Press, 1999.

Chaikin, Andrew. *A Man on the Moon: The Voyages of the Apollo Astronauts.* New York: Viking, 1994.

Chang, Kenneth. "The Allure of an Outpost on the Moon." *The New York Times,* January 13, 2004.

————. "Scientists Chip Away at Mysteries of the Moon." *The New York Times,* August 8, 2006.

————. "Scientists Find Deeper Meaning for Moon Rumblings." *The New York Times,* February 15, 2005.

Chertok, Boris. *Rockets and People: Creating a Rocket Industry.* The NASA History Series, Washington: NASA Office of External Relations, June 2006.

Chilton, Robert G. "Apollo Spacecraft Control Systems." Houston: Manned Spacecraft Center, June, 1965.

"Cold War Fallout Shelters," http://www.u-s-history.com/pages/h3706.html.

Coledan, Stefano. "Slowly Crumbling, NASA Landmarks May Face the Bulldozer." *The New York Times,* February 28, 2006.

"The Collider Calamity." *Scientific American,* March 2006.

Collins, Martin, editor. *After Sputnik: 50 Years of the Space Age.* New York: Smithsonian/Collins, 2007.

Collins, Michael. *Carrying the Fire: An Astronaut's Journeys.* New York: Farrar, Strauss & Giroux, 1974.

"A Comparison of Soviet and U.S. Defense Activities, 1973–87." Directorate of Intelligence. McLean, Virginia: The Central Intelligence Agency, CIA Archives, July 1, 1988.

Compton, William David. *Where No Man Has Gone Before: A History of Apollo Lunar Exploration Missions.* Washington: National Aeronautics and Space Administration Office of Management, Scientific and Technical Information Division, 1989.

Conway, Chris. "The Basics: Russia, Outer Space and the Profit Motive." *The New York Times,* November 26, 2006.

Constable, George, Bob Somerville, and Neil Armstrong. *A Century of Innovation: Twenty Engineering Achievements That Transformed Our Lives.* Washington: Joseph Henry Press, 2003.

Cooper, Henry S. F. *Moon Rocks.* New York: Dial Press, 1970. (Previously published as Cooper, Henry S. F., Jr. "Letter from the Space Center." *The New Yorker,* July 12, 1969–July 17, 1971.)

"A Correction." *The New York Times,* July 17, 1969.

Cortright, Edgar M., editor. *Apollo Expeditions to the Moon*. NASA SP-350, Washington: Scientific and Technical Information Office, 1975, http://history.msfc.nasa.gov/saturn _apollo/giant.html.

Craig, William H., Brigadier General, DOD Representative, Caribbean Survey Group. "Memorandum for Brigadier General Edward G. Lansdale, USAF, Assistant to the Secretary of Defense. Subject: Ideas in Support of Project." February 2, 1962. Archives, NASA Headquarters, Washington, D.C.

"The Crisis USSR/Cuba." Central Intelligence Agency Memorandum, McLean, Virginia: The Central Intelligence Agency, CIA Archives, October 25, 1962.

Crouch, Donald S. "Design Study for Lunar Exploration Hand Tools." Martin Company Report No. ER 14052. Houston: NASA Manned Spacecraft Center, December 1965.

"The Cruise of the Vostok." *Time*, April 21, 1961.

"The Cuban Missile Crisis Declassified." Gunther Klein, Stefan Brauburger, and Guido Knopp, writers and producers. Itaga Filmproduktion GMBH/ZDF, 2002.

David, Leonard. "CIA and NASA Linked During Cold War." SPACE.com, http://www .space.com/news/wsc_cia_1014.html.

Davis, Douglas. "Astronauts in New Orbits." *Newsweek*, April 1, 1974.

Dawson, Virginia P. and Mark D. Bowles. *Taming Liquid Hydrogen: The Centaur Upper Stage Rocket, 1958–2002*. The NASA History Series. Washington: NASA Office of External Relations, 2004.

———. *Realizing the Dream of Flight: Biographical Essays in Honor of the Centennial of Flight, 1903–2003*. The NASA History Series. Washington: NASA Office of External Relations, 2005.

Day, Dwayne. "Brothers in Arms: The CIA and the American Civilian Space Program, 1958–1968." *The World Space Conference*, October 2002.

Day, Dwayne, ed. "Cold War Military Space History: Programmes." *Space Chronicle*. Supplement 1, 2006.

Day, Dwayne A. "The Cold War in Space." *The Space Review*, July 31, 2006, http://www .thespacereview.com/article/671/1.

"Deke Slayton." *Legends of Airpower*. (Show 306, Russ Hodge, executive producer. 2006.)

"Department of Defense Space Program: An Executive Overview for FY 1998–2003." Washington: The Department of Defense, March 1997, http://www.fas.org/spp/military/ program/sp97/index.html.

Der Bing, William. NASA Manned Spacecraft Center Oral History Project. Houston, Texas, January 2, 1970.

Dewan, Shaila. "Huntsville Journal: When the Germans, and Their Rockets, Came to Town." *The New York Times*, December 31, 2007.

Diamond, Edwin. "The Dark Side of the Moonshot Coverage." *Columbia Journalism Review*, Fall 1969.

Disher, John H. NASA Oral History Program, NASA Headquarters, Washington, D.C., January 27, 1967.

Dobbs, Michael. *One Minute to Midnight: Kennedy, Khrushchev, and Castro on the Brink of Nuclear War*. New York: Alfred A. Knopf, 2008.

"Do We Have the 'Right Stuff' to Put an Astronaut on Mars?" *Sixty Minutes*, April 6, 2008.

Dyson, Freeman. "Rocket Man." *The New York Review of Books*, January 17, 2008.

Easterbrook, Gregg. "How NASA Screwed Up (and Four Ways to Fix It)." *Wired*, July 2007.

———. "Moon Baseless: NASA Can't Explain Why We Need a Lunar Colony." *Slate*, December 8, 2006.

———. "The Sky Is Falling." *The Atlantic*, June 2008.

Edgerton, David. *The Shock of the Old: Technology and Global History Since 1900*. New York: Oxford, 2007.

Eisenhower, Dwight D. "Farewell Address to the Nation." January 17, 1961.

Eisenhower, Dwight D. Budget Message to Congress, 1961. The Dwight D. Eisenhower Library and Museum, http://www.eisenhower.utexas.edu.

Engber, Daniel. "What's a Launch Window?" *Slate*, July 18, 2005.

———. "A Space Tourist's Itinerary: Check on the Stem Cells, Call Nelson Mandela. . . ." *Slate*, October 3, 2005.

Epstein, Robert. "Buzz Aldrin: Down to Earth." *Psychology Today*, May/June 2001, http://psychologytoday.com/articles/pto-20010501-000029.html.

Ertel, Ivan D., and Mary Louise Morse. *Apollo Spacecraft: A Chronology*. Washington: NASA Scientific and Technical Information Division, Office of Technology Utilization, 1969.

Fallaci, Oriana. *If the Sun Dies*. New York: Atheneum, 1967.

Ferris, Timothy. "A New Pathway to the Stars." *The New York Times*, December 21, 2003.

Figliola, Patricia Moloney. "U.S. Military Space Programs: An Overview of Appropriations and Current Issues." *Congressional Research Service Report for Congress*, The Library of Congress, August 7, 2006.

"First Explorers on the Moon." *National Geographic*, December 1969.

"Flight Mission Rules, Apollo 11, As-506/107/LM-5, For NASA/DOD Internal Use Only, Including Appropriate Contractors." Manned Spacecraft Center, Houston, Texas, April 16, 1969.

Florman, Samuel C. *The Existential Pleasures of Engineering*. New York: St. Martin's, 1996.

For All Mankind. director. Al Reinert, November 1, 1989.

Fountain, Henry. "The Basics: Retracing One Small Step." *The New York Times*, September 25, 2005.

Franklin, Jane. *Cuba and the United States: A Chronological History*. Sydney: Ocean Press, 1966.

Fries, Sylvia Doughty. *NASA Engineers and the Age of Apollo*. Washington: National Aeronautics and Space Administration Scientific and Technical Information Program, 1992.

Frutkin, Arnold W. "Memorandum for Mr. Webb, Dr. Dryden, Dr. Seamans, Dr. Simpson, Mr. Scheer, Mr. Callaghan, Mr. Duff. Subject: 'Military vs peaceful' space activities." March 16, 1965. Archives, NASA Headquarters, Washington, D.C.

Gaddis, John Lewis. *The Cold War: A New History*. New York: Penguin Press, 2005.

Garber, Stephen J., editor. *Looking Backward, Looking Forward: Forty Years of U.S. Human Spaceflight Symposium*. NASA History Series, Washington: NASA Office of External Relations, 2002.

Garber, Stephen J. "Multiple Means to an End: A Reexamination of President Kennedy's Decision to Go to the Moon." *Quest: The History of Spaceflight Quarterly*, 7:2, Summer 1999.

Gibney, Frank B., and George J. Feldman. *The Reluctant Space-Farers: The Political and Economic Consequences of America's Space Effort*. New York: New American Library, 1965.

Goddard, Esther. "Widow of a Pioneer in Rocketry Recalls 41-Foot Flight in '26," *The New York Times*, July 17, 1969.

Goddard, Robert Hutchings. Diaries. Robert H. Goddard Library, Clark University, Worcester, Massachusetts.

Goddard, Seth, editor. "A Giant Leap for Mankind." *Life*, http://www.life.com/Life/space/giantleap.

Goldman, Eric F. "The Wrong Man From the Wrong Place at the Wrong Time." *The New York Times*, January 5, 1969.

Goldstein, Richard. "Walter M. Schirra Jr., an Original Astronaut, Dies at 84." *The New York Times*, May 4, 2007.

Goodwin, Matthew. "The Cold War and the Early Space Race." *The Cold War*, http://www.history.ac.uk/ihr/Focus/cold/articles/godwin.html.

"Government Space Budgets to Continue Growth." *SpaceDaily*, December 11, 2003, http://www.spacedaily.com/news/satellite-biz-03zzzl.html.

Graham, John F. *Space Exploration: From Talisman of the Past to Gateway for the Future*. Washington: NASA History Series, 1995.

Greene, Daniel St. Albin. "What Next After You've Walked on the Moon?" *The National Observer*, May 17, 1975.

Griffin, Michael. "The Real Reasons We Explore Space." *Air & Space*, June–July, 2007.

Grissom, Betty, and Henry Still. *Starfall*. New York: Thomas Y. Crowell, 1974.

Gugliotta, Guy. "Keepers of the Moon." *The New York Times*, July 8, 2008.

Guterl, Fred, William Underhill, and Sarah Garland. "Land of Big Science." *Newsweek*, September 15, 2008.

Halvorson, Todd. "One Giant Leap for Mankind." *Florida Today*, July 16, 1999.

Hansen, James R. *First Man: The Life of Neil A. Armstrong: The Authorized Biography*. New York: Simon & Schuster, 2005.

Harland, David M. *The First Men on the Moon: The Story of Apollo 11*. Berlin: Springer/Praxis, 2007.

Harmon, Amy. "New Economy; Even as It Gazes Toward the Stars, the Space Program Has Broad Benefits for Those Rooted to Earth." *The New York Times*, February 10, 2003.

———. "Reviving Romance With Space, Even as 'Space Age' Fades." *The New York Times*, February 4, 2003.

Harris, Kevin, "Collected Quotes of Albert Einstein," http://rescomp.stanford.edu.

Harwood, Catherine, interviewer. Interview with Frank Borman. NASA Johnson Space Center Oral History Project, Houston, Texas, August 11, 1999.

———. Interview with Don Fuqua. NASA Johnson Space Center Oral History Project, Las Cruces, New Mexico, April 13, 1999.

"Has U.S. Settled for No. 2 in Space?" *U.S. News & World Report*, October 14, 1968.

"Hatch of Soyuz Capsule Nearly Burned Up; Crew Was in Serious Danger." The Associated Press, April 22, 2008.

Hawking, Stephen. "The Final Frontier." *Cosmos*, October 2008, http://www.cosmosmagazine .com/node/2209/full.

Heppenheimer, T. A. "How America Chose Not to Beat Sputnik into Space." *American Heritage*, Winter 2004.

Herbert, Bob. "Tears for Teddy." *The New York Times*, May 24, 2008.

Hickam, Homer. "Voices: 10/4/57." *The New York Times*, September 25, 2007.

Hines, William. "Astronauts Favors from Space Reach Acapulco Bay Hotel." *The Huntsville Times*, October 13, 1972.

———. "Gemini Flight Slated to Carry Sky-Spy Gear." *The Washington Star*, August 11, 1965.

Hitt, Jack. "The Amateur Future of Space Travel." *The New York Times*, July 1, 2007.

———. "Battlefield Space." *The New York Times*, August 5, 2001.

———. Jack. "Lunar-tics." *The New York Times*, February 9, 2003.

Holcomb, J. K. "Apollo Program Flight Summary Report, Apollo Missions As-201 through Apollo 16." National Aeronautics and Space Administration Office of Manned Space Flight, Apollo Program Office. June, 1972.

Holmes, Brainerd. NASA Oral History Program, NASA Headquarters, Washington, D.C. July 30, 1968.

"Homage to the Men from the Moon." *Time*, August 22, 1969.

Home-Douglas, Pierre. "An Engineer First." *ASEE Prism*, Summer 2004.

Hornig, Donald (Science Advisor to the President). "Memorandum for the President. Subject: NASA Distortion of Where the U.S. Stands in Space." September 26, 1968. Archives, NASA Headquarters, Washington, D.C.

———. "Memorandum for the President. Subject: Assessment of the Soviet Space Program." December 2, 1968. Archives, NASA Headquarters, Washington, D.C.

Hoover, John Edgar. *Subject: Wernher Magnus Maximilian Freiherr Von Braun*. April 16, 1961. Washington: The United States Department of Justice Federal Bureau of Investigation Archives.

In the Shadow of the Moon. David Sington, director. ThinkFilm, producer, 2007.

Ivins, Molly. "Ed Who?" *The New York Times Magazine*, June 30, 1974.

Jastrow, Robert. "Moon Still Is a Generally Silent Witness," *The New York Times*, March 24, 1974.

Jenkins, Dennis R., Tony Landis, and Jay Miller. "American X-Vehicles, An Inventory: X-1 to X-50." *Monographs in Aerospace History*. no. 31, June 2003.

Johnson, Caldwell. Oral History Transcript, Manned Spacecraft Center, Houston, Texas, December 9, 1966.

Johnson, John Jr. "Propelling America into Space." *Los Angeles Times*, January 20, 2008.

Johnson, Lyndon B. "Memorandum for Dr. Hornig." Undated. Archives, NASA Headquarters, Washington, D.C.

Johnson, Sandra, interviewer. Interview with Jay H. Greene. NASA Johnson Space Center Oral History Project, Houston, Texas, November 10, 2004.

———. Interview with Paul P. Haney. NASA Johnson Space Center Oral History Project, High Rolls, New Mexico, January 20, 2003.

———. Interview with Robert H. Heselmeyer. NASA Johnson Space Center Oral History Project, Houston, Texas, November 12, 2004.

———. Interview with John W. Holland. NASA Johnson Space Center Oral History Project, Seabrook, Texas, February 19, 2004.

———. Interview with Milton A. Silveira. NASA Johnson Space Center Oral History Project, McLean, Virginia, April 18, 2006.

Jones, Eric M., editor. *Apollo 11 Lunar Surface Journal*. March 28, 2008, http://www.hq.nasa .gov/alsj/a11/a11.html

———. *Apollo 11 Technical Air-to-Ground, PAO Spacecraft Commentary, and Command Module On-Board Transcripts*, http://history.nasa.gov/alsj/a11/a11trans.html.

Kaplan, Fred. "Shooting Down the Hype: The Satellite Takedown Doesn't Prove Anything About Our Missile-Defense Capability." *Slate*, February 22, 2008.

———. "The War Room: What Robert Dallek's New Biography Doesn't Tell You About JFK and Vietnam." *Salon*, May 19, 2003.

Kaplan, Lawrence M. *Missile Defense: The First Sixty Years*. Missile Defense Agency, Department of Defense, September 27, 2006.

Kaufman, Marc. "U.S. Finds It's Getting Crowded Out There: Dominance in Space Slips as Other Nations Step Up Efforts." *The Washington Post*, July 9, 2008.

Kazan, Casey. "Is Helium 3 Exploitation China's Hidden Lunar Agenda?" *The Daily Galaxy*, July 3, 2008.

Kelly, Michelle. Oral History Transcript, Michael Collins, Houston, Texas–8 October 1997, http://history.nasa.gov/ap11-35ann/interviewspdf/collins97.pdf.

———. Interview with Alan Bean. NASA Johnson Space Center Oral History Project, Houston, Texas, June 23, 1998.

———. Interview with Jerome B. Hammack. NASA Johnson Space Center Oral History Project, Seabrook, Texas, August 14, 1997.

———. Interview with Seymour Liebergot. NASA Johnson Space Center Oral History Project, Houston, Texas, April 27, 1998.

Kennedy, Jacqueline, "Letter to Robert C. Seamans," March 14, 1964, Robert Channing Seamans, Jr., papers, MC 247, Institute Archives and Special Collections, MIT Libraries, Cambridge, MA.

Kennedy, John F. "Address Before the 18th General Assembly of the United Nations," September 20, 1963. John F. Kennedy Library and Museum Historical Resources, http://www.jfklibrary.org/Historical+Resources/Archives/Reference+Desk/Speeches/ JFK/003POF03_18thGeneralAssembly09201963.htm.

———. "Address at Rice University on the Nation's Space Effort." Houston, Texas, September 12, 1962.

———. "National Security Action Memorandum No. 271 for The Administrator, National Aeronautics and Space Administration [James Webb]. Subject: Cooperation with the USSR on Outer Space Matters. Information copies to: Chairman, National Aeronautics and Space Council; Secretary of State; Secretary of Defense; Director of Central

Intelligence; Chairman, Atomic Energy Commission; Director, National Science Foundation; Special Assistant to the President for Science and Technology; Director, Bureau of the Budget; Director, U.S. Information Agency." November 12, 1963. Archives, NASA Headquarters, Washington, D.C.

———. "Special Message to the Congress on Urgent National Needs." Delivered to a joint session of Congress, May 25, 1961, http://www.jfklibrary.org/Historical+Resources/Archives/Reference+Desk/Speeches/JFK/003POF03NationalNeeds05251961.htm

Khanna, Parag. "Waving Goodbye to Hegemony." *The New York Times*, January 27, 2008.

Kincaide, William C. "Development of the Apollo Portable Life Support System." Houston: Manned Spacecraft Center, February 3, 1964.

Klerkx, Greg. "The Citizen Astronaut." *The New York Times*, January 17, 2004.

Kolbert, Elizabeth. "Crash Course." *The New Yorker*, May 14, 2007.

Korspeter, Caroline. "Moon Advice from Grandma." *The Akron Beacon Journal*, July 13, 1969.

Kraft, Chris. *Flight: My Life in Mission Control*. New York: Dutton, 2001.

Kramer, F. E., D. B. Twedell, and W. J. A. Walton, Jr. "Apollo-11 Lunar Sample Information Catalogue (Revised)." Houston: NASA Lyndon B. Johnson Space Center, February 1977.

Kranz, Gene. *Failure Is Not an Option*. New York: Simon & Schuster, 2000.

Krepon, Michael and Michael Katz-Hyman. "An Arms Race in Space Isn't the Problem." *Space News*, February 12, 2007.

"Krushchev Accepted Joint Moon Mission Offer." United Press International, October 2, 1997.

Kushner, David. "One Giant Screwup for Mankind." *Wired*, January 2007.

Launius, Roger. "Apollo: A Retrospective Analysis." *Monographs in Aerospace History*, Number 3, NASA SP-2004-4503. July, 2004.

———. *Apollo at Twenty-Five*. Washington: NASA, NASA-CR-193380, 1994.

———. "Interpreting the Moon Landings: Project Apollo and the Historians." *History and Technology*, vol. 22, no. 3, September 2006.

———. "Kennedy's Space Policy Reconsidered: A Post-Cold War Perspective." *Air Power History*, Winter 2003.

———. "Perceptions of Apollo: Myth, Nostalgia, Memory or All of the Above?" *Space Policy*, sciencedirect.com, April 9, 2005.

Launius, Roger, interviewer. "Bumper 8—50th Anniversary of the First Launch on Cape Canaveral—Mrs. Elizabeth M. (Carlton) Bain's Oral History." September 25, 2000. Indialantic, Florida, September 25, 2000.

Launius, Roger and Howard McCurdy. "Robots and Humans in Space Flight: Technology, Evolution, and Interplanetary Travel." *Technology in Society* 29, 2007, pp. 271–282.

Launius, Roger and Howard McCurdy, editors. *Spaceflight and the Myth of Presidential Leadership*. Urbana: University of Illinois Press, 1997.

Lawrie, Alan, and Robert Godwin. *Saturn V: The Complete Manufacturing and Test Records Plus Supplemental Material*. Burlington, Ontario: Apogee Books, 2005.

Leary, Warren E. "Lockheed Wins Job of Building Next Spaceship." *The New York Times*, September 1, 2006.

———. "NASA Plans Permanent Moon Base." *The New York Times*, December 5, 2006.

———. "To the Moon, Alice! (Use Your Internet Connection, Dear)." *The New York Times*, November 22, 2005.

Leffler, Melvyn P. "The Cold War: What do We Now Know?" *The American Historical Review*, vol. 104, no. 2. (Apr., 1999), pp. 501–24.

"LEM Guidance, Navigation and Control Subsystem, Course No. 30315, 30415, Study Guide for Training Purposes Only." March 1, 1966. NASA Historical Reference Collection, NASA Headquarters, Washington, D.C.

Levine, Arnold S. *Managing NASA in the Apollo Era*. Washington: NASA Scientific and Technical Information Branch, 1982.

Lepore, Jill. "Our Own Devices: Does Technology Drive History?" *The New Yorker,* May 12, 2008.

Light, Michael. *Full Moon.* New York: Alfred A. Knopf, 1999.

Lindbergh, Charles A. *The Wartime Diaries of Charles A. Lindbergh.* New York: Harcourt Brace Jovanovich, 1970.

Lindros, Marchs, ed. *The Soviet Manned Lunar Program.* Federation of American Scientists, 1997, http://www.fas.org/spp/eprint/lindroos_moon1.htm.

Lipton, Eric. "Administration Trying for Spy Satellites Again." *The New York Times,* September 18, 2008.

Liwshitz, M. "The Origin of Light Flashes Observed by Apollo Astronauts." Washington: Bellcom, September 30, 1971.

Logsdon, John M. *The Decision to Go to the Moon: Project Apollo and the National Interest.* Cambridge: MIT Press, 1970.

Logsdon, John M., Dwayne A. Day, and Roger D. Launius. *Exploring the Unknown: Selected Documents in the History of the U.S. Civilian Space Program.* The NASA History Series, Washington: NASA History Office, 1996.

Logsdon, John M., moderator. "Legislative Origins of the National Aeronautics and Space Act of 1958, Proceedings of an Oral History Workshop Conducted April 3, 1992." *Monographs in Aerospace History,* number 8, Washington: Office of Policy and Plans, NASA History Office, 1996.

———. "Managing the Moon Program: Lessons Learned from Project Apollo. Proceedings of an Oral History Workshop Conducted July 21, 1989." *Monographs in Aerospace History,* number 14, July 1999.

Logsdon, John, "Testimony," Senate Science, Technology, and Space Hearing: International Space Exploration Program, Source: Senate Committee on Commerce, Science, and Transportation, International Space Exploration Program, April 27 2004, http://www.spaceref.com/news/viewsr.html?pid=12688.

Loory, Stuart H. "House Rebuffs Kennedy's U.S.-Red Moon Trip in Limiting Space Funds." *The New York Herald Tribune,* October 11, 1963.

Low, George M. NASA Oral History Program, NASA Headquarters, Washington, D.C. May 1, 1964.

———. Special Notes and Daily Reports. Archives, NASA Headquarters, Washington, D.C.

Lowman, Paul D., Jr. "The Apollo Program: Was it Worth It?" *World Resources: The Forensic Quarterly,* vol. 49, August 1975.

Lucas, Paul. "Shadows of the Soviet Space Race." *Strange Horizons,* May 3, 2004.

Mackenzie, G. Calvin, and Robert Weisbrot. *The Liberal Hour: Washington and the Politics of Change in the 1960s.* New York: Penguin Press, 2008.

Mailer, Norman. *Of a Fire on the Moon.* Boston: Little, Brown and Company, 1969.

Makara, Michael. "One Giant Leap for Mankind: The 35th Anniversary of Apollo 11." http://history.nasa.gov/ap11-35ann.

"Making the Moon." *Astrobiology Magazine,* June 2, 2002, http://www.astrobio.net/news/article178.html.

Mallick, Donald L., with Peter W. Merlin. *The Smell of Kerosene: A Test Pilot's Odyssey.* NASA SP 4108, NASA History Series, Washington: NASA, 2003.

Massachusetts Institute of Technology Instrumentation Laboratory. "Apollo Guidance and Navigation." June 1965.

McCarthy, Natasha. "The Wisdom of Engineers." *The Philosopher's Magazine,* Second Quarter 2008.

McCaw, Lt. Col. Flip, and Dr. James G. Roche. "The U.S. Air Force Transformation Flight Plan." HQ USAF/XPXC, November 2003.

McDonald, Doyle, interviewer. Interview with Guenter F. Wendt. NASA Johnson Space Center Oral History Project, Coral Gables, Florida, January 16, 1998.

McDougall, Walter A. *. . . the Heavens and the Earth: A Political History of the Space Age.* New York: Basic Books, 1985.

McGrath, Matt. "France Plans Revolution in Space." BBC, July 1, 2008.

McNamara, Robert S., Defense Secretary. "Is Russia Slowing Down in Arms Race?" *U.S. News & World Report*, April 12, 1965.

Michel, Jean. *Dora: The Nazi Concentration Camp Where Modern Space Technology Was Born and 30,000 Prisoners Died.* New York: Holt, Rinehart, and Winston, 1979.

"Moon for Sale." *Big Science.* BBC, 2007.

Moore, Kevin. "A Brief History of Aircraft Flight Simulation." *Fifty Years of Flight Simulation Conference Proceedings*, April 1979, http://homepage.ntlworld.com/bleep/SimHistl.html

Moore, Mike. *Twilight War: The Folly of U.S. Space Dominance.* Washington: The Independent Institute, 2008.

Muir, Hazel. "New Clues to Birth of the Moon." *New Scientist*, August 15, 2001.

Murray, Charles, and Catherine Bly Cox. *Apollo.* New York: Simon and Schuster, 2002.

Muson, Howard. "Comedown from the Moon—What Has Happened to the Astronauts." *The New York Times Magazine*, December 3, 1972.

Myers, Steven Lee. "Look Out Below. The Arms Race in Space May Be On." *The New York Times*, March 9, 2008.

"NASA Support Manual: Apollo Spacecraft Familiarization." SID 62-435. Downey: NAA, Inc, December 1, 1966.

"NASA's Dirty Secret: Moon Dust." *Science Daily*, September 29, 2008, www.sciencedaily.com/releases/2008/09/080924191552.htm.

"National Intelligence Estimate: The Soviet Space Program." Number 11-1-62. Directorate of Intelligence. McLean, Virginia: The Central Intelligence Agency, CIA Archives, December 5, 1962.

"National Intelligence Estimate: The Soviet Space Program." Number 11-1-65. Directorate of Intelligence. McLean, Virginia: The Central Intelligence Agency, CIA Archives, January 27, 1965.

"National Intelligence Estimate: The Soviet Space Program." Number 11-1-67. Directorate of Intelligence. McLean, Virginia: The Central Intelligence Agency, CIA Archives, March 2, 1967.

Neal, Roy, interviewer. Interview with Alan B. Shepard, Jr., NASA Johnson Space Center Oral History Project, Pebble Beach, California, February 20, 1998.

"Neil Armstrong." *Sixty Minutes*, November 6, 2005.

"Neil Armstrong Proves to Be Very Much an Earthling," *Chicago Daily News*, August 11, 1977.

Neufeld, Michael J. *Von Braun: Dreamer of Space, Engineer of War,* New York: Knopf, 2007.

"New Rocket Crash Alarms Kazakhs." BBC News, September 7, 2007.

Newman, Robert E. Tenth Anniversary of Apollo 11 Moon Landing Press Conference. Houston: NASA Manned Spacecraft Center, July 20, 1979.

Nuclear Weapon Database: United States Arsenal. December 26, 2007, http://www.cdi.org/nuclear/database/usnukes.html.

Nugent, Benjamin. *American Nerd: The Story of My People.* New York: Scribner, 2008.

Oberg, James. "The Dozen Space Weapons Myths." *The Space Review*, March 12, 2007.

Oberg, James. *UFOs and Outer Space Mysteries.* Virginia Beach: Donning Press, 1982.

Ordway, Frederick I., III, and Mitchell R. Sharpe. *The Rocket Team: From the V-2 to the Saturn Moon Rocket.* Cambridge: The MIT Press, 1979.

"The Origin of the Moon." Planetary Science Institute, http://www.psi.edu/projects/moon/moon.html.

Orloff, Richard W. *Apollo by the Numbers: A Statistical Reference.* Washington: NASA History Division, 2000.

Osgood, Kenneth. *Total Cold War: Eisenhower's Secret Propaganda Battle at Home and Abroad.* Lawrence: University Press of Kansas, 2006.

Overbye, Dennis. "At NASA, Science Sharply Shifts Course." *The New York Times*, April 27, 2004.

———. "Kissing the Earth Goodbye in About 7.59 Billion Years." *The New York Times*, March 11, 2008.

Packer, George. "The Fall of Conservatism." *The New Yorker,* May 26, 2008.

Paine, Thomas O. *In this Decade: Mission to the Moon.* Washington: NASA Headquarters, 1972.

Park, Robert L. "Scorched Earth." *The New York Times,* January 15, 2006.

Perlstein, Rick. "Getting Past the '60s? It's Not Going to Happen." *The Washington Post,* February 3, 2008.

———. *Nixonland.* New York: Scribner, 2008.

Petroski, Henry. *To Engineer Is Human: The Role of Failure in Successful Design.* New York: Vintage, 1992.

Pettit, Don. "Space Chronicles." NASA Human Space Flight Web site, http://spaceflight.nasa .gov/station/crew/exp6/spacechronicles1.html.

Pincus, Walter. "Plans by U.S. to Dominate Space Raising Concerns." *The Washington Post,* March 29, 2005.

Piszkiewicz, Dennis. *Wernher Von Braun: The Man Who Sold the Moon.* Westport: Praeger, 1999.

Plait, Phil. *Bad Astronomy.* New York: John Wiley and Sons, 2002.

Platoff, Anne M., Hernandez Engineering. "Where No Flag Has Gone Before: Political and Technical Aspects of Placing a Flag on the Moon." *NASA Contractor Report 188251,* August, 1993.

Powers, Col. John A. "Shorty." Public Affairs Officer. Oral History Transcript. Johnson Space Center, November 9, 1968.

President's Report to Congress on U.S. Space Programs. December, 1961.

Public Papers of the Presidents of the United States, John F. Kennedy, 1961. Washington: Government Printing Office, 1962.

"Race to the Moon," *American Experience.* http://www.pbs.org/wgbh/amex/moon.

"Remembering Apollo 11: The 30th Anniversary Data Archive CD-ROM." NASA History Office, Office of Policy and Plans, Washington, July 1999.

"Report of Conclusion of Powers Trial." USSR International Affairs. McLean, Virginia: The Central Intelligence Agency, CIA Archives, August 22, 1960.

"Report on the Covert Activities of the Central Intelligence Agency." McLean, Virginia: The Central Intelligence Agency, CIA Archives, September 30, 1954.

Rogers, Adam. "A Conversation with Shana Dale, Deputy Administrator of NASA." *Wired,* May 21, 2007.

Rollins, Paul, interviewer. Interview with Jack Kinzler. NASA Johnson Space Center Oral History Project, Seabrook, Texas, January 16, 1998.

Romero, Simon. "In Houston, the Euphoria Vanishes." *The New York Times,* July 29, 2005.

Rose, Kenneth D. *One Nation Underground: The Fallout Shelter in American Culture.* New York: New York University Press, 2001.

Rosenbaum, Ron. "The Return of the Magic Bullet." *Slate,* June 1, 2007.

Ross-Nazzal, Jennifer, interviewer. Interview with Joseph P. Allen. NASA Johnson Space Center Oral History Project, Houston, Texas, January 28, 2003.

———. Interview with Floyd V. Bennett. NASA Johnson Space Center Oral History Project, Houston, Texas, October 22, 2003.

———. Interview with John O. Creighton. NASA Johnson Space Center Oral History Project, Seattle, Washington, May 3, 2004.

———. Interview with Jack Funk. NASA Johnson Space Center Oral History Project, Houston, Texas, October 30, 2003.

———. Interview with Alan B. Kehlet. NASA Johnson Space Center Oral History Project, San Jose, California, September 30, 2005.

———. Interview with Raymond E. Melton. NASA Johnson Space Center Oral History Project, Las Cruces, New Mexico, January 22, 2003.

———. Interview with Michael A. Reynolds. NASA Johnson Space Center Oral History Project, McKinney, Texas, March 2, 2005.

————. Interview with Frank H. Samonski. NASA Johnson Space Center Oral History Project, Houston, Texas, December 30, 2002.

Rusnak, Kevin M., interviewer. Interview with John W. Aaron. NASA Johnson Space Center Oral History Project, Houston, Texas, January 18, 2000.

————. Interview with Arnold D. Aldrich. NASA Johnson Space Center Oral History Project, Houston, Texas, June 24, 2000.

————. Interview with Donald D. Arabian. NASA Johnson Space Center Oral History Project, Cape Canaveral, Florida, February 3, 2000.

————. Interview with Peter J. Armitage. NASA Johnson Space Center Oral History Project, Houston, Texas, August 20, 2001.

————. Interview with James M. Beggs. NASA Johnson Space Center Oral History Project, Cutchogue, New York, September 19, 2000.

————. Interview with Robert L. Carlton. NASA Johnson Space Center Oral History Project, Houston, Texas, March 29, 2001.

————. Interview with Charles L. Dumis. NASA Johnson Space Center Oral History Project, Houston, Texas, March 1, 2002.

————. Interview with William B. Easter. NASA Johnson Space Center Oral History Project, Wrightsville Beach, North Carolina, May 3, 2000.

————. Interview with Edward I. Fendell. NASA Johnson Space Center Oral History Project, Houston, Texas, October 19, 2000.

————. Interview with John R. "Jack" Garman. NASA Johnson Space Center Oral History Project, Houston, Texas, March 27, 2001.

————. Interview with Joseph G. Gavin, Jr. NASA Johnson Space Center Oral History Project, Amherst, Massachusetts, January 10, 2003.

————. Interview with Charles R. Haines. NASA Johnson Space Center Oral History Project, Houston, Texas, November 7, 2000.

————. Interview with Thomas J. Kelly. NASA Johnson Space Center Oral History Project, Bethesda, Maryland, March 7, 2002.

————. Interview with John S. Llewellyn, Jr. NASA Johnson Space Center Oral History Project, Houston, Texas, January 9, 2001.

————. Interview with Grady E. McCright and Rob R. Tillett. NASA Johnson Space Center Oral History Project, Las Cruces, New Mexico, March 8, 2000.

————. Interview with Joseph E. Mechelay. NASA Johnson Space Center Oral History Project, Houston, Texas, June 16, 2000.

————. Interview with William R. Pogue. NASA Johnson Space Center Oral History Project, Houston, Texas, July 17, 2000.

————. Interview with Rodney G. Rose. NASA Johnson Space Center Oral History Project, Wimberley, Texas, November 8, 1999.

————. Interview with John C. Stonesifer. NASA Johnson Space Center Oral History Project, Houston, Texas, March 21, 2001.

————. Interview with Kenneth A. Young. NASA Johnson Space Center Oral History Project, Houston, Texas, June 6, 2001.

Saletan, William. "Evolving Predators." *Slate*, September 16, 2008.

————. "Lords of the Rings: Life Arrives on a Moon of Saturn." *Slate*, January 19, 2005.

————. "Strip Moon: Mining the Moon to Fund a Space Base." *Slate*, December 5, 2006.

Sato, Rebecca. "Space Euphoria: Do Our Brains Change When We Travel in Outer Space?" *The Daily Galaxy,*

"Saturn V Flight Manual." MSFC-MAN_503, Huntsville: NASA, November 1, 1968.

Saturn V News Reference. August, 1967, http://history.msfc.nasa.gov/saturn_apollo/saturnv_press_kit.html.

Sawyer, Kathy. "The Quiet Man on the Moon." *The Washington Post,* July 18, 1999.

Scheer, Julian. "What President Nixon Didn't Know," http://www.space.com/news/a11_plaque.html.

Schwartz, John. "Destination Is the Space Station, but Many Experts Ask What For." *The New York Times,* December 5, 2006.

———. "Private Company Plans $100 Million Tour Around the Moon." *The New York Times,* August 10, 2005.

———. "Manned Private Craft Reaches Space in a Milestone for Flight." *The New York Times,* June 22, 2004.

———. "Satellite Spotters Glimpse Secrets, and Tell Them." *The New York Times,* February 5, 2008.

———. "Thrillionaires: The New Space Capitalists." *The New York Times,* June 14, 2005.

Schwartz, John, and Simon Norfolk. "Trajectories: The Rise of Rockets in the Post-Space-Age Age." *The New York Times,* May 11, 2008.

Schweickart, Russell L. "The Sky Is Falling. Really." *The New York Times,* March 16, 2007.

Scott, Caroline. "Relative Values: Buzz Aldrin and His Son Andy." *The Sunday Times* (London), October 7, 2007.

Seamans, Robert C., Jr. NASA Oral History Program, NASA Headquarters, Washington, D.C., March 27, 1964.

———. "Project Apollo: The Tough Decisions." *Monographs in Aerospace History,* number 37, National Aeronautics and Space Administration Office of External Relations History Division, 2005.

Secrets of the Moon Landings. Nicole Teusch, director. National Geographic Channel, 2007.

Shachtman, Noah. "Pentagon Preps for War in Space." *Wired,* February 20, 2004.

Shah, Anup. "Militarization and Weaponization of Outer Space." *Global Issues,* January 21, 2007, http://www.globalissues.org/Geopolitics/ArmsControl/Space.asp.

Shapin, Steven. "What Else Is New?" *The New Yorker,* May 14, 2007.

Shawcross, Paul, OMB. "The U.S. Government Space Budget." October 17, 2006.

Shea, Joseph F. NASA Oral History Program, NASA Headquarters, Waltham, Massachusetts, January 12, 1972.

Sherrod, Robert. "Conversation with Brig. Gen. Michael Collins USAFR, director, Air and Space Museum, Smithsonian Institution, former astronaut (Apollo 11), 9 Sept. 1974, with RS. (Telephone)." http://history.nasa.gov/ap11-35ann/interviewspdf/collinssept74 .pdf.

———. "Interview with Col. Edwin E. Aldrin, Jr., USAF, NASA Hq. Washington, by Robert Sherrod, 4 December 1970, 1330-1599." http://history.nasa.gov/ap11-35ann.

———. "Interview (luncheon) with Colonel Collins, Assistant Secretary of State for Public Affairs, with RS, Federal City Club, 3 September 1970." http://history.nasa.gov/ap11 -35ann/interviewspdf/collinssept70.pdf.

———. "Interview with Deke Slayton, Director of Flight Crew Operations." Manned Spacecraft Center, Houston, July 27, 1972.

———. "Untitled Manuscript on the History of NASA." The Sherrod Archives, NASA Headquarters, Washington, D.C.

"Shot Down: U2 Spyplane," *Man Moment Machine.* Michael McInerney, producer/director. Edelman Productions, 2005.

Siddiqi, Asif A. *The Soviet Space Race With Apollo.* Gainesville: University Press of Florida, 2003.

Slade, Jim, interviewer. Interview with Charles J. Donlan. NASA Johnson Space Center Oral History Project, Washington, D.C, April 27, 1998.

———. Interview with Maxime A. Faget. NASA Johnson Space Center Oral History Project, Houston, Texas, June 18–19, 1997.

"A Small, Belated Step for Grammarians." *The New York Times,* October 3, 2006.

Smith, Andrew. *Moondust: In Search of the Men Who Fell to Earth.* New York: Fourth Estate, 2005.

Smith, Bromley. "Memo for Mr. Moyers." June 1, 1964. Archives, NASA Headquarters, Washington, D.C.

Smith, Morgan. "Can You Survive in Space without a Spacesuit?" *Salon,* August 2, 2007.

Snider, Arthur J. "Aldrin to be First Man on Moon." *New Orleans Times-Picayune,* February 27, 1969.

Sorensen, Ted. *Counselor.* New York: Harper, 2008.

———. *Kennedy.* New York: Harper & Row, 1965.

"The Soviet Bioastronautics Research Program." Office of Scientific Intelligence. McLean, Virginia: The Central Intelligence Agency, CIA Archives, February 22, 1962.

"Soviet Capabilities for Clandestine Attack Against the U.S. with Weapons of Mass Destruction and the Vulnerability of the U.S. to Suck Attack (mid-1951 to mid-1952)." *National Intelligence Estimate.* McLean, Virginia: The Central Intelligence Agency, CIA Archives, September 4, 1951.

"Soviet Satellite Sends U.S. Into a Tizzy." *Life,* vol. 43, no. 16, October 14, 1957.

"Soviets Take a New Look at Space Spending." Directorate of Intelligence. Intelligence Memorandum, McLean, Virginia: The Central Intelligence Agency, CIA Archives, October 1, 1970.

"Spaceflight: The Soviet Race to the Moon." http://www.centennialofflight.gov/essay/SPACE FLIGHT/soviet_lunar/SP21.htm.

Sparks, Heather. "Study of Lunar Soil Confirms Moon's Origins." SPACE.com, October 25, 2001.

Spinoff 2006. Publications and Graphics Department, NASA Center for AeroSpace Information (CASI). Washington: Government Printing Office, 2006.

Spires, David N. *Beyond Horizons: A Half Century of Air Force Space Leadership.* Washington: Air Force Special Command, Air University Press, Government Printing Office, 1998.

"Sputnik Declassified." *Nova.* Lone Wolf Documentary Group, WGBH/Boston, 2007.

Sterling, Bruce. "The New Cold War." *Wired,* May 2003.

Stockton, Sharon. "Engineering Power: Hoover, Rand, Pound, and the Heroic Architect." *American Literature,* vol. 72, no. 4, December 2000, pp. 813–41.

Stone, Ron, interviewer. Interview with James A. Lovell, Jr. NASA Johnson Space Center Oral History Project, Houston, Texas, May 25, 2009.

Stuart, Lettice. "A Disney Creation for Visitors to the Johnson Space Center." *The New York Times,* May 14, 1989.

Sudakov, Dmitry, translator. "Why Did the USSR Lose the Moon Race?" *Pravda,* December 3, 2002.

Svoboda, Elizabeth. "Who Owns the Moon?" *Salon,* January 19, 2008.

Swanson, Glen, editor. *Before This Decade Is Out: Personal Reflections on the Apollo Program.* NASA History Series, Washington: NASA Office of Policy and Plans, 1999.

Swanson, Glen, interviewer. Interview with Geneva B. Barnes. NASA Johnson Space Center Oral History Project, Washington, D.C., March 26, 1999.

Taubman, Philip. "In Death of Spy Satellite Program, Lofty Plans and Unrealistic Bids." *The New York Times,* November 11, 2007.

Thompson, Nicholas. "A War Best Served Cold." *The New York Times,* July 30, 2007.

Thrall, Nathan, and Jesse James Wilkins. "Kennedy Talked, Khrushchev Triumphed." *The New York Times,* May 22, 2008.

Tindall, Howard W. NASA Manned Spacecraft Center Oral History Project. Houston, Texas, December 16, 1966.

"To the Moon." *Nova,* 1999.

Tomayk, James E. *Computers Take Flight: A History of NASA's Pioneering Digital Fly-by-Wire Project.* NASA SP-2000-4224, Washington: NASA History Office, 2000.

Toth, Robert C. "House Opposes Joint Moon Trip; Votes NASA Fund; Kennedy Proposal for U.S. and Soviet Cooperation is Rebuffed, 125-110, Issue Is Not Debated." *The New York Times,* October 10, 1963.

Tucker, Tom. "Touchdown: The Development of Propulsion Controlled Aircraft at NASA Dryden." *Monographs in Aerospace History,* no. 16, Washington: Office of Policy and Plans, NASA History Office, 1999.

United States House of Representatives. "The Congressional Record." September 16, 1969.
The Universe. Tony Long, writer, producer, and director. 2008.
"US and Soviet Space Programs: Comparative Size." Directorate of Intelligence. Intelligence
    Memorandum. McLean, Virginia: The Central Intelligence Agency, CIA Archives,
    March 1, 1966.
Verne, Jules. *De la Terre à la Lune. From the Earth to the Moon. Direct in Ninety-Seven Hours
    and Twenty Minutes: And a Trip Round It.* Louis Mercier, Louis and King, Eleanor E,
    translators. New York: Scribner, Armstrong & Company, 1874.
Vick, Charles P. "CIA/CIO Declassifies N1-L3 Details." Global Security.org, http://www
    .globalsecurity.org/intell/library/imint/4_n1_1.htm.
"Vienna Meeting between the President and Chairman Khrushchev," June 3, 1961. John F.
    Kennedy Library, Cambridge, MA.
Vladimirov, Leonid. *The Russian Space Bluff.* New York: The Dial Press, 1972.
von Braun, Wernher. "To the Vice President of the United States," April 29, 1961.
Walsh, John. "The Astronaut's Mind." *Esquire,* September 1970.
"War in Space: Can America Win It?" *Space World,* February 1962.
Ward, Doug, interviewer. "Charles M. Duke, Jr." NASA Johnson Space Center Oral History
    Project, Houston, Texas, March 12, 1999.
———. Interview with Gerald D. "Gerry" Griffin. NASA Johnson Space Center Oral History
    Project, Houston, Texas, March 12, 1999.
———. Interview with Fred W. Haise, Jr. NASA Johnson Space Center Oral History Project,
    Houston, Texas, March 23, 1999.
———. Interview with James A. McDivitt. NASA Johnson Space Center Oral History Proj-
    ect, Elk Lake, Michigan, June 29, 1999.
"Washington Goes to the Moon." WAMU American University Radio. Washington, D.C.,
    May 24, 2001.
Wayne, Leslie. "A Bold Plan to Go Where Men Have Gone Before." *The New York Times,*
    February 5, 2006.
———. "With Shuttle Program, an Intricate and Integral Partnership for NASA." *The New
    York Times,* August 3, 2005.
Webb, James E. "Memorandum for Mr. Callaghan—AC." November 6, 1963. Archives,
    NASA Headquarters, Washington, D.C.
———. "Letter to President Johnson," October 1, 1968. Archives, NASA Headquarters,
    Washington, D.C.
———. "Memorandum for the President. Subject: *Reader's Digest* Article on Space." August
    9, 1963. Archives, NASA Headquarters, Washington, D.C.
———. "Letter to The Honorable Robert S. McNamara, Secretary of Defense." January 16,
    1963. Archives, NASA Headquarters, Washington, D.C.
Weiner, Tim. "Air Force Seeks Bush's Approval for Space Weapons Programs." *The New York
    Times,* May 18, 2005.
———. *Legacy of Ashes: The History of the CIA.* New York: Doubleday, 2007.
Welsh, E.C. "Letter to President Johnson," September 30, 1968. Archives, NASA Headquar-
    ters, Washington, D.C.
———. "Memorandum To: Mr. Andrew Hatcher, The White House." November 1, 1963.
    Archives, NASA Headquarters, Washington, D.C.
"Wernher von Braun and the V-2 Rocket." *Man Moment Machine.* Michael McInerney, pro-
    ducer/director. Edelman Productions, 2005.
"What Did the CIA Know and When Did They Know It?" http://www.astronautix.com/
    articles/whanowit.htm.
Whelon, Albert D., and Sidney N. Graybeal. "Intelligence for the Space Race." *Studies in In-
    telligence.* McLean, Virginia: The Central Intelligence Agency, CIA Archives, Fall
    1961.
When We Left Earth: The NASA Missions, Richard Dale and Bill Howard, executive produc-
    ers. Dangerous Films, 2008.

Wilford, John Noble. "Earthbound: Our Future in Space Is Already History." *The New York Times,* February 9, 2003.

———. "With Fear and Wonder in Its Wake, Sputnik Lifted Us into the Future." *The New York Times,* September 25, 2007.

Wilson, Chris. "What's the Point of Putting Weapons in Orbit?" *Slate,* Feb. 13, 2008.

Wright, Quincy. "Legal Aspects of the U-2 Incident." *The American Journal of International Law* 836, 849 (1960).

Wright, Rebecca, interviewer. Interview with Richard H. Battin. NASA Johnson Space Center Oral History Project, Lexington, Massachusetts, April 18, 2000.

———. Interview with Harold D. Beck. NASA Johnson Space Center Oral History Project, Houston, Texas, December 9, 2004.

———. Interview with Charles F. Deiterich. NASA Johnson Space Center Oral History Project, Bertram, Texas, February 28, 2006.

———. Interview with James W. Head III. NASA Johnson Space Center Oral History Project, Providence, Rhode Island, June 6, 2002.

———. Interview with John D. Hodge. NASA Johnson Space Center Oral History Project, Great Falls, Virginia, April 18, 1999.

———. Interview with Wayne E. Koons. NASA Johnson Space Center Oral History Project, Houston, Texas, October 15, 2004.

———. Interview with Eugene F. Kranz. NASA Johnson Space Center Oral History Project, Houston, Texas, January 8, 1999.

———. Interview with Donald L. Lind. NASA Johnson Space Center Oral History Project, Houston, Texas, May 27, 2005.

———. Interview with John E. McLeaish. NASA Johnson Space Center Oral History Project, San Antonio, Texas, November 15, 2001.

———. Interview with John G. McTigue. NASA Johnson Space Center Oral History Project, San Jose, California, September 29, 2005.

———. Interview with Delores B. "Dee" O'Hara. NASA Johnson Space Center Oral History Project, Mountain View, California, April 23, 2002.

———. Interview with John W. O'Neill. NASA Johnson Space Center Oral History Project, Houston, Texas, July 12, 2001.

———. Interview with Edward L. Pavelka, Jr. NASA Johnson Space Center Oral History Project, Houston, Texas, May 9, 2001.

———. Interview with Fred Pearce. NASA Johnson Space Center Oral History Project, Houston, Texas, November 12, 2003.

———. Interview with Chris D. Perner. NASA Johnson Space Center Oral History Project, Houston, Texas, July 26, 2001.

———. Interview with William H. Rice. NASA Johnson Space Center Oral History Project, Washington, DC, March 18, 2004.

———. Interview with Russell L. "Rusty" Schweickart. NASA Johnson Space Center Oral History Project, Houston, Texas, October 19, 1999.

———. Interview with H. Morgan Smith. NASA Johnson Space Center Oral History Project, Denton, Texas, October 31, 2002.

Wolfe, Tom. *The Right Stuff.* New York: Farrar, Strauss, Giroux, 1979.

Yergin, Daniel. *Shattered Peace: The Origins of the Cold War and the National Security State.* New York: Houghton Mifflin, 1977.

Young, Anthony. "Generation Y and Lunar Disbelief." *The Space Review,* January 22, 2007, http://www.thespacereview.com/article/787/1.

Young, Hugo, Brian Silcock, and Peter Dunn. *Journey to Tranquility.* Garden City: Doubleday, 1970.

Index